Lecture Notes in Computer Science

Commenced Publication in 1973
Founding and Former Series Editors:
Gerhard Goos, Juris Hartmanis, and Jan van Leeuwen

Epaminondas Kapetanios   Vijayan Sugumaran
Myra Spiliopoulou (Eds.)

Natural Language and Information Systems

13th International Conference on Applications
of Natural Language to Information Systems, NLDB 2008
London, UK, June 24-27, 2008
Proceedings

Springer

Volume Editors

Epaminondas Kapetanios
School of Computer Science
University of Westminster
London, UK
E-mail: e.kapetanios@wmin.ac.uk

Vijayan Sugumaran
Department of Decision and Information Sciences
School of Business Administration
Oakland University
Rochester, MI, USA
E-mail: sugumara@oakland.edu

Myra Spiliopoulou
Faculty of Computer Science
Otto-von-Guericke-University Magdeburg
Magdeburg, Germany
E-mail: myra@iti.cs.uni-magdeburg.de

Library of Congress Control Number: Applied for

CR Subject Classification (1998): H.2, H.3, I.2, F.3-4, H.4, C.2

LNCS Sublibrary: SL 3 – Information Systems and Application, incl. Internet/Web
and HCI

ISSN   0302-9743
ISBN-10   3-540-69857-4 Springer Berlin Heidelberg New York
ISBN-13   978-3-540-69857-9 Springer Berlin Heidelberg New York

Springer is a part of Springer Science+Business Media

springer.com

Typesetting: Camera-ready by author, data conversion by Scientific Publishing Services, Chennai, India
Printed on acid-free paper   SPIN: 12326839   06/3180   5 4 3 2 1 0

Preface

This volume contains the papers presented at NLDB 2008, the 13th International Conference on Natural Language and Information Systems, held June 25–27, 2008. It also contains some of the best research proposals as submitted to the NLDB 2008 doctoral symposium held on June 24, 2008. The programme also includes three invited talks covering the main perspectives of the application of natural language to information systems: the way humans process, communicate and understand natural language, what are the implications and challenges towards semantic search for the new Web generation, how natural language applies to the well-established database way of querying as a means to unlock data and information for end users.

We received 68 papers as regular papers for the main conference and 14 short papers for the doctoral symposium. Each paper for the main conference was assigned four reviewers based on the preferences expressed by the Program Committee members. We ensured that every paper had at least two reviewers that expressed interest in reviewing it or indicated that they could review it. We ensured that each paper got at least three reviews. As a result, only 10% of the papers were reviewed by three reviewers.

The Conference Chair and the two Program Committee Co-chairs acted as Meta-Reviewers. Each of them took roughly 1/3 of the papers (obviously respecting conflicts of interest), for which s/he was responsible. This included studying the reviews, launching discussions and asking for clarifications whenever necessary, as well as studying the papers whenever a need for an informed additional opinion arose or when the reviewers' notes did not allow for a decision.

After the review deadline, each meta-reviewer went through the reviews of their papers and made a preliminary assessment. To the extent possible, the meta-reviewers tried to get the inconsistencies and discrepancies resolved. We used a ranking list as resulted from the weighted average scores of all papers in a scale from 1 (lowest possible) to 6 (highest possible) as computed by taking into account the reviewer's confidence as a weighting factor. The final acceptance rate counting the number of full papers according to the NLDB tradition, at least during the last three years, was 29.4%. The following decision rules were used to make the final decisions:

- Full Papers: Papers with a weighted average score of 4.0 (weak accept) or more were accepted as full papers. There were 17 such papers. One more paper was subject to conditional acceptance. This gave us room for two more papers to be accepted as full papers. We considered the papers with a weighted average score of 3.8 or above but less than 4.0 to select the two papers. Taking into account the reviewer feedback and the scores, we decided to select two papers with the least number of negative (1.0, 2.0 or 3.0) scores.
- Short Papers: Papers with a weighted average score of 3.7 or above but less than 4.0 were accepted as short papers.

- Posters: Papers with a weighted average score of 3.0 (weak reject) or above but less than 3.7 were accepted as posters.
- Doctoral Symposium Papers: The four-page long papers, which were invited from the submissions to the doctoral symposium, received 4.0 (accept) as weighted average score having both reviews as 4.0 or 5.0 in a scale of 1 (lowest possible) to 5 (highest possible)

All papers underwent proofreading, which also resulted in their classification into the following five major thematic areas:

- Natural Language Processing and Understanding
- Conceptual Modelling and Ontologies
- Information Retrieval
- Querying and Question Answering
- Document Processing and Text Mining
- Software (Requirements) Engineering and Specification

In all these thematic areas, a series of challenges together with key ideas, emerging concepts and interesting solutions were presented illustrating the state of the art in applications of natural language as information and communication means. In addition, a series of case studies ranging from criminal to biological and traffic accident analysis demonstrated the wide variety of real-world problems where application of natural language plays a key role.

Natural Language Processing and Understanding: The three full papers (pages 15–47) from this thematic area deal with elementary questions of meaningful interpretation and knowledge representation for understanding and processing natural language terms within a context. In particular, it is shown that word division into morphemes is useful for automatic description of morphological structures of languages without existing morphological models and/or morphological dictionaries.

This is important for modern information retrieval technologies, when we are dealing with unknown languages or languages without complete description, or, possibly, for some specific terminological areas, for example, medicine. A global optimization technique (implemented as a genetic algorithm) to the task of division of words into morphemes using the Spanish language as a case study is presented, i.e., the main goal is investigating the application of the un-supervised technique for determining morpheme structure of words.

The problem of Word Sense Disambiguation (WSD) is addressed for disambiguating abbreviations in Jewish Law Documents written in Hebrew, where the excessive use of abbreviations is of paramount importance for understanding of the Hebrew language as such. The same problem has been addressed within the context of natural language parsing and text understanding, where a semantic interpreter is suggested which acquires and anchors common sense knowledge about ordinary concepts.

This goes beyond WordNet, which relates concepts only by *is-a* and *part-of* links. The described acquisition methods link concepts by semantic relations expressing events, actions. etc. This also bears implications to populate WordNet noun senses with the interpreted glosses creating a general common-sense knowledge network, that could be used for all kinds of natural language understanding tasks.

Conceptual Modelling and Ontologies: The three full and one short papers (pages 48–89) from this thematic area deal with the challenging relationship between natural language terms and words as used in writing or cognition and the formal conceptualizations as used for machine understanding.

In particular, the representation at a conceptual level of the meaning of spatial propositions as an Interlingua for French and German topological prepositions is suggested by the first paper as the right way to deal with machine translation of propositions in these grammatically and semantically rich natural languages. In this case, it is shown that Machine Translation can only take place if the meaning of these propositions can be expressed in common neutral concepts, which exemplify conceptual knowledge about object classes and relations between objects. Since they become available, they are relevant for the interpretation of the spatial expressions.

The relationship between terms and words anchored in a particular natural language and specifications of conceptualizations becomes more challenging when ontologies need to be extracted from text as available in documents. To this extent, a new concept of *ontological profiles* is introduced as a more powerful semantic representation of a domain than an ontology is on its own. It links concepts with vectors of related terms from a domain vocabulary and their use in collections of documents, together with weights showing the strength of these relations in a particular context. This is crucial for semantic search, input of weights and strength of relations is text mining and not as a linguistic ontology.

Furthermore, a considerable attempt to tackle the unresolved problem of learning and extracting non-taxonomic relationships in a conceptual schema or ontology is taken via a hybrid relationship learning approach that combines a linguistic contextual analysis of concepts with co-occurrence patterns across documents. Both techniques may be used separately to extract ontology relationships, and they tend to select slightly different types of relationships, which, if used in combination, end up with relationship candidates that are significantly better than what each of the techniques can offer in isolation. This has been implemented as part of the General Architecture for Text Engineering (GATE).

Finally, the problem of extracting semantic annotation schemes by generating linguistic indicators from documents (crucial for the success of the Semantic Web) is addressed by the short paper.

Information Retrieval: The impact of this interplay between natural language and specifications of conceptualizations is illustrated by Information Retrieval techniques as addressed by the following three full and two short papers (pages 90–137). They all target at more effective and semantically relevant retrieval of documents and information in general.

In particular, an improvement of search mechanisms is suggested by the first two papers as based on more topic-oriented and -focused crawling and classification of results, in an attempt to reduce the vast amount of hits usually returned by search engines. Another approach based on query-reformulation and expansion is addressed by the next one full and one short paper, where either the query term independence as a dominating assumption for query terms suggestions in modern information retrieval is abandoned, or suggestions are made in conjunction with retrieved concepts from the retrieved documents. The last short paper illustrates the relevance of natural language with the task of retrieving geographic information and, particularly, spatial terms as a subtask of information retrieval.

Querying and Question Answering: The following four full and one short papers (pages 138–191) put more the emphasis on implications of natural language on querying databases and question answering. Given the two different approaches between data retrieval (DR) as practised by RDBMs and information retrieval (IR), with both suffering from drawbacks as posed by flexibility versus inflexibility in query answering, these papers aspire to contribute to this merging.

In particular, the first full paper addresses a more informative way of providing answers to queries by taking into consideration not only extensional semantics of retrieved entities but also intentional ones as derived by a background knowledge base. This may increase, in particular cases, the information content returned to the user.

The next full paper illustrates an attempt to combine the DR and IR systems by using predefined word similarities to determine the correlation between a keyword query (commonly used in IR) and data records stored in the inner framework of a standard RDBMS. This could be a significant enhancement of query processing in RDBMSs.

The following full paper presents three new techniques, which combine aspects of information extraction with data integration in order to better exploit the data in applications such as crime informatics, bio-informatics, road traffic accidents, where free text is a necessary means of representing information. Nevertheless, it is believed that 80% of information stored by companies is available as unstructured text. Therefore, since the explosion of predominantly textual information made available on the Web has led to the vision of a machine-tractable Semantic Web, a database like functionality replacing today's book-like Web is emerging.

The last one full and one short paper discuss key challenges and present interesting approaches for Natural Language Interfaces for Databases (NLIDBs). Despite the fact that such interfaces have been envisioned for the last two decades, none of those became really operational in any of the commercially available database management systems.

Document Processing and Text Mining: A significant challenge is also posed by natural language as a means of information and knowledge management (pages 192–251). The four full and two short papers within this thematic area illustrate this challenge when text segmentation, summarization and mining techniques are addressed and applied in order to also meet a series of real–world problems.

In particular, the first one full and two short papers discuss extraction techniques of events, news and interesting topics, in general. These techniques range from light-weight (full paper addressing global crisis monitoring) to more natural language processing oriented ones: (a) one short paper addressing descriptions of affairs or accidents in documents, e.g., murders in Tokyo and Kyoto, as based on sequences of words rather than frequencies classification techniques, (b) one short paper addressing the problem of segmenting the television news in self-contained fragments in order to identify the points in which there is a significant change in the topic of discussion. The latter is particularly useful as a cable-based television model, where there is, apparently, an emerging requirement for random access capabilities according to the RAI Centre for Research and Technological Innovation.

The following two full papers deal with the problem of reducing the amount of text or documents available by applying either text summarization or abstract extraction techniques, or by applying un-supervised (non-parametric) methods, which can be defined as the process of grouping similar documents without requiring a priori either the number of document categories or a careful initialization of the process from a human user. The latter is particularly useful for clustering of large amounts of documents, where ensemble clustering methods are being introduced as a potential solution.

The last full paper studies the answer to the question to which extent a language modelling approach could offer solutions to contribute to the problem of pattern recognition and extraction such as extracting the characteristics of offenders from the crime or offence style and profile and, therefore, circumscribe the potential, and still unknown, offenders of a crime scene.

Software (Requirements) Engineering and Specification: The last thematic area (pages 252–299), with three full and two short papers, addresses the need to design user-friendly interfaces and as correct and re-usable as possible information systems, where natural language processing techniques are brought to the desktop as a common application.

In particular, an essential hope is offered by the first full paper to make the development of human–computer interaction components easier and less error-prone by using descriptions of the components, functionality, and behavior of these interfaces that capture the semantics of the communication involved. An important impact of such a specification technique lies in avoiding the need for technical expertise.

The need to communicate the specification of information systems requirements with non-technical people is also illustrated by the following paper. It discusses automated assistance for the detection and elicitation of non-functional requirements (NFR) in natural language text at a very early stage of software requirements specification (SRS). It has the potential to bring about significant industrial savings in cost and aggravation, as well as to prevent very costly misinterpretation. This has been a response to empirical reports consistently indicating that neglecting NFRs in the requirements elicitation and analysis phase leads to project failures, or at least to considerable delays, and, consequently, an escalation in the cost of software development.

In a similar context, the following paper presents a system addressing the problem of extracting semantic business vocabulary and rules (SBVR) and their formal expression in Semantic Web Rule Language (SWRL) from free text descriptions of financial products or services, with an application in the banking industry. It is especially useful in environments where specifications are changing everyday as a reaction to market evolutions (financial, insurance industry).

The following short paper takes a similar approach by introducing an automatic translation from UML-based conceptual schemas to SBVR, where the initial UML schemas are complemented with textual expressions in OCL (Object Constraint Language). The obtained SBVR representation can be presented to the customers as a list of self-explaining natural language expressions using the predefined alternative notations to express SBVR concepts by means of English statements, either in the SBVR Structured English style or using the RuleSpeak approach as included in the SBVR standard.

The last short paper in this thematic area discusses an open service-oriented architecture (SOA) based on established standards and software engineering practices, such as W3C Web services with WSDL descriptions, SOAP client/server connections, and OWL ontologies for service meta-data, which together facilitate an easy, flexible, and extensible integration of any kind of language service into any desktop client. It illustrates an important approach to bringing NLP to desktop client applications such as e-mail clients and word processors. In contrast with the GATE, UIMA approaches, it provides a service-oriented architecture to broker any kind of language analysis service in a network-transparent way.

Last but not least, this volume also contains the invited research proposals as solicited from the doctoral symposium as well as the papers accepted for and presented at the poster session. They all fall into one or more of the thematic areas as stated above and give interesting perspectives to emerging research projects, work in progress, framework architectures and interesting case studies and applications. Some characteristic examples are:

- A DARPA-funded project that applies rules to extract features from natural language text and tries to identify tasks within e-mails. Particularly useful for busy e-mail users.
- Detection of protein–protein interaction at a sentence level, which is considered as a crucial factor that could contribute to improving coverage of interaction relation detection in biosciences.
- A feature-driven opinion summarization method for customer reviews on the Web based on identifying general features (characteristics) describing any product, product specification features and feature attributes (adjectives grading the characteristics).
- A semantic and syntactic distance-based method in topic text segmentation, where topic variations from French political discourses are used as a case study.

- A framework combining information retrieval with machine learning and (pre-)processing for named entity recognition in order to extract events from a large document collection. As a case study, the public collection of minutes of plenary sessions of the German parliament and of petitions to the German parliament is being used.
- Methodology and approach to deal with natural language interfaces (NLI) to databases where the named entity recognition and disambiguation is primarily the problem and focus (test bed: a huge database with movies in Portuguese).
- A method to automatically generate a conceptual model from the system's textual descriptions of the use case scenarios in the Spanish language. Techniques of natural language processing (NLP) and conceptual graphs are used as the basis of the method. Especially useful when large systems are developed.
- Definition and implementation of a model for knowledge representation using a conceptualization as much as possible close to the way in which the concepts are organized and expressed in human language. It is being used in order to improve the knowledge representation accuracy for document analysis.
- An ontology-based focused crawler for restricting potential search results on the Web.
- Tackling the problem of data integration by applying word sense disambiguation techniques.
- Mapping natural language into SQL.

The conference organizers are indebted to the reviewers for their engagement in a vigorous submission evaluation process. We would also like to thank cordially the members of the conference and the doctoral symposium Organizing Committees for their work and involvement in making NLDB 2008 a successful conference. Our special thanks are also given to the Pro–Vice Chancellor of the University of Westminster, Myszka Guzkowska, who kindly supported this event by hosting NLDB 2008.

June 2008

Epaminondas Kapetanios
Vijayan Sugumaran
Myra Spiliopoulous

Organization

NLDB 2008 was organized by the University of Westminster, Central London, UK.

Conference Committee

Conference Chair	Epaminondas Kapetanios (University of Westminster, UK)
	Elisabeth Métais (CNAM, Paris, France)
	Reind van de Riet (Vrije Universiteit Amsterdam, The Netherlands)
Program Chair	Vijayan Sugumaran (Oakland University, USA)
	Myra Spiliopoulou (University of Magdeburg, Germany)
Organizing Committee	Sue Black (University of Westminster, UK)
	Diana Irina Tanase (University of Westminster, UK)
	Rob Fenwick (University of Westminster, UK)
	Wendy Meister (University of Westminster, UK)
	Panagiotis Chountas (University of Westminster, UK)
Publicity Chair	Farid Meziane (University of Salford, UK)

Program Committee

Maristella Agosti	Karin Harbusch	Manolis Koubarakis
Jacky Akoka	Harmain Harmain	Georgia Koutrika
Sophia Ananiadou	Alexander Hinneburg	Mounia Lalmas
Frederic Andres	Annika Hinze	Jana Lewerenz
Akhilesh Bajaj	Helmut Horacek	Deryle Lonsdale
Mokrane Bouzeghoub	Andreas Hotho	Stéphane Lopes
Hiram Calvo	Yannis Ioannidis	Vanessa Lopez
Roger Chiang	Jing Jiang	Robert Luk
Philip Cimiano	Paul Johannesson	Heinrich Mayr
Isabelle Comyn-Wattiau	Epam. Kapetanios	Paul McFetridge
Antje Düsterhöft	Vangelis Karkaletsis	Elisabeth Metais
Günther Fliedl	Zoubida Kedad	Farid Meziane
Alexander Gelbukh	Christian Kop	Luisa Mich
Udo Hahn	Leila Kosseim	Andrés Montoyo

Ana Maria Moreno
Rafael Muñoz
Günter Neumann
Jian-Yun Nie
Moira Norrie
Pit Pichappan
Odile Piton
Alexandra Poulovassilis
Sandeep Purao
Yacine Rezgui

Jürgen Rilling
Hae-Chang Rim
Fabio Rinaldi
Markus Schaal
Grigori Sidorov
Max Silberztein
Irena Spasic
Myra Spiliopoulou
Benno Stein
Veda Storey

Vijayan Sugumaran
Bernhard Thalheim
K.P. Thirunarayan
Juan Carlos Trujillo
Luis Alfonso Ūrena
Panos Vassiliadis
Roland Wagner
Hans Weigand
Stanislaw Wrycza
Reind van de Riet

Additional Reviewers

Maria Bergholtz
You-Jin Chang
Oscar Ferrandez
Paulina Fragou
Miguel Angel García
Hayrettin Gürkök

Trivikram Immaneni
Robert Jäschke
Yulia Ledeneva
M. Teresa Martín
Fernando Martínez
Matthew Perry

Vassiliki Rentoumi
Philipp Schugerl
Dimitris Vogiatzis
Dean Williams

Doctoral Symposium Reviewers

Anastassia Angelopoulou
  (Organizing Committee)
Mokrane Bouzeghoub
Alexander Gelbukh
Georgia Koutrika
  (Organizing Committee)

Mounia Lalmas
Elisabeth Metais
Farid Meziane
Andres Montoyo
Konstantinos Morfonios
Rafaél Muñoz

Günter Neumann
Myra Spiliopoulou
Vijay Sugumaran
Diana Irina Tanase

Sponsoring Institutions

British Computer Society, Information Retrieval Specialist Group.

Table of Contents

Information Retrieval

Querying and Question Answering

Document Processing and Text Mining

Software (Requirements) Engineering and Specification

Conceptual Modelling and Ontologies Related Posters

Information Retrieval Related Posters

Querying and Question Answering Related Posters

Document Processing and Text Mining Related Posters

Software (Requirements) Engineering and Specification Related Posters

Doctoral Symposium Papers

Invited Papers

Sentence and Text Comprehension: Evidence from Human Language Processing

Edward Gibson

Department of Brain and Cognitive Sciences
Massachusetts Institute of Technology (MIT), USA

1 Introduction

In this presentation, I will survey research that is aimed at discovering how people process sentences and texts in natural language. In particular, I will survey evidence for the existence of several kinds of information constraints in language processing, including lexical, syntactic, world-knowledge, discourse coherence, referential and intonational information.

I will also discuss evidence relevant to the architecture of the language processing system that combines these information sources in real-time language processing, as constrained by the available working memory. I will focus on recent work in discourse coherence, which aims to discover the structure underlying texts, similar to the syntactic structure underlying sentences.

The methods that we use include (a) reading methods (self-paced reading and eye-tracking) in which reading time is the dependent measure (b) event-related potentials, in which voltage changes are measured on the scalp corresponding, corresponding to cognitive events such as language processing; (c) the visual domain method, in which eye-gaze is tracked to objects in the visual context while auditory language is processed; (d) corpus analyses; and (e) sentence production. These methods apply cross-linguistically.

E. Kapetanios, V. Sugumaran, M. Spiliopoulou (Eds.): NLDB 2008, LNCS 5039, p. 3, 2008.

Towards Semantic Search

Ricardo Baeza-Yates, Massimiliano Ciaramita,
Peter Mika, and Hugo Zaragoza

Yahoo! Research, Barcelona, Spain

Abstract. Semantic search seems to be an elusive and fuzzy target to
many researchers. One of the reasons is that the task lies in between
several areas of specialization. In this extended abstract we review some
of the ideas we have been investigating while approaching this problem.
First, we present how we understand semantic search, the Web and the
current challenges. Second, how to use shallow semantics to improve Web
search. Third, how the usage of search engines can capture the implicit
semantics encoded in the queries and actions of people. To conclude, we
discuss how these ideas can create virtuous feedback circuit for machine
learning and, ultimately, better search.

1 Introduction

From the early days of Information Retrieval (IR), researchers have tried to take
into account the richness of natural language when interpreting search queries.
Early work on natural language processing (NLP) concentrated on tokenisation
and normalisation of terms (detection of phrases, stemming, lemmatisation, etc)
and was quite successful [11]. Sense disambiguation (needed to differentiate between the different meanings of the same token) and synonym expansion (needed
to take into account the different tokens that express the same meaning) seemed
the obvious next frontier to be tackled by researchers in IR. There were a number
of tools available that made the task seem easy. These tools came in many forms,
from statistical methods to analyse distributional patterns, to expert ontologies
such as WordNet. However, despite furious interest in this topic, few advances
were made for many years, and slowly the IR field moved away: concepts like
synonymy were no longer discussed and topics such as term disambiguation for
search were mostly abandoned [5]. In lack of solid failure analysis data we can
only hypothesize why this happened and we try to do so later.

Semantic search is difficult because language, its structure and its relation to
the world and human activities, is complex and only partially understood. Embedding a *semantic model*, the implementation of some more or less principled
model of both linguistic content and background knowledge, in massive applications is further complicated by the dynamic nature of the process, involving
millions of people and transactions. Deep approaches to model meaning have
failed repeatedly, and clearly the scale of the Web does not make things easier.
In [6] we argue that semantic search has not occurred for three main reasons.
First, this integration is an extremely hard scientific problem. Second, the Web

E. Kapetanios, V. Sugumaran, M. Spiliopoulou (Eds.): NLDB 2008, LNCS 5039, pp. 4–11, 2008.

imposes hard scalability and performance restrictions. Third, there is a cultural divide between the Semantic Web (SW) and IR disciplines. Our research aims at addressing these three issues.

Arguably, part of the reason the Web and search technologies have been so successful is because the underlying language model is extremely simple, and people expected relatively little from it. Although search engines are impressive pieces of engineering they address a basic task: given a query return a ranked list of documents from an existing collection. In retrospect, the challenges that search engines have overcome, mostly have to do with scalability and speed of service, while the ranking model which has supported the explosion of search technology in the past decade is quite straightforward and has not produced major break-throughs since the formulation of the classic retrieval models [4] and the discovery of features based on links and usage. Interestingly, however, the Web has cre-ated an ecosystem where both content and queries have adapted. For example, people have generated structured encyclopedic knowledge (*e.g.*, Wikipedia) and sites dedicated to multimedia content (*e.g.*, Flickr and YouTube). At the same time users started developing novel strategies for accessing this information; such as appropriate query formulation techniques ("mammals Wikipedia" instead of just "mammals"), or invented "tags" and annotated multimedia content other-wise almost inaccessible (videos, pictures, etc.) in the classic retrieval framework because of the sparsity of the associated textual information.

Clearly the current state of affairs is not optimal. One of our lines of research, semantic search, addresses these problems. To present our vision and our early findings, we first detail the complexity of semantic search and its current context. Second, we survey some of our initial results on this problem. Finally, we mention how we can use Web mining to create a virtuous feedback circle to help our quest.

2   Problem Complexity and Its Context

Search engines are hindered by their limited understanding of user queries and the content of the Web, and therefore limited in their ways of matching the two. While search engines do a generally good job on large classes of queries (*e.g.* navigational queries), there are a number of important query types that are undeserved by keyword-based approach. Ambiguous queries are the most often cited examples. In face of ambiguity, search engines manage to mask their confusion, by (1) explicitly providing diversity (in other words, letting the user choose) and (2) relying on some notion of popularity (*e.g.* PageRank), hoping that the user is interested in the most common interpretation of the query. As an example of where this fails, consider searching for George Bush, the beer brewer. The capabilities of computational advertising, which is largely also an information retrieval problem (*i.e.* the retrieval of the matching advertisements from a fixed inventory), are clearly impacted due to the relative sparsity of the search space. Without understanding the object of the query, search engines are also unable to perform queries on descriptions of objects, where no key exists. A typical, and important example of this category is product search. For example, search engines are unable to look for "music players with at least 4GB of RAM" without understanding what a music player is, what its characteristics are, etc. Current search technology is also unable to satisfy any complex queries requiring information integration such as analysis, prediction, scheduling, etc. An example of such integration-based tasks is opinion mining regarding products or services. While there have been some successes in opinion mining with pure sentiment analysis, it is often the case that one would like to know what specific aspects of a product or service are being described in positive or negative terms. However, information integration is not possible without structured representations of content, a point which is central in research on the Semantic Web, which has focused on ways of overcoming current limitations of Web technology.

The Semantic Web is about exposing structured information on the Web in a way that its semantics is grounded in ontologies, that is, agreed-upon vocabularies. Contrary to its popular image, the SW effort is agnostic to where the data will come from, and in fact large amounts of structured data have been put online by porting databases to the Semantic Web (like DBpedia, US Census data, Eurostat data, biomedical databases, etc.), known as the Linking Open Data movement[1]. This is an appealing vision in the sense that it brings the Deep Web within reach of search engines, which they could not touch up until now. Bringing the content of databases to the Web, however, is just one part of the Semantic Web vision: while many websites on the Web are automatically generated from relational databases, much of the content that matters to users (because it has been written by other users...) is still in the form of text. (Consider for example social media in blogging, wikis and other forms of user-generated content.) The vision of what we may call the Annotated Web is to encode the semantics of textual content using the same technology that is used to make the content of databases interoperable. Again, the vision of the Annotated Web is agnostic as to where the annotations would come from, that is, whether they are produced by a human or a machine. Off-the-shelf NLP technologies have been successfully applied in the Semantic Web community. Further, with the success of microformats, and the recent standardization of RDFa by the W3C, efforts toward manual annotation seem to be slowly breaking the chicken-and-egg problem that tainted the overall Semantic Web effort (that is, whether the community should aim to develop interesting applications that would compel users to annotate their web pages or focus on creating data that will attract interesting applications). Today, metadata embedded in HTML pages using microformats or RDFa seems easy enough for users to author and compelling applications are starting to emerge. One example is Yahoo's SearchMonkey, where embedded metadata is used to enrich the search result presentation. As mentioned above, the challenge lies in the integration of the existing results in NLP with the data riches and inferential power of the Semantic Web. There are important benefits to be gained on both sides. For the field of NLP, the success of the Annotated Web promises large-scale training data sets by observing how users apply annotations in a wide range of situations and in broad domains. In turn, the Semantic Web will benefit from ever improving support for automated or semi-automated annotation.

---

[1] Linking Data: http://en.wikipedia.org/wiki/Linked_Data

However, embellishing the search interface with metadata should only be the first step. In order to move toward the situation where the machine has a satisfactory understanding of the users' need in terms of the Web's content (see [8]), the interface of the search engine will likely to incur more radical transformations. In terms of input, the users will likely spend more time building Web queries in order to better convey their intent in terms of the Web's content, which will be processed at a semantic level (by relying on both automated methods to extract metadata and exploiting human made metadata where available). They will be guided in the process by constant interpretation of their query and feedback regarding the understanding of their intent. In return, the search presentation will adapt not only to the user's immediate retrieval need, but also to the task context, helping users to perform complex tasks with machine support at every step of the process. In that sense, we believe in personalizing the task at hand and not the user. This has two additional advantages: first, there is more data per task than per user, being then able to personalize more searches; and second, we move away of privacy issues as we do not need to know the user.

3   Tackling the Problem

Using NLP-based semantics to improve search is an old dream. Why do we think we can advance the state of the art in this area? Why now? We believe that although there has been extensive work in this problem, it has not been studied at the depth and scale that is required to achieve real improvement. In addition, important changes that occurred in the last ten years, makes this study today completely different. Why at Yahoo!? Partly because we have created a multidisciplinary team spanning from IR to NLP, machine learning (ML) and the semantic Web (SW).

In our view, we think three components are crucial for this:

1. Machine learning: we need more complex models. Often, researchers have tried to improve traditional search models with one or two NLP-derived features in simple combination schemes. We hope that by using much more complex models, including many types of NLP and semantic features, we can learn the appropriate features in each context. Machine learning techniques are now off-the-shelf, but were not in the past.
2. Data: we need to study cases where large amounts of annotated or semi-annotated data are available; otherwise we cannot use machine learning techniques effectively. In the past this volume of data was not available, with the Web 2.0, it exists now.
3. Tasks: we need to design the right tasks. Too easy and NLP is only a burden; too hard and the necessary inferences are beyond current NLP techniques. Finding the right difficulty is crucial to evaluate our improvement, and this element is the only one that has not really changed.

In Barcelona we have driven our research from the three points above. This means that setting up the problem is as hard as developing solutions for it. In fact we have tried and failed several times before finding a small number of tasks on which to concentrate our research. Today, some of these are complex question answering, algorithmic advertising and entity retrieval. In each of these areas we have tried to make use of a large number of techniques from NLP, IR and ML to find potentially interesting interactions.

Crucially, the realization that in the Web content changes and users adapt, highlights the importance of learning, which brings new conceptual elements to this scenario. In the last few years machine learning, together with an increased focus for on empirical experimentation typical of hard science, has re-shaped the landscape of several disciplines including information retrieval and natural language processing. As far as the former is concerned, learning has revolutionized the way ranking models are built. Within a learning framework ranking functions can be built including hundreds of complex features, rather than around hand-tuned functions based on small sets of features such as TF-IDF or PageRank. Notice that empirical optimization of ranking models can make a crucial difference in sensitive aspects of search technology such as Web advertising, particularly with respect to learning frameworks that can make immediate use of users-feedback in the form of clicks [10]). This kind of approach has the potential to impact semantic technologies directly because of the massive feedback loop involved.

Our initial efforts to improve search are based on shallow semantics. Ciaramita and Attardi [9] have shown that it is advantageous to combine syntactic parsing and semantic tagging in state-of-the-art frameworks. In fact, as a sub-product of this research, we have shared a semantically tagged version of the Wikipedia [3]. The next step is to rank information units of varying complexity and structure; e.g., entities [17] or answers [15], based on semantic annotations.

An example of this is our research in complex ("how") questions [15]. Standard Q&A concentrates on factoid questions (e.g. "when was Picasso born") where we can hope to create query templates and retrieve the exact answer. One would hope that the type of NLP technology that can effectively answers factoid questions could be used to improve answering broader questions, and ultimately improve search systems, but this has not been the case until now. In order to study this problem, we concentrated on "how" questions (e.g. "how does a helicopter fly?") because they are close to being broad questions, while at the same time they form a linguistically coherent set. By using the Yahoo! Answers social Q&A service, we were able to collect hundreds of thousands of question and answer pairs. This effectively gave us an annotated collection of questions on which to study many NLP features using ML techniques. In particular, we could use simultaneously unsupervised (parametrised similarity functions), class-conditional learning (translation models) and discriminant models (Perceptrons) on a wide range of features: from bag of words and part-of speech tags to WordNet senses, named entities, dependency parsing and semantic-role labelling [15].

The quality of user-generated content varies drastically from excellent to abuse and spam. As the availability of such content increases, the task of identifying high-quality content in social media sites becomes increasingly important. In [1] methods for exploiting such community feedback to automatically identify high quality content are investigated. In particular they focus on Yahoo! Answers, a large community question/answering portal that is particularly rich in the amount and types of content and social interactions available in it. They show that it is possible to separate high-quality items from the rest with an accuracy close to that of humans. In the case of the Flickr folksonomy, Sigurbjornsson and van Zwol [14] have shown how to use collective knowledge to enhance image tags, they also prove that almost 80% of the tags can be semantically classified by using WordNet and Wikipedia [13]. This effectively improves image search.

As shown by the examples above, machine learning in prediction, ranking and recommendation, has provided a framework for deploying new semantic representations based directly on users feedback. Proposed components can be evaluated and, if useful, added to the ranking model, although only temporarily, until something better emerges, in a natural selective loop. Several aspects of this framework need to be further investigated and better understood. One above all: the role of people and how this technology impacts their lives and satisfies their needs. Search technology and the Web have opened new channels of communication between people and between people and machines. The integration of massive people feedback and learning could lead to evolving "semantic models", which, hopefully, would not only improve applications but our understanding of communication and intelligence as well.

4   Capturing Implicit Semantics

We can distinguish two different types of semantic sources in the Web: explicit and implicit. In the previous section we have mentioned how we have used explicit sources of semantic information that are well categorized (*e.g.* Wikipedia) or that use folksonomies (*e.g.* Flickr). Implicit sources of semantic are raw Web content, and structure, as well as human interaction in the Web (what is called nowadays the Wisdom of Crowds [16]).

The main usage source are queries and the actions following their formulation. In [7] we present a first step to infer semantic relations from query logs by defining equivalent, more specific, and related queries, which may represent an implicit folksonomy. To evaluate the quality of the results we used the Open Directory Project, showing that equivalence or specificity had precision of over 70% and 60%, respectively. For the cases that were not found in the ODP, a manually verified sample showed that the real precision was close to 100%. What happened was that the ODP was not specific enough to contain those relations. So one main challenge is how to prove the quality of semantic resources if what we can generate is larger than any other available semantic resource and every day the problem gets worse as we have more data. This shows the real power of the wisdom of crowds, as queries involve almost all Internet users.

With respect to content, the amount of implicit semantic information available is only bounded by our ability to understand natural language and its relation to knowledge. Currently, search technology is moving slowly from tokens to words and to entities of different types (person names, companies, products, locations, date...). Such simple information already poses a challenge, as we have seen in the previous sections. This is only the beginning, and we hope that many of the richer semantic structures studied by NLP can be brought to help search. For example, we foresee applications in sentiment analysis, subjectivity analysis, genre classification, etc., to have an impact in search. Another implicit source of semantic information with potential impact in search is that of linguistic time expressions [2].

5   Conclusions

As we have seen, taxonomies as well as explicit and implicit folksonomies can be used to do supervised machine learning without the need of manual intervention (or at least by drastically reducing it) to improve automatic semantic annotation. In particular, SearchMonkey [2] is a strong initiative by Yahoo! to help this process by allowing people to mash up based on result metadata. Microsearch [12] is an early example of this: you can see the metadata in the search and therefore you are encouraged to add to it [3].

By being able to generate semantic meta-information automatically, even with noise, and coupling it with the open semantic resources as we have described, we plan to create a virtuous feedback circuit. In fact, one might take all the examples already given as one stage of the circuit. Afterwards, one could feedback the results on itself, and repeat the process. Using the right conditions, every iteration should improve the output, or at least keep it adaptively up-to-date with respect to the users needs, generating a virtuous cycle, and ultimately, better semantic search, our final goal.

References

1. Agichtein, E., Castillo, C., Donato, D., Gionis, A., Mishne, G.: Finding High-Quality Content in Social Media. In: First ACM Conference on Web Search and Data Mining (WSDM 2008), Stanford (February 2008)
2. Alonso, O., Gertz, M., Baeza-Yates, R.: On the Value of Temporal Information in Information Retrieval. ACM SIGIR Forum 41(2), 35–41 (2007)
3. Atserias, J., Zaragoza, H., Ciaramita, M., Attardi, G.: Semantically Annotated Snapshot of the English Wikipedia. In: Proceedings of the 6th International Conference on Language Resources and Evaluation (LREC) (2008), http://research.yahoo.com/node/1733

---

[2] See www.techcrunch.com/2008/02/25/yahoo-announces-open-search-platform/.
[3] In this demo there is a button next to every result called "Update metadata" which gives you instant feedback of what your metadata looks like.

4. Baeza-Yates, R., Ribeiro-Neto, B.: Modern Information Retrieval. ACM Press/Addison-Wesley, England (1999)
5. Baeza-Yates, R.: Challenges in the Interaction of Natural Language Processing and Information Retrieval. In: Gelbukh, A. (ed.) CICLing 2004. LNCS, vol. 2945, pp. 445–456. Springer, Heidelberg (2004)
6. Baeza-Yates, R., Mika, P., Zaragoza, H.: Search, Web 2.0, and the Semantic Web. In: Benjamins, R. (ed.) Trends and Controversies: Near-Term Prospects for Semantic Technologies; IEEE Intelligent Systems 23 (1), 80–82 (2008)
7. Baeza-Yates, R., Tiberi, A.: Extracting Semantic Relations from Query Logs. In: ACM KDD 2007, San Jose, California, USA, pp. 76–85 (2007)
8. Baeza-Yates, R., Calderón, L., González, C.: The Intention Behind Web Queries. In: SPIRE 2006. LNCS, pp. 98–109. Springer, Glasgow, Scotland (2006)
9. Ciaramita, M., Attardi, G.: Dependency Parsing with Second- Order Feature Maps and Annotated Semantic Information. In: Proceedings of the 10th International Conference on Parsing Technology (2007)
10. Ciaramita, M., Murdock, V., Plachouras, V.: Online Learning from Click Data for Sponsored Search. In: Proceedings of WWW 2008, Beijing, China (2008)
11. Lewis, D.D., Sparck-Jones, K.: Natural Language Processing for Information Retrieval. Communications of the ACM 39(1), 92–101 (1996)
12. Mika, P.: Microsearch demo (2008), http://www.yr-bcn.es/demos/microsearch/
13. Overell, S., Sigurbjornsson, B., Zwol, R.V.: Classifying Tags using Open Content Resources (submitted for publication) (2008)
14. Sigurbjornsson, B., Zwol, R.V.: Flickr Tag Recommendation based on Collective Knowledge. In: WWW 2008, Beijing, China (2008)
15. Surdeanu, M., Ciaramita, M., Zaragoza, H.: Learning to Rank Answers on Large Online QA Collections. In: Proceedings of the 46th Annual Meeting of the Association for Computational Linguistics: Human Language Technologies (ACL-HLT) (2008)
16. Surowiecki, J.: The Wisdom of Crowds. Random House, New York (2004)
17. Zaragoza, H., Rode, H., Mika, P., Atserias, J., Ciaramita, M., Attardi, G.: Ranking Very Many Typed Entities on Wikipedia. In: CIKM 2007: Proceedings of the sixteenth ACM international conference on Information and Knowledge Management, Lisbon, Portugal (2007)

From Databases to Natural Language: The Unusual Direction

Yannis Ioannidis*

Dept. of Informatics & Telecommunications, MaD*g*IK Lab
University of Athens, Hellas (Greece)
yannis@di.uoa.gr
http://www.di.uoa.gr/~yannis

Abstract. There has been much work in the past that combines the fields of *Databases* and *Natural Language Processing*. Almost all efforts, however, have gone in one direction: given unstructured, natural-language elements (requests, text excerpts, etc.), one creates structured, database elements (queries, records, etc.). The other direction has been mostly ignored, although it is very rich in application opportunities as well as in research problems. This short paper outlines several aspects of this other direction, identifying some relevant technical challenges as well as corresponding real applications.

Keywords: Databases, natural language processing, text synthesis.

1 Introduction

Any information systems environment, including one that involves database systems, may be abstracted as having three layers: its users, its functionality (realized through some form of a management system), and its content. Natural language may play some important role in the system functionality that is related to the two end layers: in the front-end, user interactions may be expressible in natural language; in the back-end, the original database content itself may actually be in natural language. Clearly, natural language is only one alternative available for each layer: in the front-end, user interactions may be (and usually are) performed using query languages, graphical or form-based user interfaces, and other dialog modes; likewise, in the back-end, content is usually structured or semi-structured, and is stored as relational or object-relational tables, XML or RDF documents, or some other form of objects.

In the past, there has been much research work on the interplay between natural languages and the other alternatives in each case above. Most of this work is on transforming natural language queries into some formal, structured query language, such as SQL. From very early on, researchers became excited about

---

* Partially supported by the European Commission under contract FP7-ICT-215874 "PAPYRUS: Cultural and Historical Digital Libraries Dynamically Mined from News Archives".

E. Kapetanios, V. Sugumaran, M. Spiliopoulou (Eds.): NLDB 2008, LNCS 5039, pp. 12–16, 2008.

this problem, and this has continued until today; there are numerous papers that address various aspects of the problem and several systems that have been developed to perform exactly such translations. Other user interactions, beyond queries (e.g., updates or constraints), have also been investigated in the past in the same way, although much more rarely.

More recently, there has been some significant activity on the content side as well. In particular, information extraction from natural-language text has been identified as an important area, and several efforts have been devoted into obtaining partial understanding of free text and generating database records, ontology relationships, or other (semi-)structured information that can be stored and manipulated by a database.

As one may easily observe from the above, most efforts of the past that combine the fields of *databases* and *natural language processing* have gone in one direction: given unstructured, natural-language items (requests, stored text documents, etc.), one creates structured, database items (SQL queries, relational records, etc.). The other direction has been mostly ignored, although it is very rich in application opportunities as well as in research problems. The two sections below motivate why the other direction is interesting as well. They describe some real applications, identify some relevant research problems, and outline some possible solutions. The first section deals with user interactions in the front-end, while the second section deals with the database contents, whether primary data inserted by users or metadata of various forms.

2 User Interaction Elements

Consider a schema with two tables:

$$\text{EMP}(\underline{\text{eid}},\text{sal},\text{age},\textit{did}) \text{ and } \text{DEPT}(\underline{\text{did}},\text{dname},\textit{mgr}),$$

where primary keys are underlined and foreign keys are in italics. Consider someone posing the following SQL query:

> **select** e1.name **from** EMP e1, EMP e2, DPT d
> **where** e1.did = d.did **and** d.mgr = e2.eid **and** e1.sal > e2.sal

There are several reasons why having the system provide a natural language interpretation of the query may be useful. Before the query is sent for execution, it may be nice for the user to see it expressed in a way that is most familiar, as verification that the query captures correctly the intended meaning. Seeing something like "Find the names of employees who make more than their managers" for the above query will be very helpful in making sure that this was indeed the user's original intention. The more complicated the query, the more important such feedback is.

In general, any situation where explanation of queries is warranted, such natural-language interpretation may be very useful and effective. For example, when a query returns an empty answer, it is nice to know the parts of the query that are responsible for the failure. Similarly, when a query is expected to return a very large number of answers, it is useful to know the reasons, in case a rewrite would reduce the number significantly and would serve the user better.

Clearly, the same can be said about all other commands a user may give to a database system. Insertions, deletions, and updates, especially those with complicated qualifications or nested constructs, will benefit from a translation into natural language. Likewise for view definitions and integrity constraints, which borrow most of their syntax from queries. Also, although the example above was in SQL, similar arguments can be made about Relational Algebra queries, RDF queries in SPARQL or RQL, even Datalog programs, and others. One can claim that novice users may benefit by natural-language specification of even queries posed by filling out a form. Especially for large forms, where a user is likely to not know the underlying semantic connections among the fields presented in the form, a textual explanation may come in handy.

Needless to say, offering the functionality described above is not trivial for complicated queries and other commands. Part of the complexity lies with the fact that there are several alternative expressions of a query in a formal language that are equivalent, based on associativity, commutativity, and other algebraic properties of the query constructs. Capturing the query elements in the right order so that the corresponding textual expression is natural and meaningful independent of the way the user has expressed the query is not straightforward. Similarly, expressing queries with complex embeddings or aggregations is hard. For example, for the same schema above, consider the following two queries:

> **select** dname **from** EMP e, DEPT d **where** e.did=d.did **groupby** did
> **having count(distinct** sal)=1 **and count(distinct** age)=1

> **select** e.name **from** EMP e **where** e.sal ≤ **all**
> (**select** sal **from** EMP **where** did = e.did)

For a system to recognize that a good way to express the meaning of these, relatively simple queries is with phrases like "Find the names of departments whose employees all have the same salary and the same age" and "Find the names of employees with the lowest salary in their department" is nontrivial. Identifying the correct use of pronouns is one source of difficulties. Another one is related to whether or not the natural-language expression will be declarative (as in the above two examples) or procedural, i.e., whether it will just specify what the query answer should satisfy or also the actions that need to be performed for the answer to be generated. The former is always desirable, but for complicated queries, the latter may be the only reasonable approach. Identifying the complexity point where this becomes the case, however, is far from understood, and work must be done on this.

3   Database Contents

Similar issues arise in the back-end, when database contents are considered for translation into natural language. For example, consider a database that follows the small schema presented earlier and suppose that one wants to have a textual description of its contents. If the database is large, it would make sense to create a textual summary of it, otherwise, a description of its full contents. There are several situations where such translation into natural language may be useful and desirable. Creating a short company description for a business plan or a bank-loan application or collateral material for marketing are some instances. Given other appropriate schemas, one can imagine textual descriptions in several other practical cases: a short description of a museum's exhibits, possibly customized to a visitor's particular interests; a brief history of a patient's medical conditions; the highlights of a collection in a digital library, with a few sentences on the main authors in the collection; a summary of a theater play in an information portal; and others.

Whatever holds for whole databases, of course, holds for query answers as well, especially those with some nontrivial structure, i.e., entire relational databases, complex objects, etc. Textual answers are often preferred by users, whether experienced or not, as they convey the essence of the entire query answer in an immediately understandable way. Moreover, they are critical for visually impaired or similarly disabled users, as they can be read to the user through a speech recognizer.

Clearly, the idea of translating data into natural language can be extended to all other forms of primary or derived data that a database may contain. Database samples, histograms, data distribution approximations are all, in some sense, small databases and can be summarized textually as above. Describing the schema itself, its basic entities, relationships, and other conceptual primitives offered by the model it is based on, is just a special case of a database description. User profiles maintained by the system for offering personalized answers, browsing indexes, and other forms of metadata are amenable to and may benefit from natural-language translation.

As with queries and other user-interaction elements, translating database contents to natural language is far from trivial. On one hand, it is simpler from translating queries, as the extent of alternative equivalent expressions of schemas and data is much narrower than that of queries. On the other hand, it is much more complicated than translating queries, as one has to choose the appropriate schema elements and data items that need to be captured in the textual summary. Furthermore, identifying the right linguistic constructs, introducing pronouns where appropriate, and synthesizing everything to produce a natural end result is equally complex. Although there has been some recent investigation of the topic, much more work is needed to devise an approach that is comprehensive, efficient, and effective at the same time.

4   Conclusions

In this short paper, we have looked into the intersection of the Database and Natural Language Processing areas and have outlined several interesting problems that arise when one attempts to translate database elements into natural language elements, i.e., going in the opposite direction than usual. We have offered example applications that indicate the practical usefulness of the problem, have identified several categories of database elements whose translation into text would be useful, and have briefly described some of the technical challenges that need to be addressed in the future. We hope that researchers will take up this type of problems and help to push this interesting area forward.

Natural Language Processing
and Understanding

Division of Spanish Words into Morphemes
with a Genetic Algorithm*

Alexander Gelbukh[1], Grigori Sidorov[1],
Diego Lara-Reyes[1] and Liliana Chanona-Hernandez[2]

[1] Natural Language and Text Processing Laboratory,
Center for Research in Computer Science, National Polytechnic Institute,
Av. Juan Dios Batiz, s/n, Zacatenco, 07738, Mexico City, Mexico
www.Gelbukh.com, www.cic.ipn.mx/~sidorov
[2] ESIME Zacatenco, National Polytechnic Institute,
Zacatenco, 07738, Mexico City, Mexico

Abstract. We discuss an unsupervised technique for determining morpheme
structure of words in an inflective language, with Spanish as a case study. For this,
we use a global optimization (implemented with a genetic algorithm), while most
of the previous works are based on heuristics calculated using conditional probabilities of word parts. Thus, we deal with complete space of solutions and do not
reduce it with the risk to eliminate some correct solutions beforehand. Also, we
are working at the derivative level as contrasted with the more traditional grammatical level interested only in flexions. The algorithm works as follows. The input data is a wordlist built on the base of a large dictionary or corpus in the given
language and the output data is the same wordlist with each word divided into
morphemes. First, we build a redundant list of all strings that might possibly be
prefixes, suffixes, and stems of the words in the wordlist. Then, we detect possible
paradigms in this set and filter out all items from the lists of possible prefixes and
suffixes (though not stems) that do not participate in such paradigms. Finally, a
subset of those lists of possible prefixes, stems, and suffixes is chosen using the
genetic algorithm. The fitness function is based on the ideas of minimum length
description, i.e. we choose the minimum number of elements that are necessary
for covering all the words. The obtained subset is used for dividing the words
from the wordlist. Algorithm parameters are presented. Preliminary evaluation of
the experimental results for a dictionary of Spanish is given.

1 Introduction

We present an application of a global optimization technique (implemented as a
genetic algorithm) to the task of division of words into morphemes using Spanish
language as a case study, i.e., our main goal is investigating the application of the
unsupervised technique for determining morpheme structure of words.

Word division into morphemes is useful for automatic description of morphological
structures of languages without existing morphological models and/or morphological

---

* Work done under partial support of Mexican Government (CONACYT, SNI) and National
Polytechnic Institute, Mexico (SIP, COFAA, PIFI).

E. Kapetanios, V. Sugumaran, M. Spiliopoulou (Eds.): NLDB 2008, LNCS 5039, pp. 19–26, 2008.

dictionaries. It is important for modern information retrieval technologies, when we are dealing with unknown languages or languages without complete description [5], or, possibly, for some specific terminological areas, say, medicine.

The prevalent approach is presented in [3] and it is implemented in *Linguistica* system. Variations of this method are described in [4], [7], and [1]. The main idea of this approach is to use heuristics for reduction of the algorithm search space. There are two principal heuristics. The first heuristics is related to the construction of the initial set of potential morphemes, which is based on their conditional probabilities. This procedure is iterative. For each word, one division with maximum weight is selected from all possible divisions. For calculation of these weights, the conditional probabilities are used under certain empirical assumptions that are not theoretically justified, for example, the assumption of Boltzmann distribution of probabilities is assumed. The procedure is repeated until the maximum weights are achieved. Then the minimum description length (MDL) technique is applied as a second heuristics to the set of possible signatures (potential paradigms) for their improvement and debugging.

There were two competitions of automatic division into morphemes. In 2005 [9], the abovementioned methods were applied to division of words in various languages (English, Finnish, Turkish); though Spanish was not considered. In 2007 [8], the task was changed to finding the aspects of morpheme meaning.

The idea of detection of repetitive sets of non-stem morphemes (potential paradigms) is vital for all methods. Its' result is also called *signature*, we still prefer more linguistic term *paradigm*; in our case, it is *derivational paradigm*. These terms refer to the fact that stems can be concatenated with sets of morphemes, and these sets are repeated across the vocabulary, for example, *high, highly, highness* and *bright, brightly, brightness* share the set {∅, *-ly*, *-ness*}. It is really surprising that usage of these sets greatly reduces the search space for all types of algorithms.

The idea of our approach is avoid using heuristics and apply a search of the global optimum in the space of all possible solutions. In our case, we implement it as a genetic algorithm. Thus, we deal with complete space of solutions and do not reduce it with the risk to eliminate some correct solutions beforehand. On the other hand, we are working at the derivative level, as contrasted with the more traditional grammatical level, where the interest is centered on flexions only. Spanish language has rather simple morphological structure, so no more than three possible word parts are considered.

There were attempts to apply genetic algorithm to this problem [2], [6]. Still, the results were not very promising because none of these methods took into account possible repetition of sets of morphemes (paradigms). In this paper, we modify the method adding this possibility.

The rest of the paper has the following structure. First, we describe the algorithm, then its' parameters and preliminary experimental results are presented, finally, conclusions are drawn and future work is discussed.

2 Algorithm

In this section we present description of the algorithm and discuss its' parameters used in the experiments.

2.1 Algorithm Description

The general scheme of the algorithm is presented in Fig. 1.

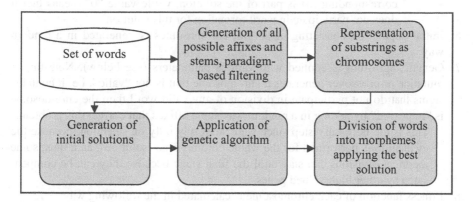

Fig. 1. Algorithm scheme

Detailed description of algorithm is as follows:

1. For each word, we detect all possible prefixes including the empty string ∅, starting from the first letter and finishing at the penultimate letter. The strings are added to the list of possible prefixes without repetitions along with their frequencies.
2. We prepare the list of possible stems, starting from the first letter to the last letter. The stem list contains unique elements with their corresponding frequencies and does not include the empty string ∅.
3. The list of possible suffixes is prepares taking the substrings starting from the last letter till the second letter from the beginning. This list contains the empty string ∅ and also does not permit repetitions. It also includes frequencies of suffixes.
4. All possible paradigms are detected for stems, i.e., the sets of morphemes that are repeated several times for various stems. For example, some paradigms are {NULL, *-ism, -iz, -idad*}, {NULL, *-ant, -acion*}. We filter out all non-stem substrings (potential morphemes) that are not part of some stem paradigm that contains more than one element (these sets can be considered also as the paradigms with only one element, but they are not really a paradigm, but a unique combination). Also, we filter out morphemes that belong only to paradigms with low frequency. In our case, we used values of frequencies of two and three.
5. Chromosomes (individuals) are formed according to the following rules:
    a. They are binary in a sense that their genes are represented as "0" and "1", thus, the chromosome is a sequence of genes.
    b. Each gene (binary position in chromosome) corresponds to a string from one of the lists,

    c. Thus, the length of the chromosome is the sum of the number of elements in the three lists.

    d. In the chromosome, value "1" means that the corresponding string of the corresponding list is part of the solution, while value "0" means that it does not participate in word formation for this solution.

6. Initial population consisting of several chromosomes is generated in a random way.

7. Genetic algorithm is applied with various parameters (see below). Note that if mutation or crossover generates a chromosome that is not "valid", i.e. it contains stems that do not participate in division of any valid word then the chromosome is "improved" by adding in a random way some affixes that correct this problem.

   This is the necessary step since we cannot rely only on selection, because the evaluation is performed for the whole chromosome (solution), and, thus, some incorrect divisions can stay until the final stage because they can be compensated by other high valued genes.

8. Fitness function of each chromosome is calculated in the following way:

    a. The number of the genes used in the solution (non-zero genes) is as small as possible;

    b. The combination of substrings used in the solution from all three lists covers major number of words from the initial vocabulary. All possible combinations of prefixes, suffixes and stems presented in the given chromosome are verified.

9. When the algorithm finishes, the best solution is used to divide words from the initial vocabulary into morphemes.

Traditional parameters of the genetic algorithm that should be mentioned are:

1. Selection procedures. Genetic operator of selection is executed using the tournament scheme, namely, for two randomly chosen individuals their fitness is compared and the best individual wins. This guarantees that the individuals are competing and that the better ones survive.

2. Crossover. This genetic operator is implemented using as its' parameter a number of blocks on the basis of which the chromosomes exchange their genetic information for creation of new individuals. The places for crossover are chosen randomly.

3. Replacement of individuals by the created ones. We used the elitist replacement scheme, when the best individuals are always conserved in the population.

4. Mutation. This genetic operator is important because it allows for creation of the new possibilities in the space of solutions. It changes values of the randomly chosen genes with certain probability.

2.2 Algorithm Parameters

We conducted several experiments with genetic algorithm and compared their results. We found out that the best results are obtained using three different sets of parameters consequently in three passes of the algorithm. Application of several passes is a common way of usage of genetic algorithms, when at each pass a specific purpose should be achieved.

We experimented within the following ranges of parameters, see Table 1.

Table 1. Ranges of parameters for the genetic algorithm

	Minimal values	Maximal values
Population size	50	5,000
Replacement	20%	100%
Mutation	Starting from 20%, reducing according to number of generations	Starting from 90%, reducing according to number of generations
Crossover	1	20
Generations	50	10,000

The following sets of parameters for the genetic algorithm give the best results applied in the consecutive manner (three passes) for population of 200 individuals; see Tables 2, 3 and 4.

Table 2. Parameters for first pass

Replacement	60%
Mutation	30%, reducing according to number of generations
Crossover	5
Generations	10,000

Table 3. Parameters for second pass

Replacement	80%
Mutation	80%, reducing according to number of generations
Crossover	20
Generations	7,000

Table 4. Parameters for third pass

Replacement	40%
Mutation	20%, reducing according to number of generations
Crossover	4
Generations	6,000

The purpose of each pass is different. At the first pass, the population is prepared and regularized. At the second pass, the population is shaken at the maximum grade. Finally, at the third pass, the populations should stabilize, for example, mutation rate is significantly reduced.

3 Experimental Results

We used as input data a Spanish dictionary that contained more than 20,000 head words. We ignored auxiliary words and adverbs, thus, working only with nouns, verbs and adjectives. Since we are interested in derivative morphology, we cut off their flexions, for example, *trabajar → trabaj- (to work), rojo → roj- (red)*, etc. Also, we treated the stressed vowels indistinctly to the non-stressed vowels because they have purely orthographic function in Spanish. The final input list contained 16,849 unique non-flexional words.

For the moment, for being able to perform a comparison with *Linguistica* system (J. Goldsmith, [3]), we made the experiments for suffixes only, though the algorithms permits simultaneous treatment of suffixes and prefixes.

The following results were obtained while preparing the lists of initial strings for the algorithm. We found 7,747 derivational paradigms, from which 6,472 contained more than one element. Thus, the paradigms that contained exactly one element were ignored. It is worth mentioning that of these 6,472: 1,852 paradigms contained zero string. Also we used a threshold for paradigm repetition, namely, we ignored paradigms that repeated less than three times, i.e., they exist for three words or less. There were 5,535 paradigms with frequency equal to 1; 404 with frequency equal to 2; and 171 with frequency equal to 3, etc. Finally, only 372 paradigms left for processing in the algorithm. They were used for filtering, i.e., only suffixes and stems that participated in them are used for representation of chromosomes. Finally, the algorithm worked with 17,085 stems and 136 suffixes. It is a really important reduction taking into account that initial number of stems was more than 44,000 and the same number of suffixes was more than 15,000.

We compared our results with the output produced by *Linguistica* system, giving it the same input. Unfortunately, we do not have the golden standard of our data for automatic evaluation. Manual evaluation of results shows that our system produced comparable division of words during the evaluation procedure described below.

The version of *Linguistica* that we have, accepts only 5,000 words as input. So we gave to both systems the same input: first 5,000 words from our dictionary. This number should be large enough because both systems use unsupervised learning. Then we evaluate manually the obtained divisions of the first hundred words. *Linguistica* has 87% of precision, while our system produces 84%. The systems have errors in different words. For example, *Linguistica* did not find the suffix *–mient(o)* that was rather frequent in the list and was detected by our system. Obviously, these values of precision correspond only to a preliminary estimation. These values may seem too high because the baseline of the existing systems is around 60%, but let us remember that we are dealing with derivative morphology and we are working with the dictionary, not with the corpus. The exact evaluation remains as a task for future work.

4 Conclusions and Future Work

We described an application of an optimization technique, namely, genetic algorithm, to the problem of division of word into morphemes. Our primary concern was derivative morphology and we made our experiments for Spanish language as a case study.

In the algorithm, the chromosome is constructed from a set of all possible substrings for stems and affixes filtered in a special way. The genes in the chromosomes are binary, when "1" means the presence of the element (stem or affix), and "0" corresponds to its' absence. We used as the filter the presence of a substring in paradigm with frequency greater than three. Also we ignored the paradigms that consisted of only one element.

Traditional genetic algorithm operators are applied to processing of chromosomes (=individuals) in populations.

Our results are comparable with more traditional techniques based on heuristics with calculations of conditional probabilities, but our method uses all space of solutions and do not filter out some possibly correct solutions beforehand.

For the moment, we are filtering out all non-stem substrings (potential morphemes) that are not part of some stem paradigm. In future, we would like to try different treatment of these paradigms, namely, including the paradigms in evaluation of fitness function or including them somehow directly in chromosomes. Also, we plan trying simultaneous treatment of suffixes and prefixes as parts of paradigms.

The future work is also related with performing the exact evaluation, though we are not aware of the golden standard for Spanish derivational morphology. We plan to develop this standard and use it for automatic evaluation.

References

[1] Baroni, M., Matiasek, J., Trost, H.: Unsupervised discovery of morphologically related words based on orthographic and semantic similarity. In: ACL Workshop on Morphological and Phonological Learning (2002)

[2] Gelbukh, A., Alexandrov, M., Han, S.: Detecting Inflection Patterns in Natural Language by Minimization of Morphological Model. In: Sanfeliu, A., Martínez Trinidad, J.F., Carrasco Ochoa, J.A. (eds.) CIARP 2004. LNCS, vol. 3287, pp. 432–438. Springer, Heidelberg (2004)

[3] Goldsmith, J.: Unsupervised Learning of the Morphology of a Natural Language. Computational Linguistics 27(2), 153–198 (2001)

[4] Creutz, M.: Unsupervised Segmentation of Words Using Prior Distributions of Morph Length and Frequency. In: Proceedings of the 41st Annual Meeting of the Association for Computational Linguistics (ACL 2003), Sapporo, Japan, pp. 280–287 (2003)

[5] Haahr, P., Baker, S.: Making search better in Catalonia, Estonia, and everywhere else. Google blog (2007), http://googleblog.blogspot.com/2008/03/making-search-better-in-catalonia.htm

[6] Kazakov, D.: Unsupervised learning of naïve morphology with genetic algorithms. In: Workshop Notes of the ECML/MLnet Workshop on Empirical Learning of Natural Language Processing Tasks. Prague, Czech Republic, pp. 105–112 (1997)

[7] Rehman, K., Hussain, I.: Unsupervised Morphemes Segmentation. In: Pascal Morphochallenge, 5p. (2005)

[8] Pascal Morphochallenge (2007), http://www.cis.hut.fi/morphochallenge2007/

[9] Pascal Morphochallenge (2005), http://www.cis.hut.fi/morphochallenge2005/

Abbreviation Disambiguation: Experiments with Various Variants of the One Sense per Discourse Hypothesis

Yaakov HaCohen-Kerner, Ariel Kass, and Ariel Peretz

Department of Computer Science, Jerusalem College of Technology (Machon Lev)
21 Havaad Haleumi St., P.O.B. 16031, 91160 Jerusalem, Israel
kerner@jct.ac.il, {ariel.kass,relperetz}@gmail.com

Abstract. Abbreviations are very common and are widely used in both written and spoken language. However, they are not always explicitly defined and in many cases they are ambiguous. In this research, we present a process that attempts to solve the problem of abbreviation ambiguity. Various features have been explored, including context-related methods and statistical methods. The application domain is Jewish Law documents written in Hebrew, which are known to be rich in ambiguous abbreviations. Various variants of the one sense per discourse hypothesis (by varying the scope of discourse) have been implemented. Several common machine learning methods have been tested to find a successful integration of these variants. The best results have been achieved by SVM, with 96.09% accuracy.

1 Introduction

One of the attractive research sub-domains in natural language processing (NLP) is the word sense disambiguation (WSD) problem [13]. Word sense disambiguation is the task of assigning to each occurrence of an ambiguous word in a text one of its possible senses. A sub-domain of WSD is resolution of ambiguous abbreviations.

Abbreviations are widely used either in writing or talking in many languages. At times, disambiguation of an abbreviation is critical to the understanding of the sentence where it appears [14]. Abbreviation is a letter or sequence of letters, which is a shortened form of a word or a sequence of words. The word or sequence of words is called a long form of an abbreviation. Abbreviation disambiguation means to choose the correct long form (sense) while depending on its context. In general, the process of abbreviation disambiguation is composed of two steps: finding all possible relevant senses and selecting the most correct sense. The various researches attempted to create human-like computational and decision processes for specific datasets.

In NLP systems, research usually focuses on either creating or using grammatical tools (e.g.: tokenizer and tagger), which detect different parts of sentences and the grammatical function and usage of different words from the immediate context in the text. These systems are built uniquely for a specific language or datasets.

Jewish Law documents written in Hebrew are known to be rich in ambiguous abbreviations. Therefore, these documents can serve as an excellent test-bed for development of models for disambiguation of ambiguous abbreviations. In addition, Hebrew is the mother tongue of the authors of this paper.

E. Kapetanios, V. Sugumaran, M. Spiliopoulou (Eds.): NLDB 2008, LNCS 5039, pp. 27–39, 2008.
© Springer-Verlag Berlin Heidelberg 2008

As opposed to the documents investigated in previous systems, Jewish Law documents usually do not contain the long form of abbreviations in the same discourse. That is, the relevant possible senses are not found in the text. They are manually entered to a dictionary built in the pre-processing stage.

Therefore, the abbreviations are regarded as more difficult to disambiguate. To determine the accuracy of the system, all the instances of the ambiguous abbreviations were solved beforehand. Some of them were based partially on published solutions and some of them were manually solved by experienced readers.

In this research, it is assumed that all abbreviations are correct. The main goal is to ease the understanding of passages in the text, by correctly and accurately solving the long form of contained abbreviations, thus shifting the focus to the text and not to the abbreviations. The secondary goal, in case the main goal is not achieved, is to propose a reduced set of long forms for the abbreviations, based on several grading methods.

Various kinds of people (e.g.: people who learn Hebrew in general, immigrants, children and people who need to read and learn documents related to a specific professional domain) need help to disambiguate abbreviations correctly. They experience great difficulty understanding the meaning of the running text. Many times they do not know what the possible extensions are or part of them and/or they cannot find the most correct extension.

This research defines features, as well as experiments with different implementations of the one sense per discourse hypothesis (by varying the scope of discourse). The developed process considers other languages and does not define pre-execution assumptions. The only limitation is the input itself: the languages of the different text documents and the man-made solution database inputted during the learning process, limit the datasets of documents that may be solved by the resulting disambiguation system.

Characterization of contextual and statistical attributes of abbreviations and experimentation with learning methods on these characteristics was performed. Exploitation of connections between the attributes may be used to significantly improve the performances of a disambiguation system.

The system designed in this research, preserves its portability between languages and domains because it does not use any NLP sub-system (e.g.: tokenizer and tagger). In this matter, the system is not limited to any specific language or dataset. The system is only limited by the different inputs used during the system's learning stage and the set of abbreviations defined.

This paper is organized as follows: Section 2 gives background concerning abbreviations in general, abbreviations in Hebrew, abbreviation disambiguation and previous systems dealing with automatic disambiguation of abbreviations. Section 3 describes the features for disambiguation of Hebrew abbreviations. Section 4 presents the various implementations of the one sense per discourse hypothesis (by varying the scope of discourse). Section 5 describes the supervised machine learning experiments that have been carried out. Section 6 concludes, summarizes and proposes future directions for research.

2   Background

Abbreviations are very common and are widely used in both written and spoken language. However, they are not always explicitly defined and in many cases they are ambiguous. Abbreviation disambiguation means to choose the correct long form depending on its context. Disambiguation of abbreviations is critical for correct understanding of not only abbreviations themselves but also of the entire text.

An automatic abbreviation dictionary, called S-RAD, has been created automatically for biomedicine by Adar [2]. Liu et al. [15] made statistical studies concerning three-letter abbreviations in MEDLINE[1] abstracts. They conclude that these abbreviations are highly ambiguous and automatic disambiguation can be achieved for abbreviations that occur frequently with their definitions.

Other statistics-based implementations were built by Okazaki and Ananiadou. In [17] they build an abbreviation dictionary using a term recognition approach, achieving 99% precision and 82-95% recall on a corpus that emulates MEDLINE. In [18] they presented a method for clustering long forms in biomedical text for disambiguation of acronyms.

2.1 Abbreviations in Hebrew

Hebrew in general is very rich in its vocabulary of abbreviations. Almost all of the abbreviations in Hebrew are easy to discover by the occurrence of a double-quote character in a token. The number of Hebrew abbreviations is about 17,000 [3] not including unique professional abbreviations, relatively high comparing to 40,000 lexical entries in the Hebrew language. About 35% of them are ambiguous, that is, about 6000 abbreviations have more than one possible long form. Moreover, Jewish Law articles written in Hebrew-Aramaic include a high rate of abbreviations.

The causes for the use of abbreviations in the Hebrew language vary. Some may be attributed to historical and religious origins: poverty, religious principle of short recitation to pupils, religious restriction of erasing complete quotes from the bible, shorthand and mnemonic devices.

Two examples of sentences in Hebrew that include ambiguous abbreviations are presented below (Hebrew is read from right to left). Their translation to English (based on [11]) is also presented where **X** presents the discussed abbreviation. The explanation for the correct sense of these abbreviations is also given.

(1) "כ"כ קדושה לו אין לצרכו בביתו מדרש הקובע יחיד אבל ...",

which means in English: "... However, if someone established a place for studying Torah in his home for his own use, it does not have **X** holiness."

כ"כ (**X**) has 3 different senses within the checked Jewish Law articles. The correct sense is "so much". This sense can be found according to the word "However" which indicates that the sense of the abbreviation is not as holy as a public place for such studies, which is mentioned at the beginning of the sentence.

(2) "בנחת אם כי במרוצה הזמירות א"א",

which means in English: "**X** the songs rapidly, but at leisure".

א"א (**X**) has 8 different senses within the checked Jewish Law articles. The correct sense is "One should not say". This sense can be found according to the words "but at", which indicate that the sense of **X** is something negative.

---

[1] MEDLINE is a bibliographical database with references to journal articles in medicine, nursing and dentistry.

2.2 Abbreviation Disambiguation

The one sense per collocation hypothesis was introduced by Yarowsky [28]. This hypothesis states that natural languages tend to use consistent spoken and written styles. Based on this hypothesis, many terms repeat themselves with the same meaning in all their occurrences. Within the context of solving the long form of abbreviations, it may be assumed that authors tend to use the same words in the vicinity of a specific long form of an abbreviation. The words may be reused as indicators of the proper solution of an additional unknown abbreviation with the same words in its vicinity. This is the basis for all contextual features defined in this research.

The one sense per discourse hypothesis was introduced by Gale et al. [7]. This hypothesis assumes that in natural languages, there is a tendency of an author to be consistent in the same discourse or article. Based on this hypothesis, if in a specific discourse, an otherwise ambiguous phrase or term has a specific meaning, any other subsequent instance of this phrase or term will have the same specific meaning. Within the context of solving the long form of abbreviations, it may be assumed that authors tend to use an abbreviation with a specific long form throughout a discourse. However, this hypothesis does not always apply to all domains.

Pustejovsky et al. [24] developed an abbreviation disambiguation system named ACROMED, using a weight scheme called ATC used as the standard in the SMART database [26]. The dataset used for this system, contained 10 medical abstracts from MEDLINE. The dataset contained additional 42 abstracts, used to validate the system's accuracy. This second set, did not contain any possible long form of the abbreviation. The system solved a single abbreviation, SRF, which had 10 different solutions in the dataset. The system achieved 97.62% accuracy.

Pakhomov [21] developed an automatic abbreviation disambiguation system for medical abstracts, using the Maximum Entropy machine learning (ML) method. The dataset contained about 10,000 medical abstracts, from the Miu Medical Institute database. The system solved 2 sets of abbreviations: (1) 6 ambiguous abbreviations, by setting a single solution for each article. These abbreviations had an average of 9 different senses in the dataset and (2) a set of 69 abbreviations in the dataset. Each abbreviation in this set had between 100 and 1,000 instances in the dataset.

This system was implemented based on 2 methods - "context of 2 words before and after the abbreviation" as well as "context of paragraph containing the abbreviation". For the first set of abbreviations, the system achieved 89.14% and 89.66% accuracy for each method. For the second set, the system achieved 89.12% and 89.02% accuracy for each method.

Yu et al. [30] developed an automatic abbreviation disambiguation system for medical abstracts, using LIBSVM [5]. The system was implemented based on the method "context of 2 words before and after the abbreviation". In addition, this system used the one sense per discourse hypothesis. The system achieved about 84% and 87% accuracy, in two separate datasets. This was an improvement of 2%, compared to the system results without relying on the above hypothesis.

Gaudan et al. [8] developed an automatic abbreviation disambiguation system for human gene symbols in biological literature, using SVM. The dataset contained the entire MEDLINE abstracts database (August 2004). The system solved a set of 1851 ambiguous abbreviations. These abbreviations had an average of 3.4 different senses in the dataset.

Their system was implemented based on the method "most similar article context". The contextual terms used for the disambiguation were extracted using the C-value algorithm [6], a method combining linguistical (adjective–noun patterns) and statistical features. Each article contained a single sense for each abbreviation, relying on the one sense per discourse hypothesis. The system achieved 98.5% accuracy. However, the researched abbreviations were only those that have been defined more than 40 times in the whole of MEDLINE.

HaCohen-Kerner et al. [12] developed an automatic abbreviation disambiguation system for Jewish Law documents written in Hebrew, without the use of any ML method. The system combined six different methods. The first three methods focused on unique characteristics of each abbreviation, e.g.: common words, prefixes and suffixes. The following two methods were statistical methods based on general grammar and literary knowledge of Hebrew. The final method was a Hebrew specific method that used of the numerical value (explanation in [12]) of the abbreviation. The system solved 50 abbreviations with an average of 2.3 different senses in the dataset. The system achieved about 60% accuracy.

Pakhomov et al. [22] developed an automatic abbreviation disambiguation system for medical abstracts, using C5.0 and Maximum Entropy. The dataset contained about 1.7 million medical notes, from the Miu Medical Institute database, search results from the internet and abstracts from MEDLINE. The system solved 8 ambiguous abbreviations. These abbreviations had an average of 9 different senses in the dataset.

The system was implemented in two parts: (1) only medical notes from the Miu data, (2) combinations of the different subsets of the dataset. The first part was implemented based on the "bag-of-words" method. The system achieved 93.90% and 95.80% accuracy, with the C5.0 and Maximum Entropy methods, accordingly.

The second part was implemented based on the "bag-of-words" method for the +- 20 word vicinity of the abbreviation and the "bag-of-words" method for all the text in the snippets of the internet search results. The system achieved 62.14% and 67.82% accuracy using the C5.0 and Context Vector method [23].

3 Features for Abbreviation Disambiguation

Ambiguous abbreviations have several senses, but within a specific context, such as a sentence, abbreviations usually have only one specific sense. To find the correct sense of any abbreviation, 18 different features were defined. Each feature is described as a rule and implemented as a baseline method in the developed system. These methods are divided into 3 distinct groups: Statistical attributes (2 methods), Hebrew specific attributes (1 method) and Contextual relationship attributes (15 methods). The third group is distinguished from the others, as it aspires to understand the underlying meaning of the abbreviation in its context. Methods in this group tried to answer questions like: Does the abbreviation prefix help understand the abbreviation? Does a specific word appear in the context of a specific abbreviation solution?

Statistical attributes
1.   **Writer Common Rule (WC):** The most common solution used for the specific abbreviation by the writer [29].
2.   **Dataset Common Rule (DC):** The most common solution used for the specific abbreviation in the entire dataset.

Hebrew specific attributes
3.   **Gimatria Rule (GM):** The numerical sum of the numerical values attributed to the Hebrew letters forming the abbreviation. This is known as the Gimatria of a word, as part of the field of word Numerology. It is important to state that authors tend to use abbreviations, numbers and letters as reference to chapters, pages and other resources.

Contextual relationship attributes
In all the following rules, all words and attributes of a sentence (including grammatical symbols such as period, comma, etc.) are used to determine the context of the specific abbreviation. The rules also take count of the attribute instances within the learning dataset, as a weighted decision node for the deduction process.
4.   **Prefix Counted Rule (PRC):** The selected sense is the most commonly appended sense by the prefix.
5-8. Before K (1,2,3,4)^2 Words Counted Rule (BKWC): The selected sense is the most commonly preceded sense by the K words in the sentence of the abbreviation.
9-12. After K (1,2,3,4) Words Counted Rule (AKWC): The selected sense is the most commonly succeeded sense by the K words in the sentence of the abbreviation.
13.   Before Sentence Counted Rule (BSC): The selected sense is the most commonly preceded sense by all words in the sentence of the abbreviation.
14.   After Sentence Counted Rule (ASC): The selected sense is the most commonly succeeded sense by all words in the sentence of the specific abbreviation instance.
15.   All Sentence Counted Rule (AllSC): The selected sense is the most commonly surrounded sense by all words in the sentence of the specific abbreviation instance.
16.   Before Article Counted Rule (BAC): The selected sense is the most commonly preceded sense by all words in the article of the specific abbreviation instance.
17.   After Article Counted Rule (AAC): The selected sense is the most commonly succeeded sense by all words in the article of the specific abbreviation instance.
18.   All Article Counted Rule (AllAC): The selected sense is the most commonly surrounded sense by all words in the article of the specific abbreviation instance.

4   Implementing the One Sense per Discourse Hypothesis

As mentioned above, the basic assumption of the one sense per discourse hypothesis is that there exists at least one solvable abbreviation in the discourse and that the sense of that abbreviation is the same for all the instances of this abbreviation in the

---

^{2} The value of k has been set at 4 to relate to 4 words before the abbreviation and 4 words after it. This is because 9 is accepted as the number of items that the average person is able to remember without apparent effort, according to the cognitive rule "7±2" [16].

discourse. The correctness of any feature was investigated based on this hypothesis for several variants of "One Sense" (OS) based on the discussed discourse: none, a sentence, an article or the entire dataset.

1. Individual correctness (No OS) – this is the "normal" case in which the sense of each abbreviation instance is deduced based on the feature alone (i.e.: without use of "one sense").
2. Sentence correctness (osS) – the sense deduced for one instance of an abbreviation is used for all instances of the abbreviation in the sentence.
3. Article correctness (osA) – the sense deduced for one instance of an abbreviation is used for all instances of the abbreviation in the article.
4. Writer correctness (osW) – the sense deduced for one instance of an abbreviation is used for all instances of the abbreviation, used by the writer in all his articles.

In this research, the OS hypothesis was implemented in two forms. The "pure" form (with the suffix S/A/W without C) uses the sense found by the majority voting method for an abbreviation in the discourse and applies it "blindly" to all other instances.

The "combined" form (with the suffix C) tries to find the sense of the abbreviation using the discussed feature only. If the feature is unsuccessful, then we use the relevant OS variant using the majority voting method. The use of the OS hypothesis, in both forms, is only relevant for context based methods, since the solutions by other methods are static and identical from one instance to another.

Therefore, for each of the 15 context based methods, six variants of the hypothesis were implemented. This produces 90 variants, which together with the 18 features in their normal form, results in a total of 108 variants. In addition, the ML methods were experimented together with the OS per discourse hypothesis. Of the 108 possible variants, for the 18 features, the best variant for each feature was chosen. In each experiment, the next best variant is added, starting from the 2 best variants up until all 18 variants.

5 Experiments

5.1 Data Sets

The examined dataset includes texts that were written by two Jewish scholars: Rabbi Yisrael Meir HaCohen [10] and Rabbi Ovadia Yosef [19, 20]. All documents belong to the same domain: Jewish Law Documents.

The entire dataset includes 564,554 words where 114,814 of them are abbreviations instances, and 42,687 of them are ambiguous. That is, about 7.5% of the words are ambiguous abbreviations. These ambiguous abbreviations are instances of a set of 135 different abbreviations. The average number of senses for each abbreviation is 3.27. Each one of the abbreviations has between 2 to 8 relevant possible senses.

As mentioned before, the relevant possible senses usually are not found in the text. They are manually entered to a dictionary built in the pre-processing stage. Most of the correct senses of the ambiguous abbreviations in the texts written by Rabbi Yisrael Meir HaCohen are found already in [10]. The correct senses of the ambiguous abbreviations in the other texts were determined by experienced readers.

5.2 Results of the Variants of "One Sense per Discourse" Hypothesis

The results of the all OS variants, for all the features (Section 3) are presented in Table 1. These results are obtained without using any ML methods.

Table 1. The results of the OS per discourse hypothesis for all the features

# of method	Method	Accuracy percentage % for various applications of OS						
		No OS	osS	osSC	osA	osAC	osW	osWC
1	PRC	33.67	34.41	34.52	52.77	54.54	66.66	71.04
2	B1WC	56.05	56.41	56.61	67.74	71.84	72.93	82.51
3	B2WC	55.72	56.23	56.35	69	72.34	74.85	82.84
4	B3WC	60.54	60.89	61.01	72.67	75.48	75.44	82.86
5	B4WC	64.49	64.72	64.85	74.29	76.5	75.52	82.2
6	BSC	75.21	75.18	75.24	76.85	78.15	74.92	78.52
7	BAC	76	76	76	76.01	76	75.39	76
8	A1WC	**78.79**	79.01	79.21	78.72	83.81	76.32	**87.75**
9	A2WC	77.57	78.07	78.26	79.15	83.43	78.54	87.62
10	A3WC	78.64	79.11	79.28	79.61	83	78.19	85.8
11	A4WC	75.44	79.28	79.5	79.41	82.42	78.01	84.99
12	ASC	78.59	78.61	78.62	78.25	78.94	77.37	79.04
13	AAC	75.44	75.44	75.44	75.34	75.44	77.28	75.44
14	AllSC	77.97	77.97	77.97	77.9	78.02	77.22	78.04
15	AllAC	74.12	74.12	74.12	74.12	74.12	76.93	74.12
16	GM	46.82	46.82	46.82	46.82	46.82	46.82	46.82
17	WC	**82.84**	82.84	82.84	82.84	82.84	82.84	82.84
18	DC	78.34	78.34	78.34	78.34	78.34	78.34	78.34

On the one hand, the two best pure features were WC (most common solution in the dataset) with 82.84% accuracy and A1WC (the most commonly succeeded sense by the word that comes after the abbreviation instance) with 78.79% accuracy. The first finding shows that about 83% of the abbreviations have the same sense in the whole dataset. The second finding shows that about 79% of the abbreviations can be solved by the first word that comes after the abbreviation.

On the other hand, the PRC (abbreviation prefix) and GM (numerical value) features achieved the worst results, with 33.67% accuracy and 46.82% accuracy, accordingly. The analysis of these results revealed that for most cases these two methods did not produce any solution and only the minority of the solutions was actually wrong. By reviewing the abbreviation instances for which the methods do not produce a solution, it is apparent that this is simply because the instances did not have a prefix or that the instance was not properly formed, based on the rules of numerology in the Hebrew Language. This proves that these attributes of the abbreviations are not sufficient properly solve ambiguous abbreviations, but may improve the results of the other methods defined by the system.

Furthermore, the results of the GM feature prove that almost 50% of all abbreviation instances represent a numerical value, not a sequence of words. It may be necessary to augment the text domain to properly deduce clearer conclusions concerning the attributes of abbreviations in the Hebrew Language.

From the results of the contextual relationship attributes, it is clear that the features, based on the context that comes after the abbreviation instance, i.e. A1WC, A2WC, A3WC, A4WC (1-4 words immediately after the abbreviation), ASC (all the words after the abbreviation in the sentence) and AAC (all the words after the abbreviation in the text), achieve considerably better results than their counterparts, based on the context that comes before the abbreviation context. Furthermore, the results for both sets of methods do not fall below 75% accuracy. It may be deduced that there is a strong relationship between the different solutions, for each abbreviation, and the surrounding words, forming a related context for each instance. These results suggest that each abbreviation has a stronger relationship to the words after a specific instance, than to all other combinations of words.

The merging of the features, as implemented by the AllSC and AllAC methods, did not contribute to the system results, but actually reduced the accuracy rates achieved by each individual methods set. This may imply that the relationship to the words before or after each specific abbreviation instance is stronger than the relationship to the combined word context of the same instance.

Finally, an improvement within the set of methods, concerning words before the abbreviation is visible. This improvement is achieved by adding the number of words included in the contextual relationship consideration. On the other hand, no such improvement or reduction in accuracy levels is apparent for the set of methods, concerning words after the abbreviation. This may imply that the strong relationship between a specific abbreviation instance and the words after it is only relevant for the first word after the instance. This may further imply that the relationship between the abbreviation instance and the words before the instance strengthens as a result of increasing the contextual relationship range.

As for the use of the OS hypothesis, the A1WC_osWC method variant achieves the best result with 87.75% accuracy. Many feature variants did not achieve substantially better results than the results achieved by the normal form of the features. This may be attributed to the fact that many abbreviation instances may be unique to the different discourses, therefore not able to benefit from the use of OS, in both of its forms.

Almost every feature has at least one variant that achieves a substantial improvement in results compared the results achieved by the method in its normal form. The average relative improvement is about 18%.

For all features, except for BAC, the best variant uses the OS implementation with the discourse defined as the entire dataset. This may be attributed to the similarity of the different articles defined in the dataset. This is supported by the fact that the best feature, in its normal form, is the WC method.

In addition, for all but three methods (BAC, AAC, AllAC), the best variant used the combined form of the OS implementation. This is intuitively understandable, since "blindly" overwriting probably erases many successes of the feature in its normal form.

5.3 The Results of the Supervised ML Methods

Several well-known supervised ML methods have been selected: artificial neural networks (ANN) [1], Naïve Bayes (NB) [9], Support Vector Machines (SVM) [4] and

J48 (an improved variant of the C4.5 decision tree induction [25]). These methods have been applied with default values and no feature normalization mainly using Weka[3] [27]. Tuning is left for future research. To test the accuracy of the models, 10-fold cross-validation was used. The results of these supervised ML methods, combining the best 18 feature variants (Table 1) incrementally, as explained above, are shown in Table 2.

Table 2. The results while using the best variant for each feature, incrementally

# of variants	Feature variants	ML method			
		ANN	NB	SVM	J48
2	A1WC_osWC + A2WC_osWC	91.56	91.40	94.29	91.94
3	+ A3WC_osWC	91.72	91.42	94.43	92.20
4	+ A4WC_osWC	91.75	91.51	94.43	92.34
5	+ B3WC_osWC	92.68	92.11	95.33	93.33
6	+ WC	92.95	**92.16**	95.71	93.54
7	+ B2WC_osWC	92.81	91.79	95.67	93.59
8	+ B1WC_osWC	92.91	91.06	95.68	93.56
9	+ B4WC_osWC	92.83	91.15	95.62	93.55
10	+ ASC_osWC	92.83	91.10	95.60	93.52
11	+ BSC_osWC	92.95	91.17	95.65	93.58
12	+ DC	92.98	91.17	95.63	93.58
13	+ AllSC_osWC	92.82	91.50	95.63	93.58
14	+ AAC_osW	92.84	91.42	95.59	93.58
15	+ AllAC_osW	93.10	91.43	95.77	93.58
16	+ BAC_osA	93.09	91.28	95.79	93.70
17	+ PRC_osWC	93.25	91.50	**96.09**	93.71
18	+ GM	**93.28**	91.52	96.02	**93.93**

From the above results it apparent that the SVM method is the most successful, with the best result of 96.09% accuracy, and a worst result of 94.29% accuracy which is better than the best result for any of the other ML methods. On the other hand, the NB method achieved the worst results, with a best result of 93.28% accuracy. Nonetheless, the use of any of the ML methods achieves better results than any one of the feature variants. The best improvement, by use of ML methods, is about 13%, from 82.84% accuracy for the best variant of any feature to 96.02% accuracy for the best variant for SVM.

The comparison of the SVM results to the results of previous systems, described in Section 2.2, shows that our system achieves relatively high accuracy percentages. However, most other developed systems researched ambiguous abbreviations in the English language, as well as different abbreviations and texts.

---

[3] Weka is a collection of machine learning algorithms programmed in Java for data mining tasks, such as: classification, regression, clustering, association rules, and visualization.

Notwithstanding, in this research, the number of abbreviations and their instances, are considerably higher than most of the other systems developed. This system is the only one that applies many variants of the OS per discourse hypothesis. In addition, we performed a comparison between the achievements of different ML methods, to the goal of achieving the best results, as opposed to the other systems that only focused on one ML method, each.

6 Conclusions, Summary and Future Work

The comparison to other system shows that our system achieves relatively high accuracy percentages. However, this comparison has several reservations: (1) Almost all the other developed systems researched ambiguous abbreviations in the English Language, as opposed to our system that researched abbreviations in the Hebrew Language. (2) The other developed systems researched a different set of abbreviations and texts. It is understandable that the average number of instances and average number of solutions, for each abbreviation, varies between the different systems. (3) It is not clear if the experiments will have similar results in other languages. In particular, the Gimatria rule (GM) is exclusive to Hebrew.

Notwithstanding, in our system, the number of abbreviations and their instances, are considerably higher than most of the other systems developed. In addition, we performed a comparison between the achievements of different ML methods, to the goal of achieving the best results, as opposed to the other systems that only focused on one method, each. In addition, most of the systems, which achieved very close results to the developed system, researched a small number of abbreviations. The system developed by Gaudan et al. [8] researches a considerable number of abbreviations, as well as achieves a high accuracy percentage, although their system uses NLP-based methods which are not language independent. Moreover, the system contains the long form of the abbreviations in the same discourse, in contrast to the proposed system.

Therefore, it is not possible to clearly deduce that the developed system in this research is better than the other developed systems and it is further not possible to deduce that the ambiguous abbreviation problem in the Hebrew Language is easier to solve than in other languages.

On the other hand, this cannot downsize, on any scale, the high accuracy percentages achieved by the developed system in this research. In addition, the system achieves about 96% accuracy, which is an improvement of about 36% compared to the basic system developed by us [12].

In summary, this is the first system for disambiguation of abbreviations in Hebrew, specifically in Jewish Law documents, that uses the OS per discourse hypothesis and ML methods.

The developed system researched a large set of abbreviations (135), more than most of the previously developed systems.

High accuracy percentages were achieved, with improvement ascribed to the use of OS hypothesis and variant combining with ML methods. These results were achieved without the use of NLP methods and are not bound to any of them. Therefore, these methods are not limited to any single language.

The developed system is adjustable to any specific type of texts, simply by changing the database of texts and abbreviations.

18 features were defined, augmented by 90 variants created by use of OS, as well as 4 different ML methods were used and compared.

Extending the implementation of the OS hypothesis to several types of discourses, as well as use of both the pure form and the combined form of the hypothesis, considerably improves the achievements of the system.

An improvement of 13% was achieved due to the use of OS and the SVM method, a greater improvement that the one presented in an additional abbreviation disambiguation system in English for medical abstracts.

Future research directions are: comparison between individual assessments of the two examined authors, i.e. training at one author and testing at the other, definition and implementation of NLP-based features and use of these methods interlaced with the already defined methods, applying additional ML methods, such as Boosting and Maximum Entropy, as well as use of additional transform kernels defined in SVM, augmenting the databases with articles from additional datasets in the Hebrew Language and in other languages, tuning the parameters of the ML methods, applying unsupervised learning algorithms, where manual solutions are not available as well as developing algorithms for detection and correction of faulty or erroneous abbreviations.

References

1. Abdi, H., Valentin, D., Edelman, B.: Neural networks. Sage, Thousand, Oaks (1999)
2. Adar, E.: S-RAD: A Simple and Robust Abbreviation Dictionary. Technical Report, HP Laboratories (2002)
3. Ashkenazi, S., Jarden, D.: Ozar Rashe Tevot: Thesaurus of Hebrew Abbreviations (in Hebrew). Kiryat Sefere LTD., Jerusalem (1994)
4. Cortes, C., Vapnik, V.: Support-Vector Networks. Machine Learning 20, 273–297 (1995)
5. Chang, C., Lin, C.: LIBSVM: a Library for Support Vector Machines. Software in Python (2001), http://www.csie.ntu.edu.tw/~cjlin/libsvm
6. Frantzi, K., Ananiadou, S.: The C value domain independent method for multiword term extraction. JNLP 6(3), 145–179 (1999)
7. Gale, W., Church, K., Yarowsky, D.: One Sense per Discourse. In: Proceedings of the 4th DARPA speech in Natural Language Workshop, pp. 233–237 (1992)
8. Gaudan, S., Kirsch, H., Rebholz-Schuhmann, D.: Resolving Abbreviations to their Senses in Medline. Bioinformatics 21(18), 3658–3664 (2005)
9. Good, I.J.: The Estimation of Probabilities: An Essay on Modern Bayesian Methods. MIT Press, Cambridge (1965)
10. Hacohen, Y.M.: Mishnah Berurah (in Hebrew). Hotzaat Leshem, Jerusalem (1995)
11. Hacohen, Y.M.: Mishnah Berurah. English Translation, Pisgah Foundation. Feldheim Publishers, Jerusalem (1990)
12. HaCohen-Kerner, Y., Kass, A., Peretz, A.: Baseline Methods for Automatic Disambiguation of Abbreviations in Jewish Law Documents. In: Vicedo, J.L., Martinez-Barco, P., Munoz, R., Noeda, M.S. (eds.) EsTAL 2004. LNCS (LNAI), vol. 3230, pp. 58–69. Springer, Heidelberg (2004)
13. Ide, N., Véronis, J.: Word Sense Disambiguation: The State of the Art. Computational Linguistics 24(1), 1–40 (1998)

14. Joint Commission on Accreditation of Healthcare Organizations: Medication errors related to potentially dangerous abbreviation. Sentinel Event Alert 23 (2001)
15. Liu, H., Aronson, A.R., Friedman, C.: A Study of Abbreviations in MEDLINE Abstracts. In: Proc AMIA Symp., pp. 464–469 (2002)
16. Miller, G.A.: The Magical Number Seven, Plus or Minus Two: Some Limits on our Capacity of Information. Psychological Science 63, 81–97 (1956)
17. Okazaki, N., Ananiadou, S.: Building an Abbreviation Dictionary using a Term Recognition Approach. Bioinformatics 22(24), 3089–3095 (2006)
18. Okazaki, N., Ananiadou, S.: Clustering Acronyms in Biomedical Text for Disambiguation. In: Proceedings of fifth international conference on Language Resources and Evaluation (LREC), pp. 959–962 (2006)
19. Ovadia, Y.: Yechave Daat (in Hebrew). Chazon Ovadia, Jerusalem (1977)
20. Ovadia, Y.: Yabia Omer (in Hebrew). Chazon Ovadia, Jerusalem (1986)
21. Pakhomov, S.: Semi-Supervised Maximum Entropy Based Approach to Acronym and Abbreviation Normalization in Medical Texts. Association for Computational Linguistics (ACL), pp. 160-167 (2002)
22. Pakhomov, S., Pedersen, T., Chute, C.G.: Abbreviation and Acronym Disambiguation in Clinical Discourse. In: American Medical Informatics Association Annual Symposium, pp. 589–593 (2005)
23. Pedersen, T., Patwardhan, S., Michelizzi, J.: WordNet: Similarity - Measuring the Relatedness of Concepts. In: Proceedings of the 9th National Conference on Artificial Intelligence, pp. 1024–1025 (2004)
24. Pustejovsky, J., Castano, J., Cochran, B., Kotecki, M., Morrell, M., Rumshisky, A.: Extraction and Disambiguation of Acronym-Meaning Pairs in Medline (unpublished manuscript) (2001)
25. Quinlan, J.R.: C4.5: Programs For Machine Learning. Morgan Kaufmann, Los Altos (1993)
26. Salton, G.: The SMART Information Retrieval System: Experiments in Automatic Document Processing. Prentice Hall, Englewood Cliffs (1971)
27. Witten, H., Frank, E.: Weka 3.4.12: Machine Learning Software in Java(2007), http://www.cs.waikato.ac.nz/~ml/weka
28. Yarowsky, D.: One Sense per Collocation. In: Proceedings of the Workshop on Human Language Technology, pp. 266–271 (1993)
29. Yu, H., Hripcsak, G., Friedman, C.: Mapping Abbreviations to Full Forms in Biomedical Articles. J. Am. Med. Inform. Assoc. 9(3), 262–272 (2002)
30. Yu, Z., Tsuruoka, Y., Tsujii, J.: Automatic Resolution of Ambiguous Abbreviations in Biomedical Texts using SVM and One Sense per Discourse Hypothesis. In: SIGIR 2003 Workshop on Text Analysis and Search for Bioinformatics (2003)

The Acquisition of Common Sense Knowledge by Being Told: An Application of NLP to Itself

Fernando Gomez

School of EECS
University of Central Florida, Orlando, FL 32816
Harris Center
gomez@eecs.ucf.edu

Abstract. This paper shows how the knowledge of a semantic interpreter can be bootstrapped for other semantic interpretation tasks. Methods are described for automatically acquiring common sense knowledge and for applying this knowledge to noun sense disambiguation. Ordinary concepts are described by several plain English sentences that are parsed and semantically interpreted. The semantic interpreted sentences are stored under these concepts to be used for semantic interpretation tasks. This paper explains the description of the concepts, the interpretation of the sentences and two algorithms for noun sense disambiguation that use the acquired knowledge.

1 Introduction

In [2], a method for defining verb predicates and an implemented algorithm [4] that resolves the verb predicate, its semantic roles and adjuncts is explained. The approach has proceeded first by defining verb predicates for WordNet verb classes [1], and then by defining verb predicates for individual verbs with a high degree of polysemy. The semantic roles of the predicates are linked to the selectional restrictions (categories in WordNet ontology for nouns) and to the grammatical relations that realize them. The selectional restrictions of the predicates are grounded on the WordNet ontology for nouns [10], whose upper level ontology has been modified and rearranged [3] by using the feedback obtained by testing the predicate definitions. The resulting work is immense: over 3000 verb predicates have been built. A corpus of over 3000 semantically interpreted sentences has been automatically created with the semantic interpreter. We have used the semantic interpreter in order to extend it in several important ways. There are several semantic tasks that the interpreter does not solve or solves them partially. One of them is noun sense disambiguation (NSD). In many cases, the selectional restrictions of the verb predicates cannot resolve the noun senses of the semantic roles, although it may narrow them down to a small set of possible senses. For instance, in a sentence such as "The batter dropped the bat" the selectional restrictions for the predicate of *drop-something-physical* select the first sense of "batter." They also rule out the third sense of "bat," (a turn batting in baseball), which has *activity* as its hypernym. But, the semantic interpreter

E. Kapetanios, V. Sugumaran, M. Spiliopoulou (Eds.): NLDB 2008, LNCS 5039, pp. 40–51, 2008.

cannot decide between the other five senses of "bat," which are all physical things, because *physical-thing* is the selectional restriction for the *theme* of *drop-something-physical*. The output of the semantic interpreter for that sentence is depicted in Figure 1. The interpreter groups the noun senses by ontological category: bat1 is baseball bat, and bat5 is a cricket bat, which have *equipment* as their upper-level ontological category. The noun senses are WordNet senses. Concepts without number senses correspond to Gomez's WordNet upper-level ontology.

```
((SUBJ ((DFART THE) (NOUN BATTER)) ((BALLPLAYER1 BATTER1)) (AGENT))
 (VERB DROPPED ((MAIN-VERB DROP DROPPED)) DROP-SOMETHING-PHYSICAL
       (DROP1 DROP2) SUPPORTED BY 2 SRS)
 (OBJ ((DFART THE) (NOUN BAT)) ((EQUIPMENT BAT1 BAT5) (PLACENTAL1 BAT2)
      (INSTRUMENTALITY BAT4 BAT6)) (THEME)))
```

Fig. 1. Output for "The batter dropped the bat"

In the sentences, "Farmers like plants," "The doctor removed the appendix in an operation," "Several demonstrators were injured in the demonstration," none of the senses of "plant," "doctor" "appendix," and "operation" can be ruled out on the basis of selectional restrictions, or by similarity in the WordNet taxonomy. However, most humans will tell us that, in those sentences, "plant" is a life form, and not a industrial plant, or an actor, etc., and that "appendix" is not the appendix of a book, but an animal body part, that "doctor" refers to a physician and not to a theologian, that "operation" is a surgical operation, and not a military operation or a business operation, etc. Humans determine these senses using commonsense knowledge which is as basic as that used in the selectional restrictions of verb predicates. The aforementioned examples do not provide much context, but sufficient to establish the senses of the nouns. The aim of this research is to let the semantic interpreter acquire this knowledge by being told and use it for NSD, prepositional attachment, discourse, etc.

The glosses in dictionaries have been used for noun sense disambiguation [6]. There have been also efforts to produce some kind of logic form transformation of WordNet glosses [11]. However, many of the glosses in WordNet and in dictionaries do not lend themselves to having knowledge extracted from them by a program, except in a superficial manner, because they require much knowledge to be understood. We understand many of the dictionary glosses because we already know a lot about the concepts expressed in them. A serious problem with the glosses is that lexicographers approach them as succinct definitions of concepts intended to capture their essential aspects. As a result of this, in many instances the language of glosses contains many intensional terms [13] that do not indicate how the words are used [15] in ordinary language. For instance, the WordNet gloss for cell2 is "the basic structural and functional unit of all organisms." The meaning of the terms "basic structural unit" and "basic functional unit" do not lend themselves to be easily acquired by a program because of their abstract content: two abstract adjectives predicated of an abstract noun, "unit." It is unlikely that a child will be introduced to the concept of cell2 that way. However, a sentence such as "All living things are made of cells" provides a better introduction to the concept of cell2 because the concept "living thing," as well as that of cell2 have well established denotations to physical objects. Hence, it becomes easy to understand sentences such as "Trees are made of cells," or compound nouns such as "plant cells" by subsumption between the concepts in the new sentences and the concepts in the defining sentence, or gloss. Of course, we will need other defining sentences, or glosses, to provide the basic aspects of the concept of cell2. Likewise, the WordNet gloss for demonstration3, "a public display of group feelings," is very good if one already knows the meaning of demonstration3. But, the gloss is not very helpful if one is acquiring the concept for the first time. Instead of that definition, we prefer glosses that indicate the events normally associated with demonstration3 e.g., "In a demonstration, humans gather in streets or plazas to protest," "Sometimes in a demonstration, people may be hurt, or physical objects damaged," "A riot may commence during a demonstration" etc. In summary, stay away from intensional terms and do not try to convey many aspects of a concept in one single sentence and assume that you are conveying these concepts to someone who does not know anything about them except their ontological classification as provided by WordNet. In this paper, we show how to provide glosses for WordNet word senses, how to parse and semantically interpret them and, then, use them for noun sense disambiguation. In the next section, we explain the definition of the glosses and their semantic interpretation. In sections 3, we deal with the relation of verb selectional restrictions and noun senses. In sections 4, and 5, we explain the algorithms for noun sense disambiguation. Sections 6 provides the testing of the algorithms, and sections 7 and 8 explain related research and conclusions, respectively.

2   Acquiring Common Sense Knowledge about Ordinary Concepts

There is no difference between a noun sense and a concept. Ambiguous words stand for various concepts, or noun senses. Thus, learning about a new sense is not different from learning about concepts. What we are describing in this paper is the acquisition of basic knowledge about concepts, and its application to an aspect of semantic interpretation, namely noun sense disambiguation. But, this knowledge can be used for other aspects of semantic interpretation, or for other applications. We have used the word "glosses" to refer to the sentences describing some of the main aspects of a concept, or noun sense. The acquisition is as simple as typing some sentences. Suppose that one wants the system to acquire some basic knowledge about pot4, "a container in which plants are cultivated." One types (acquire pot4) to lock the senses of "pot" to pot4, and one starts typing some sentences, one at a time. For instance, one may type: "People cultivate plants in a pot," "Soil is put in pots," "People plant plant parts and plants in a pot," etc. In those cases in which the system is unable to determine the senses of some of the nouns in the gloss (henceforth, GL), the user may tell the system the correct senses by typing (*refer-to sense*$_1$, *sense*$_2$...*sense*$_i$). One line fixes all senses at once. But in most cases, the system is able to determine the senses of the nouns in the glosses. For instance, out of the three glosses for pot4, the system only fails to determine the sense of "soil." For some concepts, three or four sentences are sufficient for providing the main aspects of that concept, while other concepts may need more sentences. The system parses and interprets most sentences in one or two seconds. Next, one provides glosses for the other senses of "pot." Glosses for concepts which do not denote physical objects are provided in the same way. For instance these are some of the glosses for operation7 (a surgical operation). "Doctors performed operations," "In an operation, doctors operate on humans or animals," "In an operation, some body parts are removed, or replaced," "Some diseases may be only cured by means of an operation," "Some patients may die as a result of an operation," "Most operations take place in hospitals or clinics." This set of glosses provides a frame-like or script-like type of knowledge.

3   Verb Selectional Restrictions and Noun Senses

If one enters the sentence "She ate the dates with a fork," our interpreter will select (edible-fruit1 date8), and (cutlery2 fork1). The verb predicate, the noun senses and the semantic role are all solved. But, there are many sentences for which the semantic interpreter cannot solve the noun senses based on the selectional restrictions for the semantic roles. In a framework in which the semantic interpreter is activated first to determine verb meaning (verb predicate) and semantic roles, the following cases caused by noun polysemy may occur:

- (a) The semantic interpreter resolves the verb predicate and the semantic roles, but it cannot decide on some noun senses.
- (b) The semantic interpreter is unable to narrow the verb predicates to one.
- (c) The interpreter comes up with more than one semantic role for the same grammatical relation.

The sentence, "He put the batter in the refrigerator," is a good example of case *(a)*. The verb predicate and semantic roles are solved, but the semantic interpreter cannot decide between the two senses of "batter," a baseball player or a flour mixture. That would not be the case for the sentence, "The batter cleaned/fixed the refrigerator."

```
[to-loc(xor natural_elevation1 peak5 physical-thing) (obj)
        (conveyance3) ((prep aboard))
        (xor natural_elevation1 peak5  physical-thing)((prep up to))
        (physical-thing)((prep onto into))]
```

Fig. 2. A definition of the role *to-loc* for one of the predicates of "climb"

An example of case *(b)* is the sentence, "The runners are raised in a nursery for one growing season." The interpreter comes up with the following verb predicates for "raise" in order of preference: RAISE-FARM (farm plants and/or animals), BRING-UP (to educate somebody), RAISE-SOMETHING (to lift something). The semantic roles for these verb predicates are all the same, namely, *theme, at-loc,* and *duration.* The noun senses selected for the *theme* of RAISE-FARM are runner8 (a fish) and runner5 (a stolon, a plant part). The noun senses selected for the *theme* of BRING-UP are runner1, runner2, runner3, runner4 runner6, all of which have *person* as their hypernym. The noun senses selected for the *theme* of RAISE-SOMETHING are all the 8 senses of "runner." The noun senses selected for the roles *at-loc* and *duration* are the same for all the verb predicates. The sense for "nursery" (a room for a baby, and a place to cultivate plants) cannot be resolved by the semantic interpreter in any of the predicates. "Growing season" is not ambiguous.

For an example of case *(c)* consider the sentence "A swell lurched the catamaran towards the reef." For this sentence, the semantic interpreter resolves the verb predicate for "lurch," but it cannot decide between the roles, *agent* and *inanimate-cause* (a causal agent other than a human, a social group, or an animal). For the *agent*, the interpreter selects the sense swell4 (a dandy, a person) and for the *inanimate-cause* the senses swell1 (a wave) and swell2 (a natural elevation1).

One of the difficulties in defining selectional restrictions for semantic roles is staying away from over-generalization and over-specification of the ontological categories in the selectional restrictions. Over-specification results in failing to identify some semantic roles, or over-narrowing the noun senses; while over-generalization results in not selecting between different noun senses and/or not distinguishing between verb senses, or verb predicates. Figure 2 depicts the definition of the role *to-loc* for one of the predicates of the verb "climb" when it means traveling upwards. The syntax for the roles is:

(role (slr) (grs) (slr) (grs) ... (slr) (grs))

Where *slr* stands for any number of selectional restrictions, and *grs* for any number of grammatical relations. When the semantic role is realized by prepositions, the prepositions are put in a list preceded by the word "prep." If there is more than one selectional restriction, the list must be preceded by "xor" or "xand". These entries indicate how the items in the list should be matched. In Figure 2, the first sublist contains the selectional restrictions (WordNet ontological categories) for the grammatical relation, direct object. Selectional restrictions are always matched from left to right. The entry "xor" in the sublist means that as soon as an ontological category in the sublist is matched to the head noun of the grammatical relation, the others ontological categories are not tried. Thus, for the sentence, "She climbed the hill," the system will select hill1, a natural elevation, for the *theme.* The entry "xand" in the sublist means that all ontological categories in the sublist are matched, but the senses are preferred in the order in which they are matched. Thus, if the role *to-loc* were defined using "xand," the senses selected for "hill" in "She climbed a hill" would be hill1, hill2 (mound4) and hill3 (mound1).

The definitions for the semantic roles aim at striking a balance between over-specification and over-generalization. In constructing the selectional restrictions for a semantic role using an "xor" entry, one needs to be very careful with those words whose senses range over several ontological categories in the list. Consider the sentence, "She climbed the table," and the *to-loc* role defined with an "xor." The system will select the sense table4 ("flat tableland with steep edges") which is natural elevation in WN, and will exclude the other senses of "table," which may not be correct. Suppose that one writes the following ontological categories for the *instrument* of "kill" when it means *cause-to-die*:

   (instrument(xor weaponry1 external-body-part1 physical-thing) ((prep with)))

Assume that the sentence to be interpreted is "Samson killed the lion with the arms." That definition will select the sense of arm3 (a weapon) for "arm" and exclude its other senses.

Now, consider the sentence, "The dish is made of chicken." Humans have no trouble in determining that "dish" is not dish1 (a piece of dishware) or dish5 (an antenna etc.), but dish2 (an item of prepared food). The distinction between dish1 and dish2 is a difficult one because we put food in dish1 or dish2. Similarity in the WN taxonomy does not help. Solving this problem by using selectional restrictions will require a very specific predicate, *made-of-food*. However, we can tell the system that "Dishes (dish2) are made of food," then parse and interpret the sentence, and use the interpreted sentence for deciding between the senses of "dish." The interpreted sentence can be inherited by all hyponyms of dish2. Consider the sentence "The bowl contains punch." No selectional restrictions or similarity in the WN taxonomy will help in this sentence either. However, the simple gloss "Dishes (dish1) contain food," will solve the senses of "bowl" and "punch" because glosses are inherited by all hyponyms of the synset where they are stored. Consequently, that gloss will also handle many other sentences such as "The plate contains punch," etc. Other pertinent sentences for "dish1" are "People serve food on dishes," "Dishes are made of metal, plastic, or ceramic ware," "People put dishes in dishwashers to clean them," etc. All those sentences take about a minute to teach to the system.

4 Algorithm for Determining Noun Senses by Subsumption of Verb Predicates and Semantic Roles

We have designed and implemented several algorithms to resolve noun senses based on the interpreted glosses. Some of the algorithms base their decisions only on the noun senses in the interpreted glosses, while others determine the noun senses of new sentences by using the verb predicate and the semantic roles of the interpreted glosses. For space limitations, we concentrate only on the latter algorithms.

Suppose that we have defined the following gloss for refrigerator1, a monosemous word, "People put food in refrigerators." Then, the system encounters the sentence, "He put the batter in the refrigerator." Let us refer to the sentence being interpreted as IS. As explained, the selectional restrictions for put1 cannot determine the sense of "batter" in the IS. The interpreter's output for the *theme* of the IS (the sentence being interpreted) is : *(obj ((dfart the) (noun batter)) ((ballplayer1 batter1) (concoction1 batter2)) (theme)).* The *theme* of the gloss is: *(obj ((noun food)) ((food food1)) (theme)).* The interpreter cannot decide between the two senses of "batter," and for that reason it prints both senses. The algorithm for noun sense disambiguation obtains all glosses for all noun senses in the IS (the sentence being interpreted). Then, the algorithm (see Figure 3) implemented in Lisp performs the following steps for each GL (gloss). The algorithm will establish the sense of "batter" as batter2, by verifying first that the verb predicate of a GL (gloss) subsumes the verb predicate of the IS (the sentence being interpreted) or that the verb predicate of the IS (the sentence being interpreted) and the verb predicate of the GL (gloss) belong to the same verb predicate hierarchy (a relaxed criterion).

Let LNOUNS-IS be a list containing all noun senses in the IS (the sentence being interpreted). Let L-GL be the list of all interpreted glosses for each of the noun senses in LNOUNS-IS. Let IS-VERB-PRED be the verb predicate of the IS (the sentence being interpreted).
While (L-GL is not empty) do:
Let FIRST-GL be the first gloss in L-GL. If the verb predicate of FIRST-GL subsumes the IS-VERB-PRED or they belong to the same hierarchy, append the output of **Role-Subsumption** (FIRST-GL, IS) to ANSWER. Delete FIRST-GL from L-GL.
End **While**.
Compute FINAL-ANSWER: Obtain the Longest-Sublist in ANSWER (the one with the greatest number of roles). If there is more than one list with the same number of roles in Longest-Sublist, return all noun senses in each role as the answer.
Function Role-Subsumption (FIRST-GL, IS).
Let TEP1, TEP2, TEP3, CORRESP-IS-ROLE, ANSWER, be temporary variables.
Let IS-ROLES be a list containing the roles of the IS (the sentence being interpreted).
Let ROLES-FIRST-GL be a list containing the roles of the FIRST-GL.
While (ROLES-FIRST-GL is not empty) do:
Let ROLE-GL be the first role in ROLES-FIRST-GL. Search for GL-ROLE in IS-ROLES and if it is found assign it to the variable CORRESP-IS-ROLE. If it is not found, assign NIL to CORRESP-IS-ROLE.
Assign to TEP1 the noun sense of ROLE-GL. Assign to TEP2 the noun senses of CORRESP-IS-ROLE.
Assign to TEP3 all noun senses in TEP2 subsumed by the noun sense in TEP1. Append TEP3 to ANSWER. Delete the first role from ROLES-FIRST-GL. Set TEP3 to NIL.
End **While**
Return ANSWER End **Function Role-Subsumption**

Fig. 3. Algorithm1 - Determining Noun Senses by Subsumption of Verb Predicates and Semantic Roles

If that is not the case, the algorithm discards that GL (gloss) and gets the next GL (gloss) if any. In this example, the predicate of the GL (gloss) subsumes the predicate of the IS (the sentence being interpreted).

Then, the algorithm checks if the noun sense in the semantic roles of the GL (gloss) subsumes the noun sense, or senses, in the semantic roles of the IS (the sentence being interpreted). The comparison is done *agent* of the GL (gloss) against *agent* of the IS (the sentence being interpreted), *theme* of GL (gloss) against *theme* of the IS (the sentence being interpreted), and so on and so forth. There is only a noun sense in the roles of the GL (gloss), because the noun senses of the glosses are resolved, while the roles of the IS (the sentence being interpreted) may have one or more noun senses.

Thus, for the aforementioned example we can verify that: *1)* the noun sense of the *agent* of the GL (gloss) subsumes the noun sense of the *agent* of the IS (the sentence being interpreted), *2)* the noun sense of the *theme* of the GL (gloss), namely *food1*, subsumes one of the noun senses of "batter," namely batter2, in the *theme* of the IS (the sentence being interpreted), and *3)* the noun sense of the *goal* of the GL (gloss) subsumes the noun sense of the *goal* in the IS (the sentence being interpreted).

As a result, the algorithm returns the following: *(refrigerator1 put (agent she) (theme batter2) (goal refrigerator1))*. Had the sentence been "He put the bass/shrimp in the refrigerator," the algorithm would perform likewise for "bass," and "shrimp" (the first sense of "shrimp" in WordNet 1.6 is small person). However, in the case of "bass" the algorithm will output two senses for "bass," bass4 (sea bass) and bass5 (freshwater bass) because both senses have *food* as their hypernym in WN. The algorithm outputs all the roles in the IS (the sentence being interpreted) subsumed by the roles in each GL (gloss).

Simple glosses such as "People put food in a refrigerator1/oven1/pot1/dish1 ..." go a long way in covering many noun senses for which no gloss has been defined. Consider the sentence "In the operation, the appendix was removed." Even if there is no gloss for appendix2 (a body part), the following gloss for operation7 "In an operation, some body parts are removed" will resolve not only the sense of "operation" as operation7 (surgical operation) but also the sense of "appendix" as appendix2. Likewise, the gloss "Sometimes in a demonstration, demonstrators may be hurt," will solve the sense of "demonstrator" as demonstrator3 and "demonstration" as demonstration3. The gloss for cell2, "All living things are made of cells," will resolve not only the sense of "cell" as cell2 in "Plants are made of cells," but also that of "plant" as plant2. The gloss under plant2, "Plant parts become plants," will solve the sense of "plant" as plant2 and "runner" as runner5 in "The runner became a new plant." The examples can be easily multiplied.

Algorithm1-B
This algorithm is a relaxation of algorithm1. In the cases in which the verb predicate of the GL (gloss) and the IS (the sentence being interpreted) do not belong to the same predicate hierarchy, algorithm1-B finds subsumptions only between the semantic roles of the GL (gloss) and the IS (the sentence being

Let Head-Noun-NP be the head noun of the NP and Head-Noun-PP be the head noun of the object of the PP in [NP PP].
Collect all glosses for the noun senses of Head-Noun-PP having a *goal* role, and put them in the variable GLS-GOAL.
While GLS-GOAL do
Let FIRST-GL be the first gloss in GLS-GOAL
If the head noun of the *theme* in FIRST-GL subsumes any of the senses of Head-Noun-NP and the head noun of the *goal* in FIRST-GL subsumes any of the senses of Head-Noun-PP, put all senses subsumed into the variable TEP1. Insert TEP1 into ANSWER. Set TEP1 to NIL.
Delete the first GL (gloss) from GLS-GOAL.
end **While** Return ANSWER.

Fig. 4. Algorithm2 - Resolving Noun Senses in [NP PP] segments by Reasoning with *Theme* and *Goal* Roles

interpreted). Its results are taken in account if at least two semantic roles in the IS (the sentence being interpreted) are subsumed.

5   Algorithm for Resolving Noun Senses in [NP PP] Segments by Reasoning with Theme, Goal and At-Loc Roles

This section explains how to use the interpreted glosses for resolving the senses of the head nouns of prepositional phrases modifying noun phrases. Consider the sentence, "The plants in the pot vanished," or "The batter in the refrigerator vanished." The algorithm explained in the previous section cannot help to determine the sense of "plant," or "batter" in these examples. However, the semantic roles in some of the glosses for pot4 and refrigerator1 can be used to solve the senses of "plant," "pot" and "batter" in a very decisive manner. The algorithm (see Figure 4) works as follows. Suppose that the [NP PP] to be interpreted is "The batter in the refrigerator," and that we have a gloss stored under refrigerator1 that says, "People put food in refrigerators." When a preposition that may stand for the semantic role *goal* (e.g., "into," "onto," "in," "on") follows a NP, the glosses for the noun senses of the head noun of the PP are searched and all those that have a *goal* role are collected. In our example, all glosses for "refrigerator" are searched, and those with a *goal* role are collected. Then for each GL (gloss) with a *goal* role, the algorithm verifies: *a)* if the head noun of the *theme* of the GL (gloss) subsumes any of the senses of the head noun of the NP in the [NP PP] segment ("batter" in our example) and *b)* if the *goal* of the GL (gloss) subsumes any of the senses of the head noun of the object of the PP ("refrigerator" in the example). If both *a)* and *b)* are true, the algorithm returns all noun senses subsumed by the *theme* and *goal* of the gloss. For this example, the output is (ALGORITHM2= (BATTER2 REFRIGERATOR1)).

Another version of this algorithm reasons with *theme* and *at-loc* roles. This algorithm is activated when the preposition in the [NP PP] segment may stand for an *at-loc* role, namely the prepositions "in" "on" and "at," "near," "along," "outside," etc. Suppose that we want to determine the sense of "bay" in "The ship in the bay." Let us further assume that we have stored the gloss "Vessels are found in bays" under bay1 (a body of water). This algorithm collects all glosses for the senses of "bay" that have an *at-loc* role, and is identical to the one in Figure 4 if one replaces *goal* with *at-loc*. This algorithm will be able to clearly determine the different senses of "bay" in "the ship in the bay" and "the bay (compartment) in the ship," given a gloss for bay4 (a ship compartment).

6 Testing

We have tested the algorithms (algorithm1, algorithm1-b, algorithm2) in sentences most of them taken from the (*The World Book Encyclopedia*, World Book, Inc. Chicago) and some from the BNC corpus. First, we looked into the Senseval-2 dataset which contains 29 words. Unfortunately, this data set contains few examples of cases to which the algorithms explained in this paper apply. However, many of the examples in the Senseval-2 dataset can be solved by other algorithms based on our methods, and explained in a forthcoming paper. In that paper, we show how the sense of "conductor" in the sentence "But in the twentieth century, conductors have replaced composers as the most influential people in musical life" (BNC corpus) can be solved by using only the noun senses in the interpreted glosses. The reason why the sense of "conductor" in that sentence cannot be solved by these algorithms is because the verb predicate for "replace" does not determine the sense of "conductor." The same applies to the sentence "The conductor fired the violinist." However, the situation is very different if the verb is "direct."

The purpose of this test has been to show that the algorithms explained in this paper solve noun senses which will be very hard to solve by other algorithms. We have defined and semantically interpreted 1235 glosses for 706 noun senses, or concepts. We have chosen 34 words (*conductor, striker, operation, arm, nail, table, rally, plant, chair, cell, beam, blow, dish, spring, pot, bed, ball, colony, sign, bat demonstrator, plot, crane, port, pen, star, paddle, mast, article, dam pocket, coat, bay, cabin*), and searched for sentences with those words. Some of the sentences were formed by students while we were testing the system. The sentences were parsed and semantically interpreted, and, then, the algorithms were applied. The algorithms were tested in 120 relevant sentences, identifying correctly the target noun sense in 87%. The algorithms were unable to identify the noun sense correctly, or they came up with more than one sense for the target noun in 13% of the 120 sentences. Most of the failures can be corrected by adding some basic glosses to some noun senses. Other problems are due to the semantic interpreter, or implementation. We have built a small file[1] containing 100 sentences for which the algorithms found the correct sense for the target noun. All 34 words are represented in the file. First we list the target word followed by colon, and, then, the test sentence for that word, followed by the

---

[1] www.cs.ucf.edu/~gomez/nounsenses-disambiguation

output of the algorithms. The file also contains some comments explaining the output. These are two sample sentences for the word "port":

```
(p "they make port in Portugal from grapes")
(ALGORITHM1=
 ((PORT2 MAKE-OR-CREATE-SOMETHING (THEME PORT2) (OF-STUFF GRAPE1) )))
Comment: port2 is inheriting glosses from wine1. No glosses have been
provided for port2.

(p "they brought the grapes to the port in a boat")
(ALGORITHM1= ((PORT1 BRING-THINGS (THEME GRAPE2 GRAPE1) (GOAL PORT1)
 (INSTRUMENT BOAT1) )))
```

7   Related Work and Discussion

This work falls within the knowledge-based approaches to NSD [8]. Most work on NDS has been based on semantic similarity in the WordNet ontological taxonomy [14,9]. A major distinctive aspect of our work is that the algorithms use interpreted sentences in which verb senses and noun senses are solved. Another differential aspect is that the task of NSD is linked to the overall task of semantic interpretation. The circularity between NSD and selectional restrictions is overcome by giving priority to the verb selectional restrictions, which narrow the noun senses. Then, the NSD algorithms are used to resolve the final senses. Thus, it becomes critical that the verb selectional restrictions do not select incorrect senses, or over-narrow them. As a result, the NSD algorithms are casting light on the task of defining the selectional restrictions for the verb predicates. A work that also uses semantic interpretation for acquisition is [5]. The authors present algorithms for the acquisition of linguistic knowledge and domain knowledge from texts. The learning is realized by a classifier in a terminological representation system. In contrast, in our research the acquisition occurs by being told and aims at acquiring commonsense knowledge that will permit further understanding.

The other aspect of this work is the acquisition of common sense knowledge by being told. Wordnet is a partial realization of Quillian's dream [12] of building a general semantic network of common sense concepts, because it relates concepts only by *is-a* and *part-of* links. The acquisition methods described in this paper link concepts by semantic relations expressing events, actions. etc. A good interface providing some validation of the definitions will allow students or volunteers to populate WordNet noun senses with the interpreted glosses creating a general common sense knowledge network, that could be used for all kinds of natural language understanding tasks. A major difference between this work and ConceptNet [7] is that the sentences entered in our system are parsed and semantically interpreted by determining their semantic roles and verb predicate. The semantic structures built from the user's sentences are fully disambiguated, with the main concepts linked to WordNet noun ontology, and our ontology of verb predicates. This rich knowledge structure is what permits the semantic interpreter to bootstrap its knowledge in order to deal with other semantic tasks.

8 Conclusions

We have explained methods to bootstrap the knowledge of a semantic interpreter. The semantic interpreter acquires knowledge about ordinary concepts by being told, and applies this knowledge to noun sense disambiguation. In order to show the relevance of the knowledge acquired, we have designed and tested three algorithms for noun sense disambiguation.

References

1. Fellbaum, C.: English Verbs as a Semantic Net. In: Fellbaum, C. (ed.) WordNet: An electronic Lexical Database and some of its applications, pp. 69–104. MIT Press, Cambridge (1998)
2. Gomez, F.: Building verb predicates: A computational view. In: Proceedings of the 42nd Meeting of the Association for Computational Linguistics, ACL 2004, Barcelona, Spain, pp. 351–358 (2004)
3. Gomez, F.: Semantic Interpretation and the Upper-Level Ontology of WordNet. Journal of Intelligent Systems 16(2), 93–116 (2007)
4. Gomez, F.: An algorithm for aspects of semantic interpretation using an enhanced wordnet. In: Proceedings of the 2nd North American Meeting of the North American Association for Computational Linguistics, pp. 87–94 (2001)
5. Hahn, U., Marko, K.G.: An integrated dual learner for grammars and ontologies. Data & Knowledge Engineering 42, 273–291 (2003)
6. Lesk, M.: Automatic sense disambiguation using machine readable dictionaries: How to tell a pine cone from an ice cream cone. In: Proceedings of the 1986 ACM SIGDOC Conference, Toronto, pp. 24–26 (1986)
7. Liu, H., Singh, P.: ConceptNet - a practical commonsense reasoning tool-kit. BT Technology Journal 22, 211–226 (2004)
8. Mihalcea, R.: Knowledge-based methods. In: Agirre, E., Edmonds., P. (eds.) Word Sense Disambiguation, pp. 107–127. Springer, Heidelberg (2006)
9. Mihalcea, R., Moldovan, D.: A method for word sense disambiguation of unrestricted texts. In: Proceedings of the Annual Meeting of the Association for Computational Linguistics, College Park, Maryland, pp. 152–158 (1999)
10. Miller, G.: Nouns in wordnet. In: Fellbaum, C. (ed.) WordNet: An electronic Lexical Database and some of its applications, pp. 23–46. MIT Press, Cambridge (1998)
11. Moldovan, D., Russ, V.: Logic form transformation of wordnet and its applicability to question answering. In: Proceedings of the 39th meeting of the ACL, Toulouse, France, pp. 402–409 (2001)
12. Quillian, M.: Semantic memory. In: Minsky, M. (ed.) Semantic Information Processing, pp. 216–270. MIT Press, Cambridge, Mass (1968)
13. Quine, V.: Word and Object. MIT Press, Cambridge (1960)
14. Resnik, P.: Semantic similarity in a taxonomy: An information-based measure and its application to problems of ambiguity in natural language. Journal of Artificial Intelligence Research 11, 95–130 (1999)
15. Wittgenstein, L.: Philosophical Investigations. Blackwell, Oxford (1958)

Conceptual Modelling and Ontologies

Interlingua for French and German Topological Prepositions

Djelloul Aroui and Mustapha Kamel Rahmouni

Informatic Department Es-Sénia-University, BP. 1524, El Mnaouer, 31000 Oran, Algeria
djelloul.aroui@web.de, rahmouni@mail.univ-oran.dz

Abstract. The present study focuses on the definition of an Interlingua for the French topological prepositions *"dans(in)"*, *"sur(on)"*, *"à(at)"* and their equivalent in German *"in"*, *"auf"*, *"an"* with a view to their use in French-German machine translation and second language acquisition systems. Both languages have a concept of preposition but with a different lexical use. In the same spatial situation, not always equivalent prepositions are used. The French and German prepositions can cover the description of a certain spatial situation, although its meanings can overlap themselves only partially. The choice of the appropriate preposition in a target language depends on the meaning of the topological preposition. With the second language acquisition of spatial prepositions mistakes frequently occur, because one preposition in French does not correspond necessarily to a same basic meaning of a preposition in German. Regarding the machine translation of spatial prepositions, their meanings in a target language can not be defined in the same lexical way. Therefore the machine translation and the second language acquisition of spatial expressions must be occur not directly on a language dependent, i.e. linguistic, but on a language independent level, i.e. conceptual level. The definition of an Interlingua assumes that the meaning of French and German topological prepositions can be expressed in common neutral concepts.

Keywords: Interlingua, Spatial Expression, Topological Preposition, Spatial Preposition, Spatial Representation.

1 Introduction

Our aptitude for communicating and reasoning about space is key to our abilities to navigate, give directions, and to reason analogically about other subjects [16]. One way that we describe spatial scenes is through the use of spatial prepositions. Spatial prepositions expressing spatial relations are of two kinds: local and directional. Local prepositions appear with verbs describing states or conditions, especially the verb *"to be"*. However, directional prepositions appear with verbs of motion. From the different uses of the spatial expressions we will make topological prepositions the subject of this work. We present an Interlingua for French topological prepositions *"dans(in)"*, *"sur(on)"*, *"à(at)"* and their equivalent in German *"in"*, *"auf" and "an"* with a view to use it in French-German machine translation and second language

E. Kapetanios, V. Sugumaran, M. Spiliopoulou (Eds.): NLDB 2008, LNCS 5039, pp. 55–66, 2008.
© Springer-Verlag Berlin Heidelberg 2008

acquisition systems. Expressions involving spatial prepositions in French and German convey to a hearer where one object (located object noted *LO*) is located in relation to another object (reference object noted *RO*). For example, in *the water is in the glass*, the *water* is understood to be located with reference to the *glass* in the region denoted by the preposition *in*. Consider the sentences below:

1. a. La voiture est <u>dans</u> la rue = das Auto ist *[1]<u>in</u>/<u>auf</u> der Stasse (the car is <u>on</u> the street)
    b. Les nuages sont <u>dans</u> le ciel = die Wolken sind *<u>im</u>[2]/<u>am</u>[3] Himmel
    (the clouds are <u>in</u> the sky)
    c. L'image est <u>sur</u> le mur = das Bild ist *<u>auf</u>/<u>an</u> der Wand (the picture is <u>on</u> the wall)
    d. L'homme est <u>à</u> l'ombre = der Mann ist *<u>am</u>/<u>im</u> Schatten (the man is <u>in</u> the shade)

The examples shown above demonstrate that in the same spatial situation not always equivalent prepositions are used and the description of the same spatial situation can be lexicalized differently in French and German. Therefore the word by word direct translation of the French expressions in (1) lead to a wrong use of the German prepositions: in (1a) "*in*" instead of "*auf*", in (1b) "*in*" instead of "*an*", in (1c) "*auf*" instead "*an*" and in (1d) "*an*" instead of "*in*". The choice of the appropriate preposition in a target language depends on the meaning of the topological prepositions and the relationship between *LO* and *RO*. With the second language acquisition of spatial prepositions mistakes frequently occur, because one preposition in French does not correspond necessarily to the same basic meaning of a preposition in German. Regarding the machine translation of spatial prepositions, their meanings in a target language can not be defined in the same lexical way. Therefore the machine translation and the second language acquisition of spatial expressions must occur not on a language dependent, i.e. linguistic, but on a language independent level, i.e. conceptual level. The definition of an Interlingua for French and German topological prepositions can only take place if their meaning can be expressed in common neutral concepts. On the conceptual level, conceptual knowledge's about object classes and relations between objects are available and these are relevant for the interpretation of the spatial expressions.

The issue of how language and space interact has had a long history in cognitive science research. Early theories of spatial preposition use claimed that people assigned spatial prepositions based on the geometry of a visual scene. However, more recent work has shown that the use of spatial prepositions is influenced by a variety of factors; factors such as context ([11]; [19]), functional relationships between the objects ([7]; [10]; [15]; [39]), topological [9], spatial and situational ([28]; [29]) also influence how we use prepositions in everyday language [12]. We take into consideration these factors and we examine the meaning of French and German topological prepositions in the face of (i) the topology of *RO*, (ii) the spatial characteristics of the *RO*, (iii) the functional role of the *RO* and (iv) the functional role of *LO*.

The remaining of the paper is organised as follows. In section 2 we analyse the use conditions of the French topological prepositions versus their equivalent in German.

---

[1] "*" denotes the wrong use of the preposition.

[2] "*im*" is a contraction of "*in dem*": the German preposition "*in*" and the article "*dem*".

[3] "*am*"" is a contraction of "*an dem*": the German preposition "*an*" and the article "*dem*".

Next we describe and classify objects according to their functional, topological, spatial and situational characteristics in order to interpret the French and German topological prepositions. The section 4 defines a representation of a meaning of French and German topological prepositions. The last section closed this work.

2 French Versus German Topological Prepositions

In this section we present the use conditions of the French topological prepositions and their equivalent in German.

2.1 "dans" Versus "in", "auf", "an"

The French preposition "dans" and German "in" have a same basic meaning and allow the allocation of *LO* to the interior of *RO*. They assign the place of the *LO* to the interior of the *RO*, as the *LO* place is contained totally or partly in the interior of *RO* or is part of *RO*. Which configurations between *LO* and *RO* are possible, can be taken from the knowledge of this objects and their usual interaction forms. For instance the different configurations of French "dans" and German "in" in (2) evoke different relationships between *LO* and *RO*.

2. a. L'eau est dans le vase = das Wasser ist in der Vase (the water is in the vase)
   b. Les fleurs sont dans le vase = die Blumen sind in der Vase (the flowers are in the vase)
   c. La fissure est dans le vase = der Sprung ist in der Vase (the fissure is in the vase)

The circumstances in (2) differ by the fact that a vase allowed two kinds of conceptualizations: on the one hand *vase* as a container with his intrinsic situation to contain water or/and flowers and on the other hand *vase* as a material body to contain a fissure. In addition the relationships to the vase can be totally-, partially-inclusion or part-of relation.

   In some spatial situations, different prepositions can be used in French and German (see example (1a)). In Contrast to French, where *street* is noticed as a container, it is conceptualized in German as surface and is combined with the preposition "auf". Both conceptualizations are compatible with a view of *streets* either as two-dimensional roadways or as a kind of U-shaped container that includes the buildings on both sides (see also [17]). We have the same situation for the concept *sky*: in French we use the preposition "dans" with the conceptualization of *sky* as a three dimensional containers but in German we use the proposition "an" with the conceptualization of a sky as surface (see example (1b)). This kind of expressions is understood as idiomatical use, whereby it should be decided for a certain *RO* conceptualization in each spatial situation.

2.2 "sur" versus "auf" and "an"

Where French use "sur" to express the contact of *LO* to all possible surfaces of *RO* (examples in (3a-b)) German makes a two-fold distinction between "auf" and "an", where "_auf_" is used for a top surface (3a) whereas "_an_" is used for a vertical and bottom surface (3b-c).

3. a. le vase est <u>sur</u> la table = die Vase ist <u>auf</u> dem Tisch (the vase is <u>on</u> the table)
   b. le tableau est <u>sur</u> le mur =das Gemälde ist <u>an</u> der Wand (the painting is <u>on</u> the wall)
   c. la mouche est <u>sur</u> le plafond = die Fliege ist <u>an</u> der Decke (the fly is <u>on</u> the ceiling)
   d. le lustre est <u>au</u>[4] plafond = die Lampe ist <u>an</u> der Decke (the lustre is <u>on</u> the ceiling)

The spatial perception of the horizontal surface is therefore closely connected with the functional perception of the supporting surface. In contrast to the horizontal surface, a vertical or a bottom surface appear then only as supporting surface, if any kind of attachment is available. Vandaloise [38] speaks of the activity and/or passivity of support. The relationship between the _LO_ and _RO_ is considered as active if the functional relationship is guaranteed alone by the entity involved (3c) and as passive if the relationship is materialized only by using a fixation (3d).

2.3 "_à_" versus "_an_" and "_in_"

The French preposition "_à_" corresponds to the German "_an_" or "_in_": In contrast to the German preposition "_an_", which expresses a purely spatial relation, i.e. an allocation of _LO_ to the proximal contact external region of _RO_, the French preposition "_à_" can express either a spatial relation as in (4a-e) or a functional relation as in (4f-g).

4. a. le touriste est <u>à</u> la plage = Der Tourist ist <u>am</u> Strand (the tourist is <u>at</u> the beach)
   b. la voiture est <u>au</u> croisement = das Auto is <u>an</u> der Kreuzung (the car is <u>at</u> the crossroads)
   c. Claudia est <u>à</u> Paris = Claudia ist <u>in</u> Paris (Claudia is <u>in</u> Paris)
   d. Pépé est <u>au</u> Canada = Opa ist <u>in</u> Kanada (grandpa is <u>in</u> Canada)
   e. Le vieillard est <u>à</u> l'ombre = der alte Mann ist <u>im</u> Schatten (the old man is <u>in</u> the shade)
   f. Mémé est <u>à</u> l'église = Oma ist <u>in</u> der Kirche (grandma is <u>in</u> the church)
   g. Christin est <u>à</u> l'université = Christin ist <u>an</u> der Universität (Christin is <u>at</u> the university)

The expression in (4f-g) can be interpreted in such a way that the person worship respectively studies or teaches at the university. In this case the preposition is regarded more functionally than spatially. There are also different conventions for the use conditions of the preposition "_à_" with names of locations like in (4d) defined in [1], which can be denoted in German only by the preposition "_in_". In the example presented in (3d), the fixation of _the lustre_ in _the ceiling_ implicates a passive relationship and so the use of the preposition "_à_".

3 Object Characteristics and Object Classifications

As aforementioned, the spatial, situational, topological and functional characteristics of objects play a decisive role in the interpretation of the spatial expressions. In this section, we describe and classify objects according to these characteristics.

---

[4] "_au_" is a contraction of "_à le_": the preposition "_à_" and the article "_le_".

3.1 Topological Characteristics of Objects

The spatial relations expressed by topological prepositions are primarily a relation between units of space. Each object possesses an inside, a surface and an external region adjacent to this object [31]. The proximal external region of an object x is the region, which is affected by its influence. According to [20] the proximal region (noted $Prox(x)$) called also region of interation by Miller/Johnson-Laird [32] consists on one hand of the space occupied by x ($Place(x)$) and the perceptual relevant surfaces of x ($Surf(x)$) and on the other hand of the external region of x ($Ext(x)$). For two dimensional objects $Surf(x)$ is identical with $Place(x)$. $Ext^c(x)$ is the proximal contact to the external region of x and allows the contact to the surfaces of x and constitutes the external region of x. The following relation schemas are defined:

5. a. $Place(x) \subseteq Prox(x)$ where "$\subseteq$" is the spatial part-of-relation.
    b. $Ext(x) = Prox(x) - Place(x)$ where "$-$" is the difference between the regions.
    c. $Surf(x) = Ext^c(x) - Ext(x)$

The proximal region of an Object x is not an object but a distal concept [35]. Furthermore, only the place occupied by an object x, his dimension and his perceptual surfaces are relevant for the topological classification of objects. We differentiate between objects with a material occupied space ($Place(x, MAT)$) and an empty space ($Place(x, EMPTY)$) and objects with both concepts ($Place(x, MAT\_EMPTY)$) (see also [6]). The first group consists of three and/or two-dimensional objects and designates massive objects ($Obj(x, MASSIVE)$, e.g. wall) and/or surface objects ($Obj(x, AREA)$, e.g. ceiling). The most of solid objects have top, bottom and sides surfaces ($Surf(x,TOP)$, $Surf(x, BOTTOM)$ and $Surf(x, VERT)$, e.g. desks, computers). Some surfaces can not be perceptually relevant, e.g. the sky is perceived only from his bottom surface. The second and the last group refer respectively to cavities ($Hollow(x, Space)$, e.g. the tunnel) and hollow bodies ($Hollow(x, BODY)$, e.g. the wastebasket), which can be closed ($Closed(x)$) and/or open ($Open(x)$). The cavities can be also bounded or unbounded. The characteristic "bounded" is particularly relevant for the determination of the use conditions of the French preposition "à" and German "an". Generally, it must be differentiated between interior and external "bounded". The characteristic of the interior "bounded" applies only to cavities, whose existence is fixed within a materially occupied space. The cavities are either bounded ($Bounded(x, INT)$ - "INT" for Interior) like a tunnel, vaguely bounded ($Vague-Bounded(x, INT)$) like a valley or unbounded ($Unbounded(x, INT)$) like a galaxy, which presents itself to us only from the inside. All other spatial entities, which are not cavities, are externally bounded ($Bounded(x, EXT)$ - "EXT" for exterior).

3.2 Spatial and Situational Characteristics of Objects

The shape and the size of an object define its spatial characteristics. According to [42] the perception of the shape occurs on the object parts such as surfaces, edges and corners. The surface of three-dimensional objects is structured (with exception of spherical objects) into individual surfaces limited by edges. Two-dimensional objects have logically only one surface. One dimensional object is defined as line. According to [41], the search procedure of the *LO* is facilitated, if the *RO* is bigger than the *LO*, whereby the size is measured according to the fast and successfully search. Li [30] assumes that an object is qualified as *RO* if it is comparatively bigger then *LO*. According to [1], the choice of the *RO* is not only influenced by its size (*Dim(x,1)*, *Dim(x,2)* or *Dim(x,3)* for 1, 2 or 3 dimension of an object *x*) but also by his prominent and salient characteristics. An object *x* is prominent (*Prominent(x)* e.g. the town-hall), if it dominates visually by its form and visibility or also by its function. An object *x* is salient (*Salient(x)* e.g. the bus-stop), if it represents a particularly distinguished point of reference. In addition to the spatial characteristics the situational characteristics of objects are also implicated in the description of the object, since top-, bottom-, and adjacent-regions as well as top, bottom and side surfaces of objects are always established relatively to their position in the space. An object can have a fixed position (*Fixed(x)*) e.g. the traffic light) and it is immobile or a canonical situation (*Canonical(x)* e.g. the car), which is coupled with mobility, or a flexible position (*Flexible(x)* e.g. the stone) without any fixed orientation or finally a dependent position (*Dependent(x)* e.g. the upper-arm), which is associated with a part-of relation to another object.

3.3 Functional Characteristics of Objects

The knowledge about the shape of objects is closely connected with the knowledge about their functions. A table serves to support objects and therefore has a top surface. And because it has this shape, the support function can be assigned to it. Objects like a bowl and a tray can have a similar shape, but an apple is always *in* the bowl and *on* the tray. This is to be attributed to the fact that a bowl is a container object and a tray a support object. Thus, container objects are predestinated to allocate the interior space and support objects the surface. Hollow bodies with closed or sidewise opened delimitation have a prototypical interior. This type of object is represented for example by building, rooms, means of transport, bags, bottles, glasses, vases, etc... In addition, an interior can be formed by a horizontal spatial entity, if the surface exhibits a clear delimitation. This is the case of locations like urban districts, settlements or larger geopolitical units. Those are things, which are permanently parts of objects such as ceiling. Vandaloise [38] hold a view, that some spatial expressions define a purely functional relation between the involved objects. He assumes that the French preposition "*dans*" expresses a "container/contained" (*Contained(x,y)* i.e. *x* is contained in *y*, this contenance can be partial or total) and the French preposition "*sur*" a "supporter/supported" (*Supported(x,y)*: *x* is supported by *y*) relation. There is however examples for the use of "*dans*" and "*sur*", which can not be interpreted functionally (6).

6. a. l'avion est <u>dans</u> le nuage = das Flugzeug ist <u>in</u> den Wolken (the airplane is <u>in</u> the clouds)
  b. le village est <u>sur</u> la frontière = das Dorf ist <u>an</u> der Grenze (the village is <u>at</u> the border)

This means that spatial expressions can not be determined purely functionally but with the combination of functional, topological and spatial factors. Consequently the type of the object affects the selection of the spatial preposition. Our classification of objects is dominated primarily by functional, biological or physicochemical characteristics of objects, which implicate specific spatial characteristics. Regarding the spatially relevant sorts, we based this classification on that of [6], who organized the entities in four object classes: locations, discrete objects, dependent-objects and substances. As presented above certain locations can only be used with a combination of specific topological prepositions. Alone on the basis of the knowledge about the affiliation of the designated location to the class of the lakes, rivers, mountains, island, etc., permits us to use a preposition in (7), without knowing their spatial characteristics. The Locations are divided into regions (*Loc-Region(x)*: e.g. school, settlement, province), surfaces (*Loc-Surface(x)*: e.g. continent, way), cavities of larger extent, which are related to locations (*Loc-Hollow(x)*: e.g. valley, tunnel), waters (*Loc-Water(x)*: e.g. Baltic Sea, pond) as well as geographical objects (*Loc-Geo-Obj(x)*: e.g. mountain).

7. a. Peter est sur l'île=Peter ist auf der Insel (Peter is on the island)
    b. Christin est sur le continent=Christin ist auf dem Kontinent (Christin is on the continent)
    c. Mémé est à Paris=Oma ist in Paris (grandma is in Paris)
    d. La voiture est dans le tunnel= das Auto ist im Tunnel (the car is in the tunnel)
    e. Pépé est en[5] France = Opa ist in Frankreich (grandpa is in France)

The lexical connection between the preposition and the *RO* involves an idiomatical use of certain objects with certain prepositions, i.e. *Obj(x, IDIOMAT_USE, dans)*, *Obj(x, IDIOMAT_USE, sur)*, *Obj(x, IDIOMAT_USE, à)*, *Obj(x, IDIOMAT_USE, in)*, *Obj(x, IDIOMAT_USE, auf)* and *Obj(x, IDIOMAT_USE, an)* defines the idiomatical use of an object for the specified preposition.

Discrete objects are divided into natural entities and artefacts (see [14] and [5]). Natural entities are concrete objects of natural origin, like humans (*Human(x)*), animals (*Animal(x)*) or plants (*Plant(x)*). Whereas artefacts concern concrete objects manufactured by humans, e.g. containers (*Container(x)*: e.g. bottles, glasses), buildings (*Building(x)*: e.g. schools, churches), means of transport (*Transport-Obj(x)*: e.g. buses, airplanes), articles of clothing (*Cloth(x)*: e.g. coats, towels), food (*Food(x)*: e.g. bread, cheese, cake), etc...

Dependent objects are objects, which are bound to other objects, i.e. they occur only depending on another object. Those are things, which are permanently parts of objects (*Part-Of(x)*: e.g. ceilings, legs), all kinds of cavities, which are bound to a materially occupied object area (*Hollow(x,BODY)*: e.g. hole, tear), as well as entities of the class of illustrations (*Illustration(x)*: e.g. blotch, drawing). Thereby, the spatial characterisation of such objects is always relational to the *RO* and can never take place independently from the *RO*.

Substances like liquid (*Liquid(x)* e.g. water), gaseous (*Gaseous(x)* e.g. spray), granular (*Granular(x)* e.g. sugar) assume the shape of the containers, in which they are contained. In this sense the space component of substances is to be understood as a relational size.

Referring to the object characteristics and the object classifications presented above result conceptualizations of a *ceiling*, *street* and *sky* in (8). When we assume that *street* is conceptualize as a surface and *sky* as a container, so the use of *street* with the French preposition "*dans*" and *sky* with the German preposition "*an*" are to be considered as an idiomatical use.

---

[5] The French preposition "*en*" is considered as a topological preposition. It appears particularly for the description of specific *LO-RO* constellations. It corresponds to nearly one of the meanings of the preposition "*in*". It is not considered in this work.

8. a. Conceptualization(ceiling)=Obj(ceiling, Area) ∧ Surf(ceiling, Bottom) ∧
               Ext(ceiling, Bottom) ∧ Dim(ceiling,2) ∧
               Fixed(ceiling) ∧ Bounded(ceiling, Ext) ∧ Part-Of(ceiling).
    b. Conceptualization(street)=Dim(street,2) ∧ Surf(street, TOP) ∧
               Bounded(street, TOP-EXT) ∧ Fixed(street) ∧
               Obj(street, IDIOMATICAL_USE, "_dans_")
    c. Conceptualization(sky)=Dim(sky,3) ∧ Loc-Hollow(sky) ∧ Surf(sky, BOTTOM) ∧
               Open(sky) ∧ Unbounded(sky, INT)) ∧
               Obj(sky, IDIOMATICAL_USE, "_an_")

4   Representation of a Meaning of the Spatial Prepositions

Prepositions have been studied from a variety of perspectives: geometrically oriented [3], [8] and [43], linguistically: [26], [37], [19], [36] and [21] logically based: [13], [2], [32], [34], [18], [27], [23], [24], [41], [4], [30], [22] and [33]. All authors agree on the fact that spatial prepositions do not denote like nouns an object, but rather express a relation between objects. Disagreement prevails however in the question of which nature is this relation and how it should be defined. For a geometrical approach a preposition has to be understood as geometrical relation. Clark [8] defines the preposition "_at_", "_on_" and "_in_" as one, two and three-dimensional relations. According to [8] the preposition "_at_" determines relations between points and/or lines, "_on_", relations between points, lines, and/or surfaces and "_in_" relations between points, lines, surfaces, and/or volumes. Vandeloise [40] and Herskovits [19] examined the linguistic descriptions of the spatial relations. Vandeloise [40] uses the concept *family resemblance features* in the description of the linguistic representation of spatial relationships. Herskovits [19] refers in his theory to the work of [36] and [25] and uses a notion of ideal meaning in order to associate with each spatial preposition a categorisation. In this case prototypes or ideal meanings are geometrical relations between *LO* and *RO*. For the logical approach the relation expressed by the spatial prepositions is primarily a relation between spaces.

This study refers to the logical approach, in particularly to the work of [41] based on the two-level semantics theory which tries to purge lexical entries of all details which can be inferred from general conceptual principles.

We agree with Wunderlich/Herweg [41] that the meaning of spatial prepositions is a localization relation between *LO* and a specific region concerning *RO*. This relation is defined in the language of predicate logic with λ-abstraction $\lambda y \lambda x Loc(x, PREP^*(y))$. $Loc(x, p)$ means that the place occupied from x is contained in the region p [20]. The core of the meaning of the preposition is a neighbourhood function $PREP^*$, which has to be specified for each preposition. The specification of this function takes place on the conceptual level, in which conceptual knowledge about object classes and relations between objects is used.

4.1 Formalism

French and German topological prepositions are used to describe proximal spatial relationships. The meaning of the French topological prepositions "dans", "sur", "à" and their equivalent in German "in", "auf", "an" would be specified as a two place predicate in (9).

9. a. [dans] = Loc(LO,Place(RO)) ∧ Conceptualization(LO) ∧ Conceptualization(RO)
   b.   [sur] = Loc(LO, Ext^c(RO)) ∧ Conceptualization(LO) ∧ Conceptualization(RO)
   c.   [à] = Loc(LO,Prox(RO ∧ Conceptualization(LO) ∧ Conceptualization(RO)
   d.   [in] = Loc(LO, Place(RO)) ∧ Conceptualization(LO) ∧ Conceptualization(RO)
   e.   [auf] = Loc(LO, Ext^c(RO, VERT)) ∧ Conceptualization(LO) ∧ Conceptualization(RO)
   f.   [an] = Loc(LO, Ext^c(RO)) ∧ Conceptualization(LO) ∧ Conceptualization(RO)
   with:
   - Place(RO) =Place(RO,EMPTY) ∨ Place(RO,MAT_EMPTY) ∨ Place(RO,MAT)
   - Ext^c(RO) = Ext(RO, TOP) ∨ Ext(RO, BOTTOM) ∨ Ext(RO, VERT)
   - Prox(RO) = Ext(RO,VERT) ∨ Place(RO, EMPTY)

The formalism in (9) is affected by the topology of RO and it say "LO is located in the DANS*-, SUR*-, À*-, IN*-, AUF*- or AN*-region with respect to the reference object RO". The DANS*- and IN*-region in (9a, 9d) has to be equated with the place of RO. The SUR*- and AN*-region in (9b, 9f) are defined by Ext^c(RO) as a proximal contact to the external region of RO. The proximal region of RO - Prox(RO) comes into consideration for the specification of neighbourhood function À* in (9c) because it does not take an allocation to a specific topological subspace of RO. The AUF*-region in (9e) results from the AN*-region with the restriction on a vertical region of RO. Furthermore, the meaning of these prepositions is concretized with the conceptualisation of LO and RO.

4.2 Algorithm

In order to interpret the meaning of French and German spatial expressions the following steps are necessary:

1) Monolingual processing of the spatial expressions according to the topology of RO.
2) Comparison of French and German spatial expressions according to the topology of RO. Basically all pairs of French and German spatial expressions, whose reference regions overlap one another totally or partially, can be correspond to the other.
3) Monolingual processing of the spatial expressions according to the spatial characteristics and functional roles of LO and RO.
4) Comparison of French and German spatial expressions according to the spatial characteristics and functional roles of LO and RO.

In order to explain this algorithm, we consider the examples in (10):

10. a. French: La mouche est sur le plafond (the fly is on the ceiling)
    b. German: Die Fliege ist *auf der Decke
    c. German: Die Fliege ist an der Decke (the fly is on the ceiling)

Due to substantial similarity in the use of the German preposition "an" and "auf", the French expression in (10a) can be translated into German with (10b) or (10c). The meaning of these expressions can be expressed in (11) related to the meaning representations in (9) and conceptualizations in (8a).

11. a. Loc(LO:fly,Ext^c(RO:ceiling,BOTTOM)) ∧ Conceptualization(LO:fly) ∧
    . Conceptualization(RO:ceiling)
  b. Loc(LO:fly,Ext^c(RO:ceiling,VERT)) ∧ Conceptualization(LO:fly) ∧
    . Conceptualization(RO:ceiling)
  c. Loc(LO:fly,Ext^c(RO:ceiling,BOTTOM) ∧ Conceptualization(LO:fly) ∧
    . Conceptualization(RO:ceiling)
  with Conceptualization (fly)=Animal(fly) ∧ Active(fly)

Because the ceiling has only a bottom but not a vertical surface, the translation of the French expression with (10b) violates the topology of the German "*auf*". Thus the use of the German preposition "*auf*" is inconsistent with a conceptualisation of the ceiling. Only the representation in (11c) meet conditions defined in the algorithm and can be considered as the correct translation of the French expression. In addition to the topological mapping the conceptualisations of *fly* and *ceiling* verify in both languages the functional factors to use the French preposition "*sur*" and German preposition "*an*".

Consider now the examples in (1a-b), which involve an idiomatical use of the French preposition "*dans*" with *street* and an idiomatical use of German preposition "*an*" with *sky*. The idiomatical definitions of *street* and *sky* in (8b-c) alone constrain us to choose the adequate prepositions to be use. Therefore, idiomatical spatial expressions must be learned in the way that one learns other vocabulary.

5 Conclusion

In the present study, the definition of an Interlingua for the meaning representations of the French and German topological prepositions has been presented with reference to their use in machine translation and second language acquisition systems. In the interpretation of topological prepositions, two kinds of knowledge were necessary: the spatial knowledge (structuring of the space into subspaces) and the object knowledge (object shape, object function). As was shown, certain subspaces are associated with specific object characteristics. An entity must satisfy certain conditions, so that it can be assigned to a subspace. For instance, the interior is bound to the combination of shape characteristics and container function, the surface to the support aptitude and the proximal region to the size of the reference object. The essential differences in the interpretation of topological prepositions are to be attributed to the fact that German and French do not structure the space on the same way.

References

1. Becker, A.: Lokalisierungsausdrücke im Sprachvergleich; eine lexikalischsemantische Analyse von Lokalisirieungsausdrücken im Deutschen, Englischen,Französischen und Türkischen. Niemeyer, Tübingen (1994)
2. Bennett, D.C.: Spatial and Temporal Uses of English Prepositions: An Essay in Stratificational Semantics. Longman, London (1975)

3. Bierwisch, M.: Some Semantic Universals of German Adjectivals. Foundation of Language 3 (1967)
4. Bierwisch, M., Schreuder, R.: From concepts to lexical items. In: Cognition, vol. 42, pp. 23–60 (1992)
5. Brown, P.F.: A survey of category types in natural language. In: Tso'hatzidis, pp. 17–48 (1990)
6. Buschbeck-Wolf, B.: Topologische Präpositionen und ihre Verarbeitung in der maschinellen Übersetzung. Dissertation am Institut für Maschinelle Sprachverabeitung der Universität Stuttgart (1994)
7. Carlson-Radvansky, L.A., Covey, E.S., Lattanzi, K.M.: "What" effects on "where": Functional influences on spatial relations. Psychological Science 10, 516–521 (1999)
8. Clark, H.H.: Space, Time, Semantics, and the Child. In: Moore, T.E. (ed.) Cognitive Development and the Acquisition of Language, New York, pp. 65–110. Academic Press, London (1973)
9. Cohn, A.G., Bennett, B., Gooday, J.M., Gotts, N.: RCC: a calculus for region based qualitative spatial reasoning. GeoInformatica 1, 275–316 (1997)
10. Coventry, K.R., Carmichael, R., Garrod, S.C.: Spatial prepositions, object-specific function, and task requirements. Journal of Semantics 11, 289–309 (1994)
11. Coventry, K.R.: Function, geometry, and spatial prepositions: Three experiments. Spatial Cognition and Computation 1, 145–154 (1999)
12. Coventry, K.R., Garrod, S.A.: Saying, Seeing,and Acting: The Psychological Semantics of SpatialPrepositions. Psychology Press, New York (2004)
13. Cooper, G.S.: A semantic analysis of English locative prepositions. Bolt Beranek and Newman report Nr. 1587, Springfield (1968)
14. Dahlgren, K.: Naive Semantics for Natural Language Understanding. Kluwer Academic Publishers, Dordrecht (1988)
15. Garrod, S., Ferrier, G., Campbell, S.: In and on: investigating the functional geometry of spatial prepositions. Cognition 72, 167–189 (1999)
16. Gentner, D., Imai, M., Borodisky, L.: As Time Goes By: Evidence for two systems in processing space>time metaphors. Language and Cognitive Processes 17, 537–565 (2002)
17. Grimaud, M.: Toponyns,Prepositions, and Cognitive Maps in English and French. Journal of the American Socieiy of Gcolinguistics 14, 54–76 (1988)
18. Habel, C.: Zwischen-Bericht. In: Christoph, Habel, Herweg, M., Rehkämper, K. (eds.) Raumkonzepte in Verstehensprozessen, pp. 37–69. Niemeyer, Tübingen (1989)
19. Herskovits, A.: Language and Spatial Cognition: An Interdisciplinary Atudy of the Prepositions in English. Cambridge University Press, Cambridge (1986)
20. Herweg, M.: Ansätze zu einer semantischen Beschreibung topologischer Präpositionen. In: Christoph, Habel, Herweg, M., Rehkämper, K. (eds.) Raumkonzepte in Verstehensprozessen, pp. 99–127. Niemeyer, Tübingen (1989)
21. Hottenroth, P.M.: Präpositionen und Objektkonzepte. In: Rauh, G. (ed.) Approaches to prepositions. Tübinger Beiträge zur Linguistik, vol. 358, pp. 77–108 (1991)
22. Kiparsky, P.: Remarks on Denominal Verbs. In: Alsina, A., Bresnan, J., Sells, P. (eds.) Complex Predicates, pp. 473–499. CSLI, Stanford (1997)
23. Klein, W.: Überall und nirgendwo. Subjektive und objektive Momente in der Raumreferenz. Zeitschrift für Literaturwissenschaft und Linguistik 78, 9–42 (1990)
24. Klein, W.: Raumausdrücke. Linguistische Berichte 132, 77–114 (1991)
25. Lakoff, G., Johnson, M.: Metaphors we Live by. University of Chicago Press, Chicago (1980)

26. Lakoff, G.: Categories and Cognitive Models. Cognitive Science Program Series: University of California, Berkeley, L.A.U.T.-Paper, Nr.36, Series A (1982)
27. Lang, E.: A two level approch to projective prepositions. In: Rauh, G. (Hrsg). Approches to prepositions, pp. 127–167. Tübingen (1991)
28. Lang, E.: A two-level approach to projective prepositions. In: Zelinsky-Wibbelt, C. (ed.) The semantics of prepositions: from mental processing to Natural Language processing. Mouton de Gruyter, Berlin (1993)
29. Lang, E.: Spatial Dimension Terms. In: Haspelmath, M., et al. (eds.) Language Typology and Language Universals, pp. 1251–1275. An International Handbook, Berlin (2001)
30. Li, J.: Räumliche Relationen und Objektwissen, am Beispiel an und bei. In: Studien zur deutschen Grammatik; Bd. 49, Gunter Narr Verlag Tübingen (1994)
31. Mark, D., Svorou, S., Zubin, D.: Spatial Terms and Spatial Concepts: Geographic, Cognitive, and Linguistic Perspectives. In: Proceedings of the International Geographic Information Systems (IGIS) Symposium: The Research Agenda, vol. 2 (1987)
32. Miller, G., Johnson-Laird, P.N.: Language and Perception. Cambridge University Press, Cambridge (1976)
33. Olsen, S.: Semantische und konzeptuelle Aspekte der Partikelverbbildung mit ein-. In: Olsen (ed.), pp. 9–26 (1998)
34. Pribbenow, S.: Regebasierte Interpretation lokaler Präpositionen am Beispiel von in und bei. In: Habel, et al. (Hrsg). Raumkonzepte in Verstehensprozessen, Tübingen, pp. 202–228 (1989)
35. Pribbenow, S.: Zur Verarbeitung von Lokalisierungsausdrücken in einem hybriden System. Dissertation am Fachbereich Informatik der Universität Hamburg. IBMS Report 211 (1992)
36. Rosch, E.: Human Categorization. In: Waren, N. (ed.) Advances in Cross-Cultural psychology, vol. 1. Academic Press, London (1977)
37. Talmy, L.: How Language Structures Space. In: Pick, H.L., Acredolo, L.P. (eds.) spatial Orientation, ch. 11. Plenum Press, New York (1983)
38. Vandeloise, C.: L'espace en français, Sémantique des Prépositions spatiales Ed. Seuil-Travaux en Linguistiques, Paris (1986)
39. Vandeloise, C.: Methodology and Analysis of the Preposition In. Cognitive Linguistics 5, 157–184 (1994)
40. Vandeloise, C.: The Preposition in and the relationship container/contained. LAUD, Linguistic Agency of the University of Duisbung (prepublished, 1985)
41. Wunderlich, D., Herweg, M.: Lokale und Direktionale. In: v. Stechow, A., Wunderlich, D. (eds.) Handbuch der Semantik, de Cruyter, Berlin (1991)
42. Zusne, L.: Visual Perception of Form, New York, London (1970)
43. Harwkins, B.: Semantics of English Spatial Prepositions.Ph.D. Thesis. San Diego, California: University of California. Thesis pre-published by L.A.U.D.T., Linguistic Agency of the University of Duisburg, previously Tier (1983)

Ontological Profiles as Semantic Domain Representations

Geir Solskinnsbakk and Jon Atle Gulla

Department of Computer and Information Science
Norwegian University of Science and Technology, Trondheim, Norway
{geirsols,jag}@idi.ntnu.no

Abstract. Ontologies are conceptualizations of some domain, defining its concepts and their relationships. An interesting question is *how do we relate the ontological concepts to the general vocabulary of the domain?* In this paper we present the concept of ontological profiles, what they are, how they are constructed, and how they may be used. We propose that the ontological profile is a link between the vocabulary of the domain and its conceptual specification given by the ontology. This means that the ontological profile can be tailored to a specific document collection, reflecting the vocabulary actually used. Finally we demonstrate how the ontological profile may be utilized for ontology-driven search.

Keywords: ontological profiles, ontology-driven information retrieval, query reformulation.

1 Introduction

Ontologies are now used in a wide range of applications and have been instrumental in many interoperability projects. They help applications work together by providing common vocabularies that describe all important domain concepts without being tied to particular applications in the domains.

Used as part of semantic search applications, however, the ontologies have so far had only limited success. Early semantic search engines tried to use ontology concepts and structures as controlled search vocabularies, but this was unpractical both functionally and from a usability perspective. Ontologies for query disambiguation or reformulation seem more promising, though there is a fundamental problem with comparing ontology concepts with query or document terms. Concepts are abstract notions that are not necessarily linked to a particular term. Sometimes there may be a number of terms that refer to the same concepts, and sometimes a specific term may be realizations of different concepts depending on the context. Using conceptual structures to index or retrieve document text requires that there is something bridging the conceptual and real world.

Another issue is the tailoring of ontologies to the retrieval task. Research indicates that ontologies are of little use if they are not aligned with the documents indexed by the search application. The granularity of the ontology needs to match the granularity of the document collection. While there is no need to have an elaborated ontology for a sub domain with very few documents, it

E. Kapetanios, V. Sugumaran, M. Spiliopoulou (Eds.): NLDB 2008, LNCS 5039, pp. 67–78, 2008.

is often necessary to expand ontologies in areas that are well covered by the document collection.

This paper presents an ontology enrichment approach that both bridges the conceptual and real world and ensures that the ontology is well adapted to the documents at hand. The idea is to provide contextual concept characterizations that reveal how the concepts are referred to semantically in the document collection. The characterizations come in the form of weighted terms that are all - to some extent - related to the concept itself. The ontology together with the concept characterizations are referred to as an ontological profile of the document collection.

The approach has already been used for ontology alignment, and we are now experimenting with these profiles in search and ontology learning. Our initial search prototypes display a significant improvement of search relevance, provided that the quality of the characterizations are sufficient.

The structure of the paper is as follows: Section 2 gives a short overview of related work, while Section 3 deals with defining ontological profiles. Section 4 describes how such profiles are constructed, and Section 5 shows how the ontological profile may be used as a tool for enhancing information retrieval. Finally Section 6 concludes the paper.

2   Related Work

In the last years there have been many research projects concerned with semantic search, and we will here focus on applications that employ query expansion/reformulation techniques. [1] uses WordNet to expand the user query in the geographical domain. The query is expanded by POS tagging the query and expanding proper nouns with related words (synonymy and meronymy in WordNet). [2] describes a system that employs conceptual indexing (based on WordNet) and uses a variant of LSA to add conceptually similar words to the query. [3] describes a system that represents documents as a combination of concept instances and bag-of-words. The ontology is used at query time to disambiguate the query by presenting instances to the user. In [4] a system that employs conceptual query expansion is presented. Concepts are generated based on the top ranked document from a two-word (manually generated) query. A combination of words (represented as concepts) are added to the original query. [5] presents a ontology based search for portals. The system uses ontologies to contextualize and expand the user query with related words from the ontology. The new query is entered to Google through the Google API. The system that is most similar to the one we are developing is presented in [8]. This system is also based on building concept vectors, and the main difference lies in how these are constructed together with how they are used in the query expansion process.

3   Ontological Profiles

An ontological profile is an extension of a domain ontology. The ontology is extended with semantically related terms. These terms are added as vectors for each of the concepts of the ontology. This means that in the ontological profile each concept is associated with a vector of semantically related terms (concept vector). The terms are given weights to reflect the importance of the semantic relation between the concept and the terms.

Definition of concept vector. The definition given here is adapted from [6]. Let T be the set of n terms in the document collection used for construction of the ontological profile. $t_i \in T$ denotes term i in the set of terms. Then the concept vector for concept j is defined as the vector $C_j = [w_1, w_2, \ldots, w_{n-1}]$ where each w_i denotes the semantic relatedness weight for each term t_i with respect to concept C_j.

We assume that the ontological profile is constructed on the basis of a document collection that covers the same domain as the ontology. By applying text mining techniques to the document collection we add terms that are semantically linked to the concepts of the ontology. Thus, each concept of the ontology is associated with one vector containing terms and weights that are specific to the concept. On an abstract level we may say that building an ontological profile is in fact building a weighted semantic dictionary, in which the concept vectors for each concept gives a list of terms and their weights that give an extended semantic characterization of the concept with respect to the document collection used as basis. We argue that the ontological profile, due to the construction based on a domain collection, gives a representation of the concept and its semantically linked terms that reflects how the concept is used in the language of the documents. This is an important point, since authors may have problems using a vocabulary consistent over a large domain. The concept vectors will typically contain terms that are synonymous to the concept and that are more indirect references to the concept or the use of it. One may argue that the use of thesaurus or WordNet may give much of the same information, but with one important difference. The information found in such formal sources are more general, and possibly not applicable to very specific concepts in large ontologies. Therefore we argue that the ontological profile possibly is better suited, since it is adapted to the document collection and the vocabulary used in it.

We argue that when applied to information retrieval (see Section 5) ontological profiles are generic to the search process. By this we mean that an information retrieval system that is built to utilize such profiles may be adapted to different document collections or even domains by exchanging the document collection and/or the ontology used. This is illustrated by Figure 1. We see that two different document collections are used to build two different ontological profiles for the ontology. By substituting the ontological profile the system can search more focused in a different document collection. We might also imagine that an enterprise through the years have collected a large amount of documents, and the use of the vocabulary might have changed over time. Thus by building ontological profiles for certain time spans, these ontological profiles may provide bridging between the vocabulary used at different points in time.

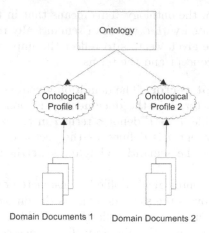

Fig. 1. The concept of ontological profiles

Terms in the concept vectors are weighted in the range [0,1] where 0 means that there is no relation between the term and the concept, while 1 designates the term as being highly related to or even synonymous with the concept.

The terms contained in the concept vectors are semantic extensions of the concept. It could possibly be argued that the terms could be added as concepts to the ontology, further specializing the ontology. However, this is not necessarily desirable. First of all, the terms are very specialized, tailored towards the vocabulary of the underlying document collection, suggesting that they are not generic to the domain, and thus would only clutter the further use of the ontology with respect to other document collections. Secondly, the purpose of constructing the ontological profile is to do a more deep semantic analysis of the document collection, finding relations that are found in the documents, but that need not be generic to the domain. The terms are not generic enough, and are to fine-grained to be used as concepts in the ontology.

The construction of these ontological profiles is based on three different aspects of the content of the documents used. The first is that we apply statistical techniques, counting the frequency of the terms in the documents. Terms that co-occur with a concept more frequent are hypothesized to be more relevant for a concept than terms that do not co-occur as frequently. The second is that we apply linguistic techniques, i.e. stemming, to collapse certain terms into a single form. The third aspect is that we use a proximity analysis of the text. The assumption that lies behind the proximity analysis is that the closer terms are found in the text, the more semantically related they are. These three aspects of the underlying document collection are the basis for the construction of ontological profiles that we suggest in Section 4.

Finally we will in this section show an example of a concept vector. We will be using the concept *Christmas Tree* from the IIP ontology [12]. The setup of the experiment can be found in Section 5.3. A *christmas tree* (see Figure 2) is by the ISO 15926 standard used in the petroleum industry defined as *"an artefact that is an assembly of pipes and piping parts, with valves and associated control equipment that is connected to the top of a wellhead and is intended for control of fluid from a well."* The top 10 terms for the concept vector are shown below:

$$C_{christmastree} = [christma_{0.71}, tree_{0.60}, valve_{0.13}, master_{0.11}, wing_{0.11},$$
$$bop_{0.08}, located_{0.08}, stack_{0.07}, choke_{0.07}, wellhead_{0.06}]$$

Note that we have applied stemming, resulting in christmas being stemmed to christma. The first two terms in the vector are the constituents of the concept name, and have also received the highest relevance score. The terms valve, and wellhead are clearly related to the concept (as we may note from both the definition and Figure 2). Master is also contained in the vector, and looking at Figure 2 we see that several valves are referred to as "master" valves. Bop (an abbreviation for blowout prevention) is certainly relevant, although it is not mentioned in the definition. This term demonstrates in a good way that terms are picked up in the process of building the ontological profile based on semantic relations. A point that is worth mentioning (although more a implementational issue) is that the vector does not contain any phrases (for instance the concept name is split into two separate terms). Adding phrases to the vectors would add even more semantics to the concept vector, and is an issue that will be addressed in the next stage of research.

4 Construction

The ontological profile is constructed on the basis of a domain relevant document collection. This assures that the connection we want between the vocabulary of the domain (at least the vocabulary in the document collection used) and the concepts of the domain is found, letting the ontological profile be a semantic characterization of the domain. A detailed description of the approach we have used for the construction of the ontological profile is found in [7], and is based on a method described by [6]. [8] describes another approach for the construction of ontological profiles (referred to as feature vectors).

The overall process of constructing the ontological profile is shown in Figure 3. The first step is to preprocess the documents used during the construction phase. During this process we remove all stop words, and stem the terms lightly (removing plural s), using the conversion $s \rightarrow \emptyset$ for all terms not terminated with ss. A light stemming algorithm is chosen to reduce the effect of decreased precision which is a problem with stronger stemming algorithms (e.g. the Porter stemming algorithm) [9].

Next, we build three separate indexes of the relevant documents, reflecting three different semantic views of the documents. In the first case, the whole document is viewed as a set of semantically related terms. The second case splits the documents into paragraphs, where each paragraph is considered a semantic entity in which the terms are closer related semantically than in the document. We have chosen to split the paragraphs at the boundary of two or

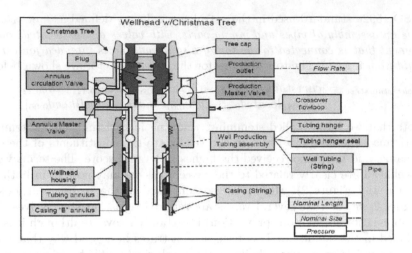

Fig. 2. Schematic view of a *christmas tree* [12]

more consecutive line breaks. The last case is where the document is split into sentences, and each sentence is considered a semantic entity. Terms found in a single sentence are considered to be even closer semantically related than in the paragraph. We have used punctuations as the boundaries between the sentences (".","!", and "?"). Thus we have formed a hierarchy of increasing semantics over the text. Once the documents have been split according to our schema we construct three separate indexes, one for whole documents, one for paragraphs, and one for sentences. The indexes are constructed based on Apache Lucene[1] and the vector space model.

The next step in the construction of the ontological profile is to assign to each of the concepts in the ontology the set of relevant documents (whole documents, paragraph documents, and sentence documents). We use the concept name as a phrase query into the three indexes, and all documents containing the phrase are assigned to the concept as relevant. Of course, using the concept name as a phrase query into the three indexes imposes a challenge; some of the concept names are artificial in their construction or are not used in the form given in the concept. This means that many of the concepts are not found during the assignment of documents to the concepts. We have not researched how to handle this, so this is a matter that needs further research.

The final step in the construction of the ontological profile is to calculate the weights for the terms assigned to the concepts. Recall that the previous step assigned all documents found to be relevant (i.e. contains the concept name as a phrase) to the concept. This means that all terms found in the relevant documents are also assigned to the concept. Having the text partitioned into three different views, we use this partitioning to boost terms that are found closer to the concept in the text. In effect this means that we give the highest weight

---

[1] http://lucene.apache.org/

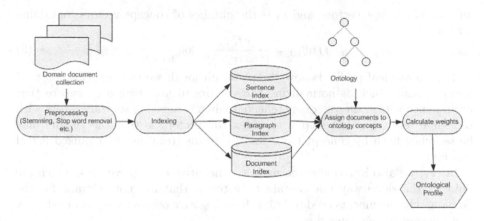

Fig. 3. The process of constructing the ontological profile

to terms that are found in the same sentence as the concept name phrase (the highest semantic coherence), terms found in the same paragraph as the concept are given lower weight than sentence - terms, and higher than document terms. The basis for the weight calculation is the term frequency for each term found in the relevant documents. Equation 1 shows the calculation, where $vf_{i,j}$ is the term frequency for term i in concept vector j, $f_{i,k}$ is the term frequency for term i in document vector k, D, P, and S are the possibly empty sets of relevant documents, paragraph documents and sentence documents assigned to j, and $\alpha = 0.1$, $\beta = 1.0$, and $\gamma = 10.0$ are the constant modifiers for documents, paragraph documents, and sentence documents, respectively. The modifiers (α, β, and γ) simply reflect the relative importance of terms found in whole documents, paragraph documents, and sentence documents. Although the absolute numerical value for these have not been researched extensively, we have found that the set of modifiers shown perform quite well.

$$vf_{i,j} = \alpha \cdot \sum_{d \in D} f_{i,d} + \beta \cdot \sum_{p \in P} f_{i,p} + \gamma \cdot \sum_{s \in S} f_{i,s} \tag{1}$$

The vectors resulting after the calculations in Equation 1 are what we refer to as the basic vectors. Applying the familiar tf*idf [10] score to the frequencies we get closer to the final representation of the vectors. The idf factor gives more importance to terms that are found in few documents across the document collection, and is used in an analogue way here. The only difference is that we now use the basic vectors as our documents, meaning that the idf factor gives higher weight to terms that are found in few concept vectors. The calculations are shown in Equation 2, where $tfidf_{i,j}$ is the tfidf score for term i in concept vector j, $vf_{i,j}$ is the term frequency for term i in concept vector j, $max(vf_{l,j})$ is the frequency of the most frequent occurring term l in concept vector j, N is the number of concept vectors, and n_i is the number of concept vectors containing term i.

$$tfidf_{i,j} = \frac{vf_{i,j}}{max(vf_{l,j})} \cdot \log \frac{N}{n_i} \tag{2}$$

The ontological profile is now complete, although we may apply some final normalization, such as normalizing the vectors to unit length to ensure that prominence within the vectors is reflected when comparing several vectors with the same term. The last step is to index the vectors so that the vectors may be searched both by concept name and by terms (resulting in a ranked list of concepts).

In [11] we also introduced the notion of negative concept vectors, which are built in the same way but contain only terms that are not relevant for the domain. The documents used to build these negative concept vectors are strictly non-relevant to the domain.

5   Ontological Profiles in Search

This section will give a description of how the ontological profile may be used in a search application. Our approach to using the ontological profile in search, is to use it as a tool for semantic reformulation of queries on top of a standard vector space based search engine (we use Lucene), using the reformulated query as a query into the index. This approach lets the system hide from the user the fact that an ontology is used, and the user is only faced with entering familiar keyword queries.

The query reformulation process is based on two steps. The first is to interpret the user query, in effect this means that we map the user query on to a set of one or more concepts in the ontology. The second step is to expand the query with semantically related terms, i.e. we use the concepts from the first step. These steps will be described in the following.

5.1   Query Interpretation

The goal of the query interpretation is to map the user query (keywords supplied by the user) on to a set of one or more concepts of the ontology. In this mapping process we use the ontological profile as a measure of the coherence between the concepts and the users query terms. We have suggested four such query interpretation techniques which produce a ranked list of concepts to choose from during the query expansion. Recall that the ontological profile contains for each concept a list of terms and their weights showing their importance for the concept. In effect we try to find the set of concepts that maximize the semantic coherence between the query and the concepts.

Simple query interpretation. The simple query interpretation is the most basic schema, mapping each query term to a single concept. For each query term we find the concept which has the highest tf*idf score for that term in its vector. The concept with the highest tf*idf score is subsequently picked as the semantic representation of the query term and picked for expansion in the next step.

Best match query interpretation. In the simple interpretation scheme we do not consider any relations between the user entered query terms. The best match approach assumes that there is in fact a relation between the user entered query terms, and tries to recognize this relationship by attempting to map the user query terms on to a single concept which has a good representation of the query terms collectively. This is done by requiring the candidate concepts to contain all query terms. The concept which maximizes the score given in Equation 3 ($ccore_c$ is the score for concept c, $t_{i,c}$ is the tf*idf score for query term i in concept c) is picked for subsequent expansion.

$$score_c = t_{0,c} + t_{1,c} + \ldots + t_{n-1,c} \tag{3}$$

Cosine similarity query interpretation. The last two interpretation schemas are attempts at disambiguating the user query. This is done by attempting to recognize relations between the query terms and the concepts they map to. There could possibly be a relation between the first concept for term 1 and the third concept for term 2, which is not recognized by the simple interpretation. In the cosine similarity approach we use the cosine similarity between concept vectors as a measure of the relationship between the concepts. The first part of the interpretation is in fact similar to the simple approach, in which for each query term we generate a ranked list of the 15 most related concepts. For each pair of concepts we calculate a score (Equation 4, cin is the concept ranked as n with respect to query term i) which takes into account both the score for each term and the relationship between the concept vectors. The pair of concepts with the highest score will be picked as the semantic representation of the query and used for subsequent expansion.

$$score_{cin,cjm} = t_{i,n} * t_{j,m} * cos\_sim(cin, cjm) \tag{4}$$

Due to complexity we have only implemented and tested the last two approaches for two query terms.

Ontology structure query interpretation. This query interpretation approach is similar to the cosine similarity interpretation, but uses a different measure for the similarity between the concepts. The similarity measure is based on the structure of the ontology, where we use the distance between the concepts as a measure of the similarity. We rely on the assumption that concepts that are close in the ontology have a higher semantic coherence than concepts that are further apart. We generate a graph structure representation of the ontology in which the concepts are nodes and the relations are edges. This graph is then traversed to find the distance for each of the concept pairs. As for the cosine similarity based approach, we generate a ranked list of the 15 most related concepts for each of the query terms. The score for each pair of concepts is calculated as shown in Equation 5 where the path function gives the distance between the concept pairs. The pair of concepts with the highest score is picked as the semantic representation of the query and used in the subsequent expansion process.

$$score_{cin,cjm} = t_{i,n} * t_{j,m} * \frac{1}{path(cin, cjm)} \tag{5}$$

5.2   Query Expansion

The query expansion process reformulates the original query by adding semantically related terms to the query. This is done by adding the top 15 terms to the original query with their weights. This means that the final query is a weighted query, and the original query terms are boosted to signal that these are the most important terms in the query. In addition we use the negative concept vectors to add for each concept the top 15 negative terms to the query. These terms are added as NOT terms, signaling that any documents containing these terms should not be included in the result set. This takes care of removing some of the noise introduced by the increased set of query terms. Finally the weighted query is fired against a vector space search engine (Lucene) and the result set is returned to the user.

5.3   Results

There are two main approaches to evaluating the usefulness of ontological profiles. The first is to evaluate it in an information retrieval system, and the second is to evaluate the concept vectors with respect to reflecting the quality of the concept vectors. First we give a short description of the evaluation of the ontological profile used in information retrieval (based on the approach described in Section 5). We have used the IIP [12] core ontology, which contains 18,675 concepts, covering the domain of oil production and drilling in subsea conditions. For the construction of the ontological profile we used the Schlumberger Oilfield Glossary [2] and successfully created concept vectors for 2,195 concepts. Our evaluation, which consisted of 7 queries and 5 test subjects, revealed that all four of our reformulation strategies performed significantly better than pure keyword search (see Figure 4, left). One surprising result was that all four of the strategies performed very equally, none of them stood out in any way with respect to query type. This suggests that the structure of the ontology is not as important as we first thought. Although the evaluation is not statistically significant, it gives an impression of the performance of our strategy. For a more thorough description of the evaluation, see [13].

As we introduce new terms to the query, we hypothesize that the reformulated queries enhance the recall of the system, as a broadened set of query terms will retrieve more documents. We have however not investigated how precision is affected by the proposed search strategy. The quality of the vectors is of critical importance for the system to perform well, as is also indicated by [8]. Further work includes tweaking of parameters and giving the system a more thorough evaluation.

The second evaluation considers the quality of the vectors. We did an evaluation (in the project management domain) in which we evaluated the quality of the relationships found by calculating the cosine similarity between the concept vectors. With relationships we mean not necessarily the relationships found in

---

[2] http://www.glossary.oilfield.slb.com

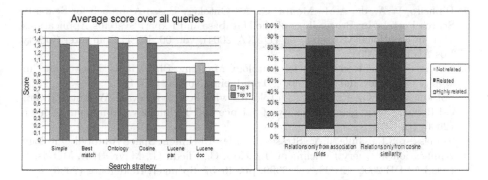

Fig. 4. Left: Evaluation of ontological profiles in search [13] Right: Evaluation of concept relations [14]

the ontology, but the relationships found among the concepts based on the concept vectors. The right part of Figure 4 show that the relationships found were generally of higher quality than those found by other means [14].

6 Conclusion

We have in this paper presented the concept of ontological profiles, and how they are constructed. We argue that they are more powerful semantic representations of a domain than an ontology is on its own. Text mining techniques are employed to extend the ontological concepts with terms that are semantically linked to the concepts, and weights showing the strength of these relations. Finally we showed how a search system based on ontological profiles may be constructed. The evaluation of the search prototype showed promising results, and we will continue research on ontological profiles for search applications.

Acknowledgment. This research was carried out as part of the IS_A project, project no. 176755, funded by the Norwegian Research Council under the VERDIKT program.

References

1. Buscaldi, D., Rosso, P., Arnal, E.S.: A wordnet-based query expansion method for geographical information retrieval. In: Working noted for CLEF workshop (2005)
2. Ozcan, R., Aslangdogan, Y.A.: Concept based information access using ontologies and latent semantic analysis. Technical report cse-2004-8, University of Texas at Arlington
3. Nagypal, G.: Improving information retrieval effectiveness by using domain knowledge stored in ontologies. In: Meersman, R., Tari, Z., Herrero, P. (eds.) OTM-WS 2005. LNCS, vol. 3762, pp. 780–789. Springer, Heidelberg (2005)
4. Weide, T.P., Grootjen, F.A.: Conceptual Query Expansion. Data & Knowledge Engineering (56), 174–193 (2006)

5. Pinheiro, W.A., de, A.M., Moura, C.: An Ontology Based-Approach for Semantic Search Portals. In: Proceedings of the Database and Expert Systems Applications, 15th International Workshop on (DEXA 2004), vol. 00. IEEE Computer Society, Los Alamitos (2004)

6. Su, X.: Semantic Enrichment for Ontology Mapping. PhD Thesis, Norwegian University of Science and Technology, Trondheim, Norway (2004)

7. Solskinnsbakk, G.: Extending Ontologies with Search-Relevant Weights. Technical Report, Norwegian University of Science and Technology, Trondheim, Norway (2006)

8. Tomassen, S.L., Gulla, J.A., Strasunskas, D.: Document Space Adapted Ontology: Application in Query Enrichment. In: Kop, C., Fliedl, G., Mayr, H.C., Métais, E. (eds.) NLDB 2006. LNCS, vol. 3999, pp. 46–57. Springer, Heidelberg (2006)

9. Frakes, W.B., Fox, C.J.: Strength and similarity of affix removal stemming algorithms. SIGIR Forum 37(1), 26–30 (2003)

10. Baeza-Yates, R., Ribeiro-Neto, B.: Modern Information Retrieval. ACM Press, New York (1999)

11. Solskinnsbakk, G.: Ontology-Driven Query Reformulation in Semantic Search. MSc Thesis, Norwegian University of Science and Technology, Trondheim, Norway (2007)

12. Gulla, J.A., Tomassen, S.L., Strasunskas, D.: Semantic interoperability in the Norwegian petroleum industry. In: Karagiannis, D., Mayer, H.C. (eds.) 5th International Conference on Information Systems Technology and its Applications (ISTA 2006). Lecture Notes in Informatics (LNI), vol. P-84, pp. 81–94. Köllen Druck Verlag GmbH, Bonn, Klagenfurt Austria (2006)

13. Solskinnsbakk, G., Gulla, J.A.: Ontological Profiles in Enterprise Search. In: EKAW 2008 - 16th International Conference on Knowledge Engineering and Knowledge Management, Acitrezza, Catania, Italy, September 29- October 3, 2008 (submitted, 2008)

14. Gulla, J.A., Brasethvik, T., Sveia Kvarv, G.: Using Association Rules to Learn Concept Relationships in Ontologies. In: ICEIS 2008: 10th International Conference on Enterprise Information Systems, Barcelona, Spain, June 12 - 16, 2008 (accepted, 2008)

A Hybrid Approach to Ontology Relationship Learning

Jon Atle Gulla and Terje Brasethvik

Department of Computer and Information Sciences,
Norwegian University of Science and Technology, Trondheim
jag@idi.ntnu.no

Abstract. Most ontology learning tools concentrate on extracting concepts and instances from text corpora. There are some recent tools that employ linguistics or data mining to uncover concept relationships, but the results are mixed. Since relationships are semantically complex notions, it seems interesting to combine approaches that address different aspects of concept relationships. In this paper we present a hybrid approach that combines the co-occurrence principle from association rules with contextual similarities from linguistics. The technique has been tested in an ontology engineering project, and the results show significant improvements over traditional techniques.

1 Introduction

Ontology engineering is a tedious and labor-intensive process that requires a wide range of skills and experiences. Except for the challenges of dealing with very complex and formal representations, the modelers need to manage and coordinate the contributions from various types of domain experts. Moreover, there are strategic issues that tend to interfer with the process and increase the costs of constructing and maintaining ontologies.

While ontologies are expensive to develop, they provide the common vocabulary that many applications need to interoperate. In spite of the costs and time needed, many projects resort to traditional ontology modeling approaches that emphasize the systematic manual assessment of the domain and gradual elaboration of model descriptions (e.g. [5, 7]).

Ontology learning tools promise to automate parts of the ontology engineering process (e.g. [11, 14, 16]). These tools process textual domain descriptions and try to come up with ontological structures with no or limited human intervention. The assumption is that the domain text reflects the terminology that should go into the ontology, and that appropriate linguistic and statistical methods may automatically extract suitable concept candidates and their relationships.

Already current ontology learning tools allow ontologies to be generated faster and with less costs than traditional modeling environments. However, manual verification and correction of the generated structures are still needed, as there are many aspects that are hard to address with automatic techniques. The best results so far are for the learning of prominent phrases, synonyms and concepts. Extracting relationships between concepts is an inherently complex task that requires a thorough understanding of how relationships are perceived. The relationships range from generalization hierarchies and part-of structures to vague and undefinable associations between concepts. Even though there are some ontology learning tools that offer relationship learning facilities, the

E. Kapetanios, V. Sugumaran, M. Spiliopoulou (Eds.): NLDB 2008, LNCS 5039, pp. 79–90, 2008.

accuracy of the relationships extracted is questionable and there has only been limited work on comparing the various approaches to relationship learning. This is unfortunate, as there are indications that several approaches may be successfully combined in multi-perspective relationship learning approaches.

In this paper we present a hybrid relationship learning approach that combines a linguistic contextual analysis of concepts with co-occurrence patterns across documents. Both techniques may be used separately to extract ontology relationships, and they tend to select slightly different types of relationships. Used in combination, though, we end up with relationship candidates that are significantly better than what each of the techniques can offer in isolation.

The paper is structured as follows. Section 2 discusses the nature of relationships in ontologies and explains why it may be useful to adopt a multi-perspective approach. Section 3 introduces association rules, and Section 4 presents a new learning approach that uses concept profiles to extract relationships. Whereas Section 5 outlines the implementation of our hybrid approach, the main conclusions of the evaluation are discussed in Section 5. The conclusions are found in Section 6.

2 Ontology Relationships

Ontology relationships provide logical links or dependencies between concepts or instances in ontologies. Whereas some form taxonomic hierarchies, there are others like synonyms and antonyms that relate concepts or terms horizontally across hierarchies. For ontologies used in enterprise applications, we tend to need at least the following types of relationships:

♦ *Hyperonymy* (is_a relationships). These are taxonomic relationships that establish abstraction hierarchies, specifying for example that *project managers* and *project members* are both company *employees*.
♦ *Meronymy* (part_of relationships). These relationships indicate that several concepts together make up another concept, and it includes both the notion of aggregation and composition from UML. Examples from the project management domain are the elements of a *project charter*, or the *people* forming a project *organization*.
♦ *Associations*. These are non-taxonomic relationships that say that one concept is logically related to another. Examples are *sponsors* financing *projects* or *ordinary project members* reporting to their *managers*.

Relationships are of fundamental importance to many Semantic Web applications. Whereas constraints and rules are sometimes used to provide accurate concept definitions, it is often the relationships that help applications relate concepts and give users and applications the flexibility they need to retrieve and exchange information.

This work addresses the learning of mostly non-taxonomic relationships. As opposed to taxonomic relationships, for which it is often possible to take advantage of external linguistic resources or the internal structure of terms to uncover hierarchical structures, non-taxonomic relationships tend to be domain-dependent and can only be understood in the context of the documentation available in the domain. There are basically two overall approaches to learning non-taxonomic relationships, the co-occurrence approach and the definitional approach:

♦ *Co-occurrence approach.* Concepts may be considered related if they consistently show up in the same context. If two concepts occur frequently in the same documents – compared to any other pair of concepts – we may assume that they semantically deal with related issues and there may be a relationship between them. The approach does not attempt to expose any semantic aspects of the concepts, and the quality depends heavily on the availability of statistical data.

♦ *Definitional approach.* Another approach is to use textual data available to characterize concepts semantically. Having generated formal semantic characterizations of all concepts, one can compute the semantic similarities of any two pair of concepts. The hypothesis is that there is a relationship between two concepts if their semantic similarity is above a certain threshold. This approach needs less statistical data and tends to capture more low-level relationships like synonyms and predicate-argument structures.

Our hybrid approach to ontology learning makes use of both a co-occurrence approach and a definitional approach, and the final results are found by intersecting the part results of the two techniques. We are using a well-known co-occurrence approach, *association rules*, to learn general high-level relationships between concepts. The technique is well understood and has proven efficient in many other ontology learning tools. For the definitional approach, we have developed an entirely new technique using *concept profiles* and cosine similarities. This technique addresses more subtle semantic issues and has been useful in generating low-level relationships between specialized domain concepts. The details of both methods are explained in the following.

3 Association Rules for Ontology Learning

Association rules come from data mining and is a technique identifying data or text elements that co-occur frequently within a dataset. They were first introduced in [1] as a technique for market basket analysis, where it was used to predict the purchasing behavior of customers. An example of such an association rule is the statement that *"90% of the transactions that purchased bread and butter also purchased milk."*
   The problem in association rules mining can be formally stated as follows:

Let I be a set of literals, called items. Let D be a set of transactions, where each transaction T is a set of items such that $T \subseteq I$. A transaction T contains X, a set of some items in I, if $X \subseteq T$.

An association rule is an implication of the form

$X \Rightarrow Y$,   where $X \subset I, Y \subset I, X \cap Y = \emptyset$

A rule $X \Rightarrow Y$ holds in the transaction set D with *confidence* c if c% of the transactions in D that contain X also contain Y. The rule $X \Rightarrow Y$ has *support s* in the transaction set D if s% of the transactions in D contain $X \cup Y$.

The approach is to generate all association rules that have support and confidence greater than user-specified minimum support and confidence levels. Implementationally, the most common algorithm for the generation of association rules is the Apriori algorithm, introduced originally in [2]. The algorithm detects all sets of items that have support greater than the minimum support. These sets are called *frequent item sets*. For every itemset l in the frequent itemset, L_k, it finds subsets of size k-1. For every subset X, it generates a rule $X \Rightarrow Y$, where $Y = l - X$. The rule is stored if the confidence

$$support(X \cup Y) / support(X)$$

is greater than or equal to the minimum confidence.

In a text mining context, association rules are used to indicate high-level relationships between concepts. Let us assume that an item set is a set of one or more concepts. If the rule $X \Rightarrow Y$ has been generated and stored, we can conclude that there is a relationship between the concepts in X and the concepts in Y. With item sets of size 1, we have rules that indicate relationships between two concepts.

In order to run association rule mining on text, we need to structure the text to mirror the setup in data mining. Following [6, 10], we consider documents – rather than sentences or paragraphs – to correspond to transactions in data mining. Since we are only interested in extracting relationships between potential concepts, we also restrict the analysis to noun phrases only. We reduce the noun phrases to their base forms, so that *project plans* and *project plan* count as the same term and only include noun phrases that have a certain prominence in the document set. We then have documents as item sets and lemmatized prominent noun phrases as items and can run a standard association rules analysis to suggest relationships between the prominent noun phrases.

4 Concept Profiles for Ontology Learning

Concept profiles are semantic characterizations of concepts in terms of words and phrases used to refer to the concept in a particular domain. They share similarities with synsets in WordNet, though the profiles are larger and include also words that are more distantly related to the concept. Weights are used to indicate to what extent a profile's word is central to the semantics of the corresponding concept. The concept profile for a project manager may include terms like [manager$_{1.0}$, leader$_{1.0}$, lead$_{0.8}$, charge$_{0.6}$, management$_{0.4}$, report$_{0.3}$, etc]. It does not only include noun phrases, but any term that is used to address semantic issues related to the concept.

Technically, the profiles are vectors of weighted terms related to the concept. The weights are calculated on the basis of a proximity analysis of sentences, paragraphs and documents in the domain. The idea is that terms used in direct proximity of a concept should be semantically closer to the concept than terms used farther away. Terms that are consistently used in the same phrases as a specific concept should semantically capture parts of the concept's meaning itself. The computation of concept vectors is done as follows:

Let j be a concept and $\vec{v}_j$ be concept j's profile. The term frequency for term i in the concept profile $\vec{v}_j$ of concept j is calculated as

$$vf_{i,j} = \alpha \cdot \sum_{d \in D} f_{i,d} + \beta \cdot \sum_{p \in P} f_{i,p} + \gamma \cdot \sum_{s \in S} f_{i,s}$$

where D is the possibly empty set of relevant documents including j
$\quad$ P is the possibly empty set of relevant paragraphs including j
$\quad$ S is the possibly empty set of relevant sentences including j
$\quad$ $f_{i,d}$ is the frequency of term i in document d
$\quad$ $f_{i,p}$ is the frequency of term i in paragraph p
$\quad$ $f_{i,s}$ is the frequency of term i in sentence s
$\quad$ $\alpha < \beta < \gamma$ (values used in experiment: α=0.1 , β=1.0 , γ=10.0)

The concept profile of concept j is a vector $\vec{v}_j = [w_{1,j}..., w_{k,j}]$, where $w_{i,j}$ is the weight of term i and is equal to the following *tf.icf* score:

$$tficf_{i,j} = \frac{vf_{i,j}}{max(vf_{l,j})} \cdot \log \frac{N}{n_i}$$

where $tficf_{i,j}$ is the weight of term i in the concept profile vector of concept j
$\quad$ $max(vf_{l,j})$ is the frequency of the most frequently occuring term l in vector $\vec{v}_j$
$\quad$ N is the number of concept profiles
$\quad$ n_i is the number of concept profiles containing term i

---

When all concepts are described in terms of concept profiles, we may calculate the relatedness between concepts using the cosine formula

$$cos(\vec{x}, \vec{y}) = \frac{\sum_{i=1}^{n} x_i y_i}{\sqrt{\sum_{i=1}^{n} x_i^2} \sqrt{\sum_{i=1}^{n} y_i^2}}$$

where $\vec{x}$ and $\vec{y}$ are concept profiles and x_i is the weight of the ith term of profile $\vec{x}$. If the cosine similarity is above a certain threshold, we conclude that there is a relationship between the concepts. The set of all cosine similarities above this threshold for all concepts pairs in the ontology is the system's suggested list of relationships in the domain. More details about this method are found in [17].

5 Learning a Project Management Ontology

Our ontology learning approach requires two separate chains of analysis.

The association rules chain is built as an extension to the GATE environment from the University of Sheffield[1] [8]. *General Architecture for Text Engineering* (GATE) is an open source Java framework for text mining. It contains an architecture and a development environment that allows new components to be added and integrated with existing ones. An analysis with GATE typically consists of a chain of components that one by one goes through the text and annotates it with information that will be used by later components. With our own components for association rules added, we built the analysis chain shown in the lower part of Figure 1. The analysis is run on a repository of documents representative to the project management domain. Tokenization, sentence splitting, parts-of-speech tagging, lemmatization and noun phrase extraction are all carried out with standard GATE components. Whereas the GATE components work on individual documents, we developed our own modules for association rules that pulled the individual files together, extracted prominent noun phrases as keywords, and suggested relationships between these phrases.

The *noun phrase extractor* from GATE identifies noun phrases in the text of the form Noun (Noun)*, i.e. phrases that consist of consecutive nouns. This means that a phrase like *very large databases* will not be recognized, since *very* is an adverb and *large* is an adjective, whereas *project cost plan* is a perfectly recognized phrase. Not all recognized noun phrases are suitable concepts, though. Our *noun phrase indexer* removes stopwords and extracts and counts the frequencies of noun phrases in the document set. A normalized term-frequency score (tf score) is used to select those prominent noun phrases that are most likely to be concepts in the domain. These are carried over to the association rules component that produces association rules between the prominent noun phrases (concepts) found by the previous component.

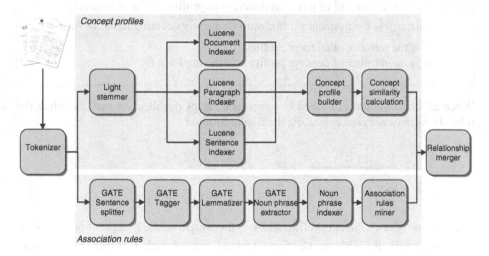

Fig. 1. Process for extraction relationships using concept profiles and association rules

The second chain of analysis is based on concept profiles and is shown in the upper part of the figure. After removing any trailing "s" in the stemming part, the Lucene search platform is used to construct indices at the level of documents, paragraphs, and

[1] http://gate.ac.uk/

sentences. Components in Java are then used to compute concept profiles for concepts extracted with tf.idf scores from the document collection, and then to calculate similarities between concepts using the cosine similarity formula.

Both chains of analysis produce sets of ranked concept relationships. The relationship merger at the end of the analysis simply selects those relationships that are suggested by both techniques.

The relationship learning approach was tested on the project management discipline PMI in Statoil, Norway's biggest petroleum company. The domain is documented by a project management manual, the PMBOK[2], which contains about 50.600 words (tokens) divided into 12 chapter. In total, a 76 document large collection was uploaded and used for the generation of relationship candidates.

6 Evaluation

Evaluating ontology relationships learning systems is notoriously difficult, as there are potential relationships between all concepts and only subjective judgment can tell the

Related concepts only from association rules		Related concepts only from cosine similarity		Related concepts from both methods	
project management team	R	cost management	HR	activity	R
management team	R	cost baseline	HR	assumption	NR
organization	R	actual cost	HR	control	R
product	HR	schedule	R	cost estimate	HR
information	R	project schedule	R	performance	R
tool	R	earn value	R	process	R
project team	R	staff	R	project	HR
application area	R	project staff	R	project management	R
risk analysis	R	milestone	NR	project objective	R
result	R	plan value	R	project plan	R
risk	R	stakeholder	HR	quality	R
resource	R	project deliverable	R	scope	R
consequence	R	ev	NR	scope statement	R
estimate	R	earn value management	R		
phase	NR	management	R		
probability	R	scope definition	NR		
action	R	scope management	R		
analysis	R	customer	R		
seller HR	HR	sponsor	R		
		project management IS	R		
		constraint	R		
		project manager	R		
		project plan development	R		
		procurement management	NR		
		project plan execution	NR		
		quality management	R		
		work breakdown structure	R		

Fig. 2. Relationships suggested for the concept *Cost*. The average scores are NR (not related), R (related) or HR (highly related).

---

[2] Project Management Institute. A Guide to the Project: Management Body of Knowledge (PMBOK), 2000.

important ones from the others. More interesting is perhaps the comparative evaluation of the two different approaches, as well as the comparison with the hybrid approach.

The two approaches did not assume the same set of concepts, as they used slightly different techniques for extracting concept candidates. The tf.idf-based approach with concept profiles produced 196 concepts, while the tf-based approach with association rules gave us 142 concepts. 111 concepts, or 56.7 % of the 196, are shared by the two approaches. Of the relationships generated by the two approaches, slightly more than 50% were confirmed by both.

A group of 4 people with strong project management skills were asked individually and independently to rate the relationships only suggested by the association rule approach, the relationships only suggested by the concept profile approach, and the relationships suggested by both approaches. Each person rated each relationship as *not related* (these two concepts are not related), *related* (there is probably a relationship between the two concepts) or *highly related* (there is definitely a relationship between these two concepts). An average score for each relationship was afterwards calculated on the basis of the individual scores. Figure 2 shows the related concepts suggested for the ontology concept *Cost* for the three groups, as well as their average scores.

Adding the results for all concepts together, we can compare the quality of relationships for the three groups. As shown in Figure 3, association rules and concept profiles tend to produce the same share of good relationships (score *Related* and *Highly related* added together). The two methods suggested 82% and 86% good relationships, respectively, which is a fairly good result for such a small document collection. It should be noted, though, that this does not mean that they necessarily suggest the same relationships. The share of very good relationships is worth a closer inspection. Whereas the association rules method only generated 7% very good relationships, the concept profile method reached an impressive 24%.

A possible explanation for this difference lies in the mechanics of association rules and cosine similarity. For an association rule to be generated, the corresponding concepts need to occur is a wide range of documents. This will typically be the case for very general concepts and their rather general relationships. The cosine similarity

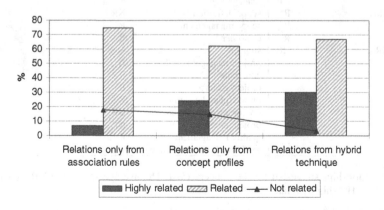

Fig. 3. Evaluation results for three categories of relationships

method, on the other hand, makes use of tf.idf to characterize concepts by their differences to other concepts, and the relationships based on cosine similarities will be based on these discriminating concept vectors. The relationships get more specialized and precise and are easier to recognize as very good relationships. This may also explain why the association rule method had a larger share of normally good relationships (75%) than the cosine similarity method (65%).

The hybrid approach is worth a closer inspection. The individual methods carry significant noise, but our results indicate that this noise is dramatically reduced when we only keep the results that are common to both methods. In total, 97% of the relationships suggested by the hybrid method were rated as good relationships by the test group (right columns in Figure 3). 30% were considered very good relationships. This suggests that the two subapproaches – although comparable in quality – are fundamentally different with their own weaknesses and strengths. As far as association rules and concept profiles are concerned, our research indicates that the hybrid approach displays substantially better results than each individual approach.

7 Related Work

There are today several ontology learning environments that include facilities for relationship learning (see Figure 4). Association rules are already in use in some of these systems [4].

Systems	Terms	Synonyms	Concepts	Hierarchy	Relations	Other
Text2Onto	X	clusters	X	X	X	X
HASTI	X			X	X	X
OntoBasis		clusters	clusters			
OntoLT/ RelExt	X			X	X	
CBC/DIRT		clusters	clusters			
DOOBLE		X			X	
ASIUM		clusters	clusters	X	X	
OntoLearn	X	X	X	X	X	
ATRACT	X	clusters	clusters			

Fig. 4. Relationships learning in current systems

Our approach to association rules is similar to what can be found in other ontology learning tools. The accuracy of ontology relationship learning is still not satisfactory and suffers from both overgeneration and uncertainty. So far, the techniques have also failed in coming up with good labels for these relationships. Text2Onto uses an intermediate structure, the Probabilistic Ontology Model (POM), that allows them to incrementally learn concepts and relationships [3].

Our approach with association rules draws on many of the ideas advocated by Haddad et al. [10]. In their work they also use documents as transactions and focus on noun phrases as the carriers or meanings and the objects of analysis. A similar approach is taken in [15].

Another interesting application of association rules is presented in Delgado et al. [6]. Their idea is to use association rules to refine vague queries to search engine applications. After the search engine has processed the initial query, their system weights the words in the retrieved documents with tf.idf and extracts an initial set of prominent keywords. Stopwords are removed and the remaining keywords are stemmed. Representing the stemmed keywords of each document as a transaction, their system is able to derive association rules that relate the initial query terms with other terms that can be added as a refined query.

Association rules have also been applied in web news monitoring systems. Ingvaldsen et al. [12] incorporate association rules and latent semantic analysis in a system that extracts the most popular news from RSS feeds and identifies important relationships between companies, products and people.

The concept profile approach is not used by any other existing ontology learning tool. A variant of the method has however been used in mapping concepts from one ontology to another [18]. The calculation of concept profiles does not take proximity into account in that system, but its similarity measures are slightly more complex with additional input from WordNet.

Our research is now focused on the integration of different relationship learning approaches. The combination of association rules and concept profiles is promising, and the results are significantly better than for each individual technique. To our knowledge, there are no tools available today that successfully combine co-occurrence approaches with definitional approaches in concept relationship learning.

8 Conclusions

This paper presented a hybrid ontology relationship learning approach that makes use of association rules and concept profiles to identify relationships between concepts. The approach is implemented as a text mining analysis system that makes use of both GATE components and the Lucene indexing system. New components can easily be added to this architecture, as the approach is gradually refined with additional techniques.

Association rules provide a powerful and straight-forward method for extracting possible ontology relationships from domain text. The relationships extracted may be both taxonomic and non-taxonomic, though it is difficult to use the analysis alone to decide on the nature of the relationships. Of the relationships extracted for the project management domain, about 82% were considered valid relationships by a test group with previous experience in project management.

However, association rules do not perform better than methods based on concept profiles and cosine similarity calculations. In fact, the concept profile approach seems to generate more precise relationships than association rules. The methods are however complementary of nature, since association rules tend to focus on general relationships between high-level concepts and cosine similarity approaches focus on specialized relationships among low-level concepts.

The hybrid ontology relationship learning approach indicates that it may be useful to combine approaches that address different aspects of ontology relationships. This leads to a substantial reduction of noise, and we can single out more very good relationships. We are now investigating to what extent several relationship learning strategies can be combined in an incremental learning strategy. An extension of the POM structure from Text2Onto may be useful in this respect, though we need to carry over more than just probability measures when these techniques are applied sequentially. The whole set of uncertainties, possibly supported by evidence in terms of vectors or raw calculation, need to come together in such a comprehensive ontology learning framework.

References

1. Agrawal, R., Imielinski, T., Swami, A.N.: Mining Association Rules between Sets of Items in large Databases. In: Proceedings of the 1993 ACM SIGMOD International Conference on Management of Data (1993)
2. Agrawal, R., Srikant, R.: Fast Algorithms for Mining Association Rules. In: Proceedings of the 20th International Conference on Very Large Data Bases (VLDS 1994) (1994)
3. Cimiano, P., Völker, J.: Text2Onto. In: Montoyo, A., Muñoz, R., Métais, E. (eds.) NLDB 2005. LNCS, vol. 3513, pp. 227–238. Springer, Heidelberg (2005)
4. Cimiano, P., Völker, J., Studer, R.: Ontologies on Demand? A Description of the State-of-the-Art, Applications, Challenges and Trends for Ontology Learning from Text. Information, Wissenschaft und Praxis 57(6-7), 315–320 (2006)
5. Cristiani, M., Cuel, R.: A Survey on Ontology Creation Methodologies. Idea Group Publishing (2005)
6. Delgado, M., et al.: Association Rule Extraction for Text Mining. In: Andreasen, T., Motro, A., Christiansen, H., Larsen, H.L. (eds.) FQAS 2002. LNCS (LNAI), vol. 2522. Springer, Heidelberg (2002)
7. Fernandez, M., Goméz-Peréz, A., Juristo, N.: Methontology: from ontological art towards ontological engineering. In: Proceedings of the AAAI 1997 Spring Symposium Series on Ontological Engineering, pp. 33–40. Stanford, Menlo Park (1997)
8. Gaizauskas, R., et al.: GATE User Guide (1996), http://gate.ac.uk/sale/tao/index.html#x1-40001.2
9. Gulla, J.A., Borch, H.O., Ingvaldsen, J.E.: Ontology Learning for Search Applications. In: Proceedings of the 6th International Conference on Ontologies, Databases and Applications of Semantics (ODBASE 2007). Springer, Vilamoura (2007)
10. Haddad, H., Chevallet, J., Bruandet, M.: Relations between Terms Discovered by Association Rules. In: Zighed, A.D.A., Komorowski, J., Żytkow, J.M. (eds.) PKDD 2000. LNCS (LNAI), vol. 1910. Springer, Heidelberg (2000)
11. Haase, P., Völker, J.: Ontology Learning and Reasoning - Dealing with Uncertainty and Inconsistency. In: da Costa, P.C.G., et al. (eds.) Proceedings of the International Semantic Web Conference. Workshop 3: Uncertainty Reasoning for the Semantic Web (ISWC-URSW 2005), pp. 45–55. Galway (2005)
12. Ingvaldsen, J.E., et al.: Financial News Mining: Monitoring Continuous Streams of Text. In: Proceedings of the 2006 IEEE/WIC/ACM International Conference on Web Intelligence, Hong Kong, December 2006, pp. 321–324 (2006)
13. Maedche, A., Staab, S.: Semi-automatic Engineering of Ontologies from Text. In: Proceedings of the 12th Internal Conference on Software and Knowledge Engineering, Chicago (2000)
14. Navigli, R., Velardi, P.: Learning Domain Ontologies from Document Warehouses and Dedicated Web Sites. Computational Linguistics 30(2), 151–179 (2004)

15. Nørvåg, K., Eriksen, T.Ø., Skogstad, K.-I.: Mining Association Rules in Temporal Document Collections. In: Esposito, F., Raś, Z.W., Malerba, D., Semeraro, G. (eds.) ISMIS 2006. LNCS (LNAI), vol. 4203, pp. 745–754. Springer, Heidelberg (2006)
16. Sabou, M., et al.: Learning Domain Ontologies for Semantic Web Service Descriptions. Journal of Web Semantics (accepted, 2008)
17. Solskinnsbakk, G.: Ontology-Driven Query Reformulation in Semantic Search, in Department of Computer and Information Sciences. Norwegian University of Science and Technology, Trondheim (2007)
18. Xu, X., Gulla, J.A.: An information retrieval approach to ontology mapping. Data & Knowledge Engineering 58(1), 47–69 (2006)

Automating the Generation
of Semantic Annotation Tools
Using a Clustering Technique*

Vitór Souza[1], Nicola Zeni[1], Nadzeya Kiyavitskaya[1], Periklis Andritsos[1],
Luisa Mich[2], and John Mylopoulos[1]

[1] Dept. of Information Engineering and Computer Science
[2] Dept. of Computer and Management Sciences,
University of Trento, Italy

Abstract. In order to generate semantic annotations for a collection
of documents, one needs an annotation schema consisting of a semantic
model (a.k.a. ontology) along with lists of linguistic indicators (keywords
and patterns) for each concept in the ontology. The focus of this paper is
the automatic generation of the linguistic indicators for a given semantic
model and a corpus of documents. Our approach needs a small number of
user-defined seeds and bootstraps itself by exploiting a novel clustering
technique. The baseline for this work is the Cerno project [8] and the
clustering algorithm LIMBO [2]. We also present results that compare
the output of the clustering algorithm with linguistic indicators created
manually for two case studies.

1 Introduction

Semantic annotation is commonly recognized as the one of the cornerstones
of the Semantic Web. To generate domain-dependent metadata a semantic annotation system utilizes a semantic model, a.k.a. ontology along with sets of
linguistic indicators (keywords and patterns, usually constructed manually by a
domain expert) that determine what text fragments are to be annotated. This
work was conducted in the context of the Cerno [8] project to explore the applicability of some of the main ingredients of a supervised categorical clustering
algorithm LIMBO [2] for producing linguistic indicators for a given semantic
model. LIMBO was originally proposed for clustering structural information of
database tuples in relational databases. The main motivation for applying this
method is that it requires a limited amounts of training data to bootstrap the
learning algorithm and of human intervention at the initial stage. Moreover, the
distance measure employed in LIMBO works with categorical data (i.e. data that
do not have an inherent order) unlike most clustering techniques. Our primary
goal in this work is to verify whether such a lightweight approach can facilitate

---
* This work has been partially funded by the EU Commission through the SERENITY
and WEE-NET projects and by Provincia Autonoma di Trento through the STAMPS
project.

E. Kapetanios, V. Sugumaran, M. Spiliopoulou (Eds.): NLDB 2008, LNCS 5039, pp. 91–96, 2008.

the construction of an annotation schema, given a semantic model and a training set of documents. The focus of this paper is the automation of the generation of an annotation schema as a component of domain dependent annotation tools using a clustering approach. We evaluate the performance of the method on two different data sets and the related annotation schemas manually developed for the previous applications of Cerno. This paper is structured as follows. The baseline of the present work in sketched in Section 2. It introduces the semantic annotation framework Cerno and LIMBO, the clustering technique adopted. Section 3 describes how the baseline technologies were extended and shows the tool built on top of the LIMBO. Section 4 presents the setup and evaluation of two experimental case studies and summarizes the lessons learned. Section 5 recalls the related work. Finally, conclusions are drawn in Section 6.

2   Research Baseline

2.1   Cerno Semantic Annotation Framework

Cerno is a lightweight semantic annotation framework that exploits fast and scalable techniques from the software reverse engineering area. To annotate input documents, Cerno uses context-free grammars, generates a parse tree, and applies transformation rules to generate output in a target format [8]. The reader can find a detailed description of the architecture and the performance of the system in [8]. Normally, adapting Cerno to a new application domain requires a couple of weeks, because its domain dependent components have to be tuned for a given type of documents and a specific semantic model. In this work we explore the possibilities to automate the generation of such indicators for specific semantic domain. Having a set of examples, one can try to identify a set of contextual keywords describing relevant concepts using well-established statistical methods that have been proven effective in many areas. To this end, we have been experimenting with a scalable hierarchical categorical clustering algorithm called LIMBO [2].

2.2   Data Clustering with LIMBO

Data clustering [7] is a common technique for statistical data analysis and is widely used in many fields. Our approach is based on LIMBO [2], a scalable hierarchical categorical clustering algorithm that builds on the Information Bottleneck (IB) [11] framework for quantifying the relevant information preserved when clustering. The algorithm proceeds in three phases: Phase 1 constructs a cluster representative for the initial data set for efficiency purposes, Phase 2 performs the clustering on the representative and Phase 3 labels the initial input with the appropriate cluster information.

In our work we assume that apart from the initial data set, we give the algorithm as input an initial clustering. This clustering corresponds to the set of input records that contain the keywords a user indicated as seeds for the underlying semantic domain. As a consequence, we process the data in a hierarchical fashion, starting with the initial clustering of the documents and proceeding until all relevant text fragments have been identified.

The method proceeds as follows: (1) Given a clustering C, group the set of input fragments T into the corresponding clusters S_t; (2) Merge all fragments of S_t into a representative R_t; (3) Find the fragments of $S \backslash S_t$ that are closest to the representative R_t; (4) Analyze the fragments found in step 3 for new semantic annotations of the domain and add them to S_t; (5) Repeat from step 2 until stopping criteria are satisfied.

3 Generation of the Annotation Schema

To provide a user assistance in generating linguistic indicators for Cerno's semantic annotation process the LIMBO algorithm was integrated in a user-friendly tool. Having such a support will allow to quickly adapt the framework to new application domains in terms of both different annotation schemas and types of documents. This tool has a graphical user interface (GUI) developed in Java. The input to LIMBO includes: the initial data set that is then transformed into a cluster representative by the clustering algorithm and a set of documents which are used for training. The graphic interface provides a step-by-step wizard that allows the user to configure the experiment, then separates the input file into clusters, repeatedly runs the algorithm and provides the results of each run. To initialize LIMBO, on the first step of the wizard the user specifies the following input files and parameters:

- *The input document* is the original unannotated input document.
- *The clusters file* is the text file that contains the clustering information.
- *The stopping criterion* specifies the way of terminating the algorithm.
- *The parsing mode* defines how the clusters will be generated from the input document. In our case, we take n words starting at the 1st word of the sentence, then n words starting at the 2nd word, and so forth. Thus, the n-grams parsing mode generates the largest number of clusters and consequently requires longer processing times.
- *The analyzer* is the module responsible for extracting the keywords out of the input document. Currently, the prototype contains two standard analyzers for English language with and without stemming that normalize the input text.

The tool parses the input file and separates the clusters using the specified parsing mode. Then, for each concept, it marks the clusters that contain any of the keywords of the category and runs the algorithm as many times as specified in the stopping criterion. When ran a fixed number of times, the k nearest neighbors are marked at each run. When finished, the prototype shows for each concept the most relevant words in each run of the algorithm.

4   Experimental Case Studies

To verify the feasibility of the proposed approach, we applied the LIMBO-based tool on two different experiments. The stopping criterion for the clustering algorithm was set to 10 iterations with the addition of the 2 nearest-neighbors. We selected 10 as the number of iterations empirically, given that the output of the clustering algorithm remained almost unchanged after a number of iterations greater than 10. The tool was run with 8 different parsing configurations per each experiment, half of them with stemming and half without, varying the parsing mode: sentences, all punctuation marks, 3-grams and 7-grams. Number 3 for n-grams mode was chosen to account for commonly used word collocations, such as for instance "information system" or "health care cleaninghouse", and number 7 was defined as the highest upper bound for a possible number of words in collocations.

We evaluated the performance by comparing automated results to a *Gold model*, i.e. the list of indicators drawn manually by the experts, and calculating recall and precision quality measures [12].

The HIPAA experiment. In the past a Cerno adaptation to the text of the Health Insurance Portability and Accountability Act (HIPAA) was generated by manual analysis of the document [3], annotating document fragments describing rights, anti-rights, obligations, anti-obligations, and related constraints. Thus, the purpose of this experiment was to evaluate how many of these indicators can be extracted by the clustering technique. We used as input four semantic categories and several corresponding keyword-seeds. Overall, in this experiment the tool has demonstrated low recall (from 0.13 to 0.38), except for the *Condition* concept (0.75). Better results were obtained for the runs with the stemming analyzer. Among the unstemmed results, the best average score is delivered by the 3-grams parsing mode. The processing times changes depending on parsing mode. In particular, n-grams mode causes generation of a larger number of clusters from the input document, compared to other two modes, thus increasing processing times of the algorithm (average 25.5 min against 3.75 min for non n-grams runs).

The accommodation ads experiment. In our previous work [9], to annotate advertisements for accommodation in Rome drawn from an on-line newspaper, we used the annotation schema which represented the information needs of a tourist and included the concepts: *Accommodation Type, Contact, Facility, Term (of availability), Location,* and *Price*. The lists with linguistic indicators were constructed by hand from a set of examples. This experiment utilized the same input documents and categories from an earlier experiment using accommodation ads retrieved from tourism websites. Selected randomly, one third of the keywords found through the manual extraction process performed previously were included in the clusters file. This experiment has shown results of higher quality in respect to both recall and precision values (many runs above 0.6). The runs with the stemming option turned off have demonstrated higher scores.

Either with or without stemming, the best average scores were obtained for the 3-grams parsing mode.

Discussion of results. The evaluation results suggest that it is most effective to use the 3-grams parsing mode, to obtain the output of the best quality either for stemmed or non-stemmed processing. 3-grams parsing mode generates the largest number of clusters from the input text, thus essentially increasing processing times of the LIMBO algorithm.

The legal documents turned out to be more difficult for automated generation of linguistic indicators. This shortcoming is caused by the nature of the concepts of interest. Right, obligation, condition, and exception are very abstract entities and normally span relatively large text fragments, which makes it difficult to apply clustering techniques to identify appropriate contextual keywords. While short ads documents written in a very precise language and having similar structure provide a better learning environment for the LIMBO method.

Although it may seem that the Limbo-based technique does not achieve desired high recall values, some keywords found do not appear in the hand-crafted list (and thus were not counted as true positives), but were found relevant by a human judge. Therefore, we believe that LIMBO can provide a more complete approach to populate annotation schema with domain-specific indicators. Results produced by LIMBO can be a good starting point for a human expert when working with a new semantic domain. Using clustering techniques we are better able to support the generation of new annotation schemas in a systematic way. To further improve the LIMBO-based tool, we plan to provide a better guidance to the user thtough the underlying process.

5   Related Work

There are several proposals for weakly supervised methods intended to populate an ontology, a task similar to the generation of linguistic indicators. One of these is the *Class-Example* method [10] that exploits lexico-syntactic features to learn a classification rule from a seed set of terms. In contrast, the *Class-Pattern* approach [6] relies on using a set of patterns that indicate the presence of certain relationships, such as "is-a". *Class-Word* technique [5] uses contextual features to extract features in which a concept occurs.

Among the systems that use statistical techniques for populating semantic models is Ontosophie [4]. The system is based on machine learning natural language processing techniques to learn extraction rules for the concepts of a given ontology combining a shallow parsing tool called Marmot and a conceptual dictionary induction system called Crystal. The OntoPop [1] methodology strives for documents annotation and ontology population under a unified framework. In addition, it adopts two other tools: Intelligent Topic Manager for representing and managing the domain model and Insight Discoverer Extractor for extracting information from texts.

6   Conclusions and Future Work

In this work, we explore the problem of generating linguistic indicators for semantic annotation tools. The contribution of this paper consists of utilizing novel statistical clustering techniques and in particular is inspired by LIMBO [2] in order to automatically generate these indicators. Moreover, in order to allow experimenting with clustering techniques and facilitate the user's work, a tool implementing the LIMBO algorithm was developed in Java. We verified the effectiveness of the proposed technique in two different case studies.

Our future work includes further experimentation with different configurations of the clustering technique in order to improve the quality of results produced. As well, we propose to actually run semantic annotation experiments using the linguistic indicators generated by the LIMBO tool, in order to better assess their effectiveness.

References

1. Amardeilh, F.: OntoPop or how to annotate documents and populate ontologies from texts. In: Proc. of the ESWC 2006 Workshop on Mastering the Gap: From Information Extraction to Semantic Representation, Budva, Montenegro (2006)
2. Andritsos, P., Tsaparas, P., Miller, R.J., Sevcik, K.C.: LIMBO: Scalable Clustering of Categorical Data. In: Bertino, E., Christodoulakis, S., Plexousakis, D., Christophides, V., Koubarakis, M., Böhm, K., Ferrari, E. (eds.) EDBT 2004. LNCS, vol. 2992, pp. 123–146. Springer, Heidelberg (2004)
3. Breaux, T.D., Vail, M.W., Antón, A.I.: Towards regulatory compliance: Extracting rights and obligations to align requirements with regulations. In: Proc. of RE 2006, Washington, DC, USA, pp. 46–55. IEEE Computer Society, Los Alamitos (2006)
4. Celjuska, D., Vargas-Vera, M.: Ontosophie: A Semi-Automatic System for Ontology Population from Text. In: Proc. of ICON 2004, Hyderabad, India (2004)
5. Cimiano, P., Völker, J.: Towards Large-Scale, Open-Domain and Ontology-Based Named Entity Classification. In: Proceedings of RANLP 2005, pp. 166–172 (2005)
6. Hearst, M.: Automated Discovery of WordNet Relations. In: Fellbaum, C. (ed.) WordNet: An Electronic Lexical Database. MIT Press, Cambridge (1998)
7. Jardine, N., Sibson, R.: The construction of hierarchic and non-hierarchic classifications. The Computer Journal 11, 117–184 (1968)
8. Kiyavitskaya, N., Zeni, N., Mich, L., Cordy, J.R., Mylopoulos, J.: Text mining through semi automatic semantic annotation. In: Reimer, U., Karagiannis, D. (eds.) PAKM 2006. LNCS (LNAI), vol. 4333, pp. 143–154. Springer, Heidelberg (2006)
9. Kiyavitskaya, N., Zeni, N., Mich, L., Cordy, J.R., Mylopoulos, J.: Annotating Accommodation Advertisements using CERNO. In: Proc. of ENTER 2007, pp. 389–400. Springer, Wien (2007)
10. Tanev, H., Magnini, B.: Weakly Supervised Approaches for Ontology Population. In: Proc. of EACL 2006, Trento, Italy (2006)
11. Tishby, N., Pereira, F.C., Bialek, W.: The Information Bottleneck Method. In: 37th Annual Allerton Conf. on Communication, Control and Computing (1999)
12. Baeza-Yates, R., Ribeiro-Neto, B.: Modern Information Retrieval. Addison Wesley, Reading (1999)

Information Retrieval

Exploiting Multiple Features with MEMMs for Focused Web Crawling

Hongyu Liu, Evangelos Milios, and Larry Korba

National Research Council Institute for Information Technology, Canada
Faculty of Computer Science, Dalhousie University, Canada
hongyu.liu@nrc-cnrc.gc.ca, eem@cs.dal.ca, larry.korba@nrc-cnrc.gc.ca

Abstract. Focused web crawling traverses the Web to collect documents on a specific topic. This is not an easy task, since focused crawlers need to identify the next most promising link to follow based on the topic and the content and links of previously crawled pages. In this paper, we present a framework based on Maximum Entropy Markov Models(MEMMs) for an enhanced focused web crawler to take advantage of richer representations of multiple features extracted from Web pages, such as anchor text and the keywords embedded in the link URL, to represent useful context. The key idea of our approach is to treat the focused web crawling problem as a sequential task and use a combination of content analysis and link structure to capture sequential patterns leading to targets. The experimental results showed that focused crawling using MEMMs is a very competitive crawler in general over Best-First crawling on Web Data in terms of two metrics: Precision and Maximum Average Similarity.

Keywords: Focused Crawling, Web Search, Feature Selection, MEMMs.

1 Introduction

General search engines are not always sufficient to satisfy all needs. To address specialized search needs, general search engines and crawlers are being evolved, leading to personalization of search engines, for example, *MyAsk, Google Personalized Search, My Yahoo Search*; localization of search engines, for example, *Google Local, Yahoo Local, Citysearch*; topic-specific search engines and portals, for example, *Kosmix, IMDB, Scirus, Citeseer*. The success of topic-specific search tools depends on the ability to locate topic-specific pages on the Web while using limited storage and network resources. This can be achieved if the Web is explored by means of a *focused crawler*. A focused crawler is a crawler that is designed to traverse a subset of the Web for gathering only documents on a specific topic, instead of searching the whole Web exhaustively. The challenge in designing a focused crawler is to predict which links lead to target pages. Focused crawler can only use information gleaned from previously crawled pages to estimate the relevance of a newly seen URL, therefore, the effectiveness of the focused crawler depends on the accuracy of this estimation process.

E. Kapetanios, V. Sugumaran, M. Spiliopoulou (Eds.): NLDB 2008, LNCS 5039, pp. 99–110, 2008.

A variety of methods for focused crawling have been developed and focused crawling algorithms can be roughly categorized along two different dimensions: local-feature based and path based. The underlying paradigm of local-feature algorithms is to train a learner with only *local* features collected about relevant nodes *alone*(i.e., the parent pages and sibling pages). These works include Fish-Search, Shark-Search, URL Ordering[1], focused crawler[2], intelligent crawling[3], InfoSpiders[4], generic programming[5], ontology approach[6], and classification[7,8,9]. Path based algorithms include reinforcement learning[10] and Context Graph algorithm[11]. Both methods capture *longer* path information leading to targets rather than relevant nodes alone, as was the case with the local-feature based crawler. In [10], crawlers are modeled as autonomous agents to learn to choose optimal actions to achieve their goal. The Context Graph method[11] uses the text of page u to estimate the link distance from u to some target pages. Documents classified into layers closer to the target are crawled first. However, two issues remain to be addressed. One is that the assumption that all pages in a certain layer centered at a target document belong to the same topic described by a set of terms does not always hold. Second, there is no discrimination among different links on a page. Since only a fraction of out-links from a page are worth following, offering additional guidance to the crawler based on local features in the page to rule out some unimportant links can be helpful.

Our approach is to model focused crawling as a sequential task, over an underlying chain of hidden states, defined by hop distance from targets, from which the actual documents are observed. In this paper, we extend our work[12] to exploit multiple overlapping features, such as title, anchor text, and URL token, with Maximum Entropy Markov Models(MEMMs) to represent useful context including not only text content, but also linkage relations.

2   MEMMs

Maximum Entropy Markov Models(MEMMs)[13] are probabilistic sequence models that define conditional probabilities of state sequences given observation sequences. Formally, let o and s be random variables ranging over observation sequences and their corresponding state (label) sequences respectively. We use $s = s_1, s_2, ..., s_n$ and $o = o_1, o_2, ..., o_n$ for the hidden state sequence and observation sequence respectively, where s and o have the same length n. State s_t depends on observations o_t and previous state s_{t-1}. MEMMs are discriminative models that define the conditional probability of a hidden state sequence s given an observation sequence o, $p(s|o)$.

Let n be the length of the input sequence, m be the number of features. MEMMs make a first-order Markov independence assumption among states, that is, the current state depends only on the previous state and not on any earlier states, so $p(s|o)$ can be written as:

$$p(s|o) = \prod_{t=1}^{n} p(s_t|s_{t-1}, o_t) \tag{1}$$

where t ranges over input positions $1..n$. Applying the maximum entropy principle, we can rewrite $p(s_t|s_{t-1}, o_t)$ as the following:

$$p(s_t|s_{t-1}, o_t) = \frac{1}{z(o_t)} \exp(\sum_{i=1}^{m} \lambda_i f_i(s_{t-1}, s_t, o_t)) \tag{2}$$

$$z(o_t) = \sum_{s' \in S} \exp \sum_{i=1}^{m} \lambda_i f_i(s_{t-1}, s', o_t) \tag{3}$$

where, s_t is the state at position t, o_t is the observation at position t, the f_i are arbitrary features, λ_i is the weight of the feature f_i, and S indicates a set of all possible states. $z(o_t)$ is called the per-state normalizing factor.

The use of feature functions allows arbitrary, non-independent features in the observation sequence o. The weights λ are the parameters of the model. Training an MEMM involves maximizing the conditional probability, $p(s|o)$, to find the best set of feature weights $\lambda = \{\lambda_1, \lambda_2, ..., \lambda_m\}$.

3   Proposed Approach

We model focused crawling as a sequential task and learn the sequential linkage patterns along paths leading to relevant pages by using a combination of content analysis and link structure of such paths. To capture such sequential patterns, we propose to apply MEMM, where the hidden states are based on hop distance from the target and observations consist of the values of a set of pre-defined feature values of observed pages, such as anchor text, URLs, and keywords extracted from the pages.

3.1   Structure of MEMMs for Focused Crawling

Let k be the number of hidden states. The key quantities associated with MEMM are the hidden states, observations(features), and the parameters(λ). Fig. 1 shows the structure of MEMM for focused crawling.

- Hidden states: $S = \{T_{k-1}, T_{k-2}, ..., T_1, T_0\}$. The focused crawler is assumed to be in state T_i if the current page is i hops away from a target. The state T_{k-1} represents "$k-1$" or more hops to a target page.
- Observations: Collections of feature values of page sequences $O = \{page_1, page_2, page_3, ...\}$. Observable page sequences represented by a sequence of values for a set of predefined feature functions $f = \{f_1, f_2, ..., f_m\}$. m is the number of feature functions.
- Set of parameters $\lambda = \{\lambda_1, \lambda_2, ..., \lambda_m\}$, where λ_i is associated with feature function f_i.

3.2   Training MEMMs

MEMMs have parameters $\lambda = \{\lambda_1, \lambda_2, ..., \lambda_m\}$ which are the weights for each feature function $f_1, f_2, ..., f_m$. Training means estimating these parameters from

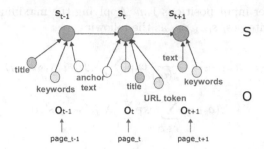

Fig. 1. Dependency structure of MEMMs on modeling a sequence of Web pages. Directed graphical model, arrow shows dependency (cause).

the training data. Given the training data D consisting of N state-observation sequences, $D = \{\mathbf{S}, \mathbf{O}\} = \{(s^j, o^j)\}_{j=1}^{N}$ (we use superscript j to represent training instances), where each $o^j = \{o_1^j, o_2^j, ..., o_n^j\}$ is a sequence of observations with length n, and each $s^j = \{s_1^j, s_2^j, ..., s_n^j\}$ is the corresponding sequence of states. The task of training MEMMs is to choose values of parameters $\{\lambda_i\}$ which maximize the log-likelihood, $L'_\lambda = \log p(\mathbf{S}|\mathbf{O})$, of the training data. We use (s^j, o^j) to represent the jth state-observation sequence from the training data set, s_t^j, o_t^j to indicate the state and observation at position t of the jth state-observation sequence respectively. The objective function can be written as:

$$L'_\lambda = \log p(\mathbf{S}|\mathbf{O}) = \sum_{j=1}^{N} \log p(s^j | o^j)$$

$$= \sum_{j=1}^{N} \left(\sum_{t=1}^{n} \sum_{i=1}^{m} \lambda_i f_i(s_{t-1}^j, s_t^j, o_t^j) \right) - \sum_{j=1}^{N} \log \prod_{t=1}^{n} \sum_{s' \in S} \exp \sum_{i} \lambda_i f_i(s_{t-1}^j, s', o_t^j)$$

To perform the optimization of L'_λ with respect to λ, we consider the gradient of the log-likelihood L'_λ and set it to zero. Parameter estimation in MEMMs uses the Limited Memory Quasi-Newton Method (L-BFGS)[14,15] to iteratively estimate the model parameters λ.

3.3   Training Data Collection

Collecting training data consists of two extraction processes: Local web graph extraction and Page sequence extraction.

Selection of Target Pages on Topics. We collect three kinds of data about the topics: keywords, descriptions and target pages. Keywords are formed by concatenating the words appearing in the different levels along the topical hierarchy directory from the top. For example, "Home, Gardening, Plants, House Plants" are extracted keywords for the topic. We also further extract important words embedded in the target pages themselves, including words from title and headers(`<title>`...`</title>`, `<h1>`...`</h1>` etc.) and keywords and descriptions from meta data(`<meta>`...`</meta>`). Descriptions are generated using the descriptive text and the anchor text in the page of the topic in the Open Directory Project(ODP)[1]. We select topics from an existing hierarchical concept index such as the ODP, and pick topics which are neither too general nor too specific. Our criteria for selecting topics is to pick the topics under 4 or 5 level of ODP topic hierarchy. An example of such a topic is *House Plants*, to be found under the ODP topic hierarchy *Home - Gardening - Plants - House Plants*.

Extraction of the Local Web Graph. In order to capture the page sequences leading to targets for training, first we construct a local Web graph to represent the content and linkage structure associated with the known targets. The process of Local Web graph extraction takes specified target pages as input, and builds the local Web graph by following the inlinks to target pages using an inlink retrieval service (as offered by the Yahoo! or Google search engines).

To construct a local Web graph, each Web page is represented by a node, and all hyperlinks between pages are added as edges between the nodes. When a new Web page is found and added to the existing graph, a new node will be created and all the hyperlinks between it and existing nodes will be added into the Web graph as edges. The local Web graph is created layer-by-layer starting from one or more user-specified target pages. Note that we only use ODP to select initial target pages on the topic, rather than using the ODP hierarchy for training. We use the inlink service from Yahoo Web API[2], which retrieves Web pages that have links to the specified Web page. Starting from layer 0 corresponding to user-specified target page(s), the graph is created layer by layer, up to layer 4.

Extraction of Page Sequences and State Sequences. Page sequences are extracted directly from the constructed local Web graph. We extract sequences in a random manner: the process starts with a randomly-picked node in layer 4, randomly selects one of its children for the next node, and repeats until a randomly-generated sequence length between 2 and 10 is reached. The following rules are considered for the page sequence extraction:

- Only extract pages from higher layers to lower layers or from the same layer, without reversing the layer order in the sequences.
- Avoiding loops in the sequence.

The complete training data includes page sequences and their corresponding layer sequences. Page sequences are referred to as observation sequences or observable input sequences, and layer sequences are referred to as hidden state sequences or state sequences. The hidden state for a page in our system is its lowest layer number.

---

[1] http://www.dmoz.org

[2] http://developer.yahoo.com/search/siteexplorer/V1/inlinkData.html

3.4   Focused Crawling

After the learning phase, the system is ready to start focused crawling on the real Web to find relevant pages based on learned parameters. The crawler utilizes a queue, which is initialized with the starting URL of the crawl, and keeps all candidate URLs ordered by their visit priority value. The crawling respects the Robot Exclusion Protocol and distributes the load over remote Web servers. The crawler downloads the page pointed to by the URL at the head of the queue, extracts all the outlinks and performs feature extraction. The predicted state for each child URL is calculated based on the current observable features and corresponding weight parameters, and the visit priority values are computed accordingly. The start page are picked randomly from the 4th level of Web graph created using backlink service for training, arranging from 2 to 6.

Efficient Inference. We now discuss two kinds of inference we are going to use in Focused Crawling stage. When the crawler sees a new page, the task of the inference is to estimate the probability that the page is in a given state s based on the values of all observed pages already visited before. We are using two major approaches to compute the probabilities in our experiments: marginal probability and the Viterbi algorithm, and they can be performed efficiently using dynamic programming.

`Marginal Probability` – The marginal probability of states at each position t in the sequence is defined as the probability of states given the observation sequence up to position t. Specifically, the forward probability, $\alpha(s,t)$ is defined as the probability of being in state s at position t given the observation sequence up to position t. The recursive steps are:

$$\alpha(s,t) = \sum_{s'} \alpha(s',t-1)\, p(s|s',o_t) \tag{4}$$

Hidden states are denoted as T_j, $j = 0..k-1$, the values $\alpha(T_j,t)$ in our focused crawling system are calculated as:

$$\alpha(T_j,t) = \sum_{j'=0}^{k-1} \alpha(T_{j'},t-1)\, \frac{1}{z(o_t)} \exp(\sum_{i=1}^{m} \lambda_i f_i(T_j, T_{j'}, o_t))$$

$$z(o_t) = \sum_{j''=0}^{k-1} \exp \sum_{i=1}^{m} \lambda_i f_i(T_{j'}, T_{j''}, o_t)$$

`The Viterbi Algorithm` – the goal is to compute the most likely hidden state sequence given the data:

$$s^* = \arg\max_s p(s|o) \tag{5}$$

$\delta(s,t)$ is defined as the best score (i.e. the score with the highest probability) over all possible configurations of the state sequence ending in state s at position t given the observations up to position t. That is

$$\delta(s,t) = \max_{s'} \delta(s',t-1)\, p(s|s',o_t) \tag{6}$$

This is the same recursive formulate as the forward values (Equation 4), except we replace sum with max.

Features and Feature Functions. Each feature function $f(s, o, t)$ is defined as a factored representation:

$$f(s, o, t) = L(s_{t-1}, s_t, t) * O(o, t) \tag{7}$$

where $L(s_{t-1}, s_t, t)$ are transition feature functions, and $O(o, t)$ are observation feature functions.

1. **Edge Features:** Transition feature functions $L(s_{t-1}, s_t, t)$ can have two forms: L_1 and L_2. We use Edge feature $L_1(s_{t-1}, s_t, t)$ to capture the possible transitions from states s_{t-1} to s_t, and $L_2(s_{t-1}, s_t, t)$ to capture the possible states at position t.

   Formally, for all $i, j = 0, 1, ..., k - 1$ so that specified $T_i, T_j \in S = \{T_{k-1}, T_{k-2}, ..., T_1, T_0\}$, we can have feature functions of the following form:

$$L_1^{(i,j)}(s_{t-1}, s_t, t) = \begin{cases} 1 \text{ if } s_{t-1} = T_i \text{ and } s_t = T_j \text{ is an allowed transition;} \\ 0 \text{ otherwise.} \end{cases}$$

$$L_2^{(i)}(s_{t-1}, s_t, t) = \begin{cases} 1 \text{ if } s_t = T_i \text{ exists;} \\ 0 \text{ otherwise.} \end{cases}$$

2. **Text Feature:** Maximal cosine similarity value between the content of a given candidate page and the set of targets. We define it as $O_1(o, t)$.
3. **Description Feature:** Cosine similarity value between the page description of a given candidate page and the target description. We define it as $O_2(o, t)$.
4. **Word Feature:** Word feature $O_w(o, t)$ identifies the keywords appearing in the page text, as described in sec. 3.3.

$$O_w(o, t) = \begin{cases} 1 \text{ if word } w \text{ appears in the current page at position } t; \\ 0 \text{ otherwise.} \end{cases}$$

We may also use the count of word w as the value of this feature, instead of the binary value.

5. **URL Token Feature:** There are two possible kinds of URLs related to the current observed page: one is the URL of the current page itself, and another one is the URL the current page is pointing to. We define two token feature functions $O_3(o, t)$ and $O_4(o, t)$ to identify if the keywords appear in the URLs.

$$O_3(o, t) = \begin{cases} 1 \text{ if any of URLs in the current page at position } t \text{ contains} \\ \quad \text{at least one keyword;} \\ 0 \text{ otherwise.} \end{cases}$$

$$O_4(o, t) = \begin{cases} 1 \text{ if the URL of the current page at position } t(\text{contained in} \\ \quad \text{the parent page}) \text{ contains at least one target keyword;} \\ 0 \text{ otherwise.} \end{cases}$$

6. Anchor Text Feature: We capture word w in the anchor surrounding text by defining two anchor features. At least 4 words are chosen from text around `<a>..</a>`.

$$O_{5,w}(o,t) = \begin{cases} 1 \text{ if word } w \text{ appears in the anchor text of the link in} \\ \quad \text{the parent page linking to current page at position } t; \\ 0 \text{ otherwise.} \end{cases}$$

$$O_{6,w}(o,t) = \begin{cases} 1 \text{ if word } w \text{ appears in the anchor text in the current} \\ \quad \text{page at position } t \text{ pointing to the page at position } t+1 \text{ ;} \\ 0 \text{ otherwise.} \end{cases}$$

4   Experiments

In this section, we conduct experiments to test our MEMM-based focused crawling approach empirically. The topics are chosen from the ODP categories, and the target pages are chosen based on the listed URLs under each category. The following table shows some information about the 10 topics for the experiments.

Topic	# of Target Pages	# of training sequences	Start Urls
Linux	19	11394	6
Biking	17	9790	2
Butterfly	17	12172	2
Hearthealthy	8	12928	2
Hockey	19	4368	2
Fitnessyoga	7	11680	3
Balletdance	7	12317	3
Skymaps	16	12073	3
Callforpapers	11	9632	3
Internetlaw	18	12210	4

4.1   Evaluation Methods

The *precision* is the percentage of the Web pages crawled that are relevant to the topic. The relevance assessment of a page p we are using is based on maximal cosine similarity to the set of target pages T compared with a confidence threshold γ. That is, if $\max_{d \in T} \cos(p, d) \geq \gamma$ then p is considered as relevant. Some topics may be sensitive to the threshold, therefore we choose the threshold values between 0.5-0.8 for general comparisons. Too high or too low threshold may result in too few or too many relevant pages based on the target pages and the start URLs, which does not provide sufficient information for comparison and for figure presentations.

The ability of the crawler to remain focused on the topical Web pages during crawling can also be measured by the average relevance of the downloaded documents[16,17,18]. In our system, since there are multiple pre-specified target pages, we used the *Maximum Average Similarity* σ.

$$\sigma = \max_{d \in T} \frac{\sum_{p \in S} \cos(p, d)}{|S|} \tag{8}$$

where T is the set of target pages, S is the set of pages crawled, $|S|$ is the number of targets.

4.2 Results

We have conducted two experiments. One is to compare MEMM-based method with different inference algorithms against Best-First Search(BFS) crawl. BFS crawl assigns priorities to all children of the current page using standard lexical cosine similarity between the content of the current page and target pages. The URL with the best score will be crawled first. Another one is to test the impact of the choice features on performance.

Comparison with Different Inference Algorithms: Viterbi Algorithm and Marginal Mode. First we compare our MEMM-based methods with all the features against BFS crawl. We find that our MEMM-based crawl significantly outperforms BFS crawl on 8 out of 10 topics. The performance of three different crawling methods, BFS crawl, MEMM-marginal crawl, and MEMM-Viterbi crawl, on the topic *Fitnessyoga* is shown in Fig. 2 (a). All three methods give very good results on this topic, however, two MEMM-based crawls still work better than BFS crawl on the number of relevant pages returned, which also is confirmed on the *Maximum Average Similarity* metric, as shown in Fig. 2 (b).

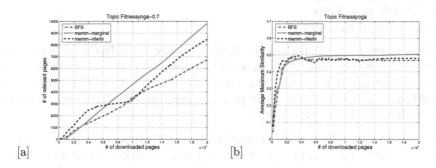

[a]   [b]

Fig. 2. Topic *Fitnessyoga*: (a) the number of relevant pages with threshold 0.7, (b) the Maximum Average Similarity

The results on topic *Linux* are shown in Fig. 3 (a). Both MEMM-marginal crawl and MEMM-Viterbi crawl also outperform BFS crawl. MEMM with marginal mode shows significant improvement over BFS crawl, while MEMM with Viterbi algorithm shows slight improvement on the number of the relevant pages. However, MEMM-Viterbi crawl gives a very close performance to MEMM-marginal

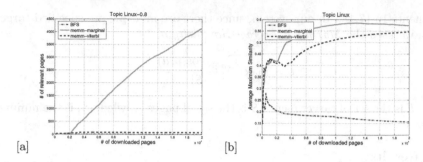

Fig. 3. Topic *Linux*: (a) the number of relevant pages with threshold 0.8, (b) the Maximum Average Similarity

crawl on the *Maximum Average Similarity*, which significantly outperforms BFS crawl as shown in Fig. 3 (b). This shows that two MEMM-based methods stay on the topic, whereas BFS method crawls away from the topic resulting in poor performance.

Compared to the results only based on MEMM-marginal and MEMM-Viterbi crawls, we found that the marginal mode outperforms the Viterbi algorithm on 7 topics out of 10. In the focused crawling problem, finding the distribution for each individual hidden state at a particular instant is more important than finding the best "string" of hidden states of each Web page along the sequence, since there may be many very unlikely paths that lead to large marginal probability. To be more specific, let us see an example. If we have

"aaa" 30% probability
"abb" 20% probability
"bab" 25% probability
"bbb" 25% probability

In this example, "aaa" is the most likely sequence, 'a' is the most likely first character, 'a' is the most likely second character, 'b' is the most likely third character, but the string "aab" has 0 probability. Therefore, as we expected, MEMM-marginal crawl shows better performance than MEMM-Viterbi crawl in most of the cases in our experiments.

Comparison with Different Features. In this section, we test the impact of the selected features on the performance. The last section demonstrated that MEMM-marginal crawl is better than MEMM-Viterbi crawl, so in this experiment, we choose to compare MEMM-marginal crawl with different features: with all features, with word feature only, and with the following features: Text feature, Description feature, URL token feature and Anchor text feature (See all features in Section 3.4). We denote them as *MEMM-marginal*, *MEMM-marginal-word*, and *MEMM-marginal-sim-meta-T* in the figures, respectively.

As the results show, MEMM-marginal crawl with all features performs consistently better than with Word feature only and with sim-meta-T features on almost all topics, except topic *Hearthealthy*, in which MEMM-marginal-word

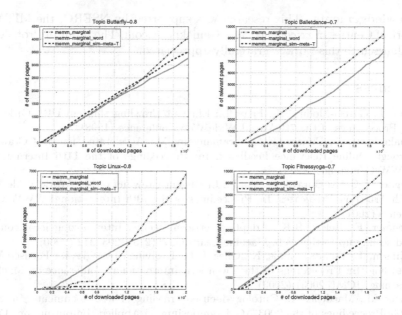

Fig. 4. Topic *Butterfly, Balletdance, Linux* and *Fitnessyoga*: Comparisons of different methods: with Word feature only, simmetaT feature only and the all features combination on the number of relevant pages within the set of downloaded pages

shows the best performance, and topic *Callforpapers*, in which MEMM-marginal-sim-meta-T shows the best. Fig. 4 shows some of the results. This confirms our approach that by using all the features, even if only some of them are present, relevant paths can be effectively identified.

5   Conclusion

Our approach is unique that we model the process of crawling by a walk along an underlying chain of hidden states, defined by hop distance from target pages, from which the actual topics of the documents are observed. When a new document is seen, prediction amounts to estimating the distance of this document from a target. In this way, good performance depends on powerful modeling of context as well as the current observations. The advantages and flexibility of MEMMs fit our approach well and are able to represent useful context. With Maximum Entropy Markov Models (MEMMs), we exploit multiple overlapping and correlated features, such as anchor text, to represent useful context and form a chain of local classifier models. We have studied the impact of different combination strategies, and the results showed that using marginal mode performs better than using Viterbi algorithm, and the crawler using the combination of all features performs consistently better than the crawler that depends on just one or some of them.

Acknowledgements. This research was supported by NSERC, the MITACS Network of Centres of Excellence, and Genieknows.com. The input of Prof. Jeannette Janssen to this work is gratefully appreciated.

References

1. Cho, J., Garcia-Molina, H., Page, L.: Efficient Crawling through URL Ordering. In: Proceedings of the 7th World Wide Web Conference (1998)
2. Chakrabarti, S., Punera, K., Subramanyam, M.: Accelerated Focused Crawling through Online Relevance Feedback. In: Proceedings of the 11th International WWW Conference (1999)
3. Aggarwal, C., Al-Garawi, F., Yu, P.: Intelligent Crawling on the World Wide Web with Arbitrary Predicates. In: Proceedings of the 10th International WWW Conference (2001)
4. Menczer, F., Belew, R.K.: Adaptive retrieval agents: Internalizing local context and scaling up to the Web. Machine Learning 39(2/3), 203–242 (2000)
5. Johnson, J., Tsioutsiouliklis, K., Giles, C.L.: Evolving Strategies for Focused Web Crawling. In: Proceedings of the Twentieth International Conference on Machine Learning (ICML 2003) (2003)
6. Ehrig, M., Maedche, A.: Ontology-focused crawling of web documents. In: SAC 2003: Proceedings of the 2003 ACM symposium on Applied computing, pp. 1174–1178. ACM, New York (2003)
7. Pant, G., Srinivasan, P.: Learning to Crawl: Comparing Classification Schemes. ACM Trans. Information Systems. 23(4) (2005)
8. Pant, G., Srinivasan, P.: Link Contexts in Classifier-Guided Topical Crawlers. IEEE Transactions on Knowledge and Data Engineering 18(1), 107–122 (2006)
9. Frnkranz, J.: Hyperlink ensembles: A case study in hypertext classification. Information Fusion 3(4), 299–312 (2002)
10. Rennie, J., McCallum, A.: Using Reinforcement Learning to Spider the Web Efficiently. In: Proceedings of the Sixteenth International Conference on Machine Learning (ICML 1999) (1999)
11. Diligenti, M., Coetzee, F., Lawrence, S., Giles, C., Gori, M.: Focused Crawling Using Context Graphs. In: Proceedings of the 26th International Conference on Very Large Databases (VLDB 2000) (2000)
12. Liu, H., Janssen, J., Milios, E.: Using hmm to learn user browsing patterns for focused web crawling. Data & Knowledge Engineering 59(2), 270–291 (2006)
13. McCallum, A., Freitag, D., Pereira, F.: Maxiumu Entropy Markov Models for Information Extraction and Segmantation. In: Proceedings of the Seventeenth International Conference on Machine Learning, pp. 591–598 (2000)
14. Nocedal, J., Wright, S.J.: Numerical Optimization. Springer, Heidelberg (1999)
15. Sha, F., Pereira, F.: Shallow Parsing with Conditional Random Fields. In: Proceedings of the 2003 Conference of the North American Chapter of the Association for Computational Linguistics on Human Language Technology, pp. 134–141 (2003)
16. Menczer, F., Pant, G., Srinivasan, P., Ruiz, M.: Evaluating Topic-Driven Web Crawlers. In: Proceedings of the 24th ACM/SIGIR Conference. Research and Development in Information Retrieval (2001)
17. Menczer, F., Pant, G., Srinivasan, P.: Topical Web Crawlers: Evaluating Adaptive Algorithms. ACM TOIT 4(4), 378–419 (2004)
18. Srinivasan, P., Menczer, F., Pant, G.: A General Evaluation Framework for Topical Crawlers. Information Retrieval 8(3), 417–447 (2005)

Ranked-Listed or Categorized Results in IR: 2 Is Better Than 1

Zheng Zhu[1], Ingemar J. Cox[2], and Mark Levene[1]

[1] School of Computer Science and Information Systems,
Birkbeck College, University of London
zheng@dcs.bbk.ac.uk, mark@dcs.bbk.ac.uk
[2] Department of Computer Science, University College London
ingemar@ieee.org

Abstract. In this paper we examine the performance of both ranked-listed and categorized results in the context of known-item search (target testing). Performance of known-item search is easy to quantify based on the number of examined documents and class descriptions. Results are reported on a subset of the Open Directory classification hierarchy, which enable us to control the error rate and investigate how performance degrades with error. Three types of simulated user model are identified together with the two operating scenarios of correct and incorrect classification. Extensive empirical testing reveals that in the ideal scenario, i.e. perfect classification by both human and machine, a category-based system significantly outperforms a ranked list for all but the best queries, i.e. queries for which the target document was initially retrieved in the top-5. When either human or machine error occurs, and the user performs a search strategy that is exclusively category based, then performance is much worse than for a ranked list. However, most interestingly, if the user follows a hybrid strategy of first looking in the expected category and then reverting to a ranked list if the target is absent, then performance can remain significantly better than for a ranked list, even with misclassification rates as high as 30%. We also observe that this hybrid strategy results in performance degradations that degrade gracefully with error rate.

1 Introduction

Search engines play a crucial role in information retrieval on the web. Given a query, search engines, such as Google, Yahoo! and Windows Live, return a ranked list of results, referred to as the result set. For many queries, the result set includes documents on a variety of topics, rather than a single topic. This variation is often due to ambiguous queries. For example, the query "Jaguar" will often return documents referring to both the car and the animal. While the user is only interested in one topic, it is not possible for the search engine to know which topic is relevant based on the query alone. Moreover, the standard ranking of the documents in the result set is independent of the topic. Thus, the rank-ordered result set has an arbitrary topic ordering. Referring to the "Jaguar" example, this means that a user must scroll through a ranked list in which many documents are not relevant.

E. Kapetanios, V. Sugumaran, M. Spiliopoulou (Eds.): NLDB 2008, LNCS 5039, pp. 111–123, 2008.

There have been several proposals [1, 2, 3] to assist the user by organising the documents in the result set into groups, all documents within a group referring to a common topic. Thus, for the query "Jaguar", a user might be shown two distinct groups of documents, one referring to the animal and the other referring to the car. A user can immediately ignore the non-relevant topic and focus his attention only on the relevant topic. For this simple example, this grouping may, on average, halve the number of documents the user must examine.

Intuitively, we would expect grouping to substantially reduce search time[4], where search time is measured by the number of documents a user must examine before finding the desired document. Although previous researchers have evaluated their prototype systems, there has been no attempt, to our knowledge, of formulating a generic user interaction model for a retrieval system, which allows the benefits of grouping to be quantified in comparison to a standard retrieval system which does not group its results.

In this paper, we attempt to quantify the benefits of grouping documents based on classification, where for demonstration and ground-truth purposes we make use of the Open Directory (dmoz)[1], a large and comprehensive human-edited directory on the web. However, we note that our experimental approach may be applied to any method of grouping documents.

The remainder of the paper is organised as follows. Section 2 reviews related work. Section 3 then describes the architecture and ranking method of the classification-based information retrieval (IR) system that have investigated. Section 4 describes the experimental methodology used to evaluate the system and Section 5 describes the experimental results. Finally, Section 6 provides a summary and discussion.

2   Related Work

This concept of grouping search results has been discussed in Hearst and Pedersen [1], where it was shown that relevant documents tend to be more similar to each other than to non-relevant documents, indicating that relevant documents can be grouped into one category. The two main methods of grouping results are clustering and classification.

Clustering methods typically extracts key phrases from the search results for grouping purposes and attach to each group a candidate cluster label [5, 3]. The search results are treated as a bag of words/phrases, which are ranked according to the statistical features that have been found.

Classification uses predefined category labels that are more meaningful to users than generated labels. Chen and Dumais [2] suggest that category search based on classification can improve search time in comparison to the traditional list-based search, where documents within a category are ranked according to their relative ranking in the original search engine results list. A user study comparing the two interfaces demonstrated the potential superiority of a

---

[1] http://www.dmoz.org

classification-based user interface that can assist the user in quickly focusing in on task-relevant information, this evaluation method is more or less similar to Krishna's work[4].

Previous research in this area has focused on evaluating the grouping of search results versus the traditional list-based method and has not considered a hybrid model; even when a hybrid model is implemented[2], there has not been, to our knowledge, a quantitive analysis of the model as considered here, where users may use either interface to optimise their search performance. Further, the effect of the error rate that can occur when grouping results on users' performance has not previously been given much attention.

3   Classification-Based Information Retrieval

We describe the architecture and ranking method of a classification-based IR system that we have been developing in this section. We assume the existence of a standard retrieval system that, given a query, returns a ranked list of documents as the result set. Web search engines such as Google, Windows Live and Yahoo! satisfy this assumption.

Given a ranked set of documents, it is necessary in our system to classify the documents into their respective classes. Figure 1 provides a conceptual view of the classification-based information retrieval system we are developing. Figure 1a depicts the ranked set of documents provided by a standard IR system. Our system classifies these documents into a number of classes, ranks the classes and then displays a ranked list of classes to the user, as depicted in Figure 1b. When a user clicks on a particular class, the ranked set of documents in this class is then displayed to the user, as shown in Figure 1c.

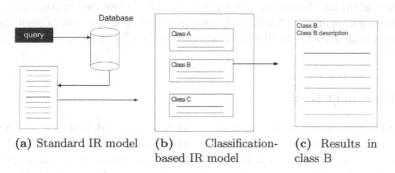

(a) Standard IR model   (b)   Classification-   (c) Results in
                              based IR model   class B

Fig. 1. Conceptual framework of a classification-based IR system

We now provide a more formal framework for our system. We assume that there are $|D|$ documents in the original result set, D, and we denote the standard IR rank of document, $d_k \in D$ as $s(d_k)$; we refer to this rank as the *scroll rank*

---

[2] http://demo.carrot2.org/demo-stable/main

(SR). For convenience of presentation we assume that the documents are ordered such that document d_k has scroll rank $s(d_k) = k$. According to the eye tracking experiment[3] and [4], the position in the list can be approximated by the time to find a result.

3.1   Class Rank

After performing classification on the original result set returned by the standard IR system, the documents are grouped into $|C|$ top-level classes. Each class, c_i, consists of a set of documents, $d_{i,j}$, where $1 \leq j \leq |c_i|$, and $|c_i|$ denotes the number of documents in class, c_i.

Each class, c_i, consists of a set of documents, $d_{i,j}$, where $1 \leq j \leq |c_i|$, and $|c_i|$ denotes the number of documents in class, c_i.

Given the set of classes, C, a rank ordering of the classes is necessary. For a document, $d_{i,j}$, let $\psi(i,j) = k$ denote the corresponding index of the same document in the initial result set as output by the standard IR system. Thus, the score associated with document $d_{i,j}$ is $s(d_{\psi(i,j)}) = s(d_k) = k$. Then, the score of class, c_i is given by

$$\phi(c_i) = -min(s(d_{\psi(i,j)})) \quad \text{for } 1 \leq j \leq |c_i|,$$

where $\phi(c_i)$ outputs the score for each class. In our case scores of each class based on the scroll rank of the document within the class. For notational convenience, we assume the classes to be ordered such that class c_i has rank i.

We believe that this simple method of ranking classes is novel and, more importantly, minimizes the affects of the ranking method on the performance of the classification-based system. Conversely, if we had developed a sophisticated ranking system for classes and documents within classes (see Section 3.2), then it becomes increasingly difficult to determine whether differences in performance compared with a standard IR system are due to classification or the new ranking algorithm.

3.2   Document Rank

Having ranked each class, it is now necessary to rank the documents, $d_{i,j}$, within each class, c_i. To do so, we assume the existence of a function, $\varphi(d_{i,j})$, that outputs a score for each document. Here we adopt one of the popular method, the scroll ranks of the documents, $s(d_k)$, as output by the standard IR system, as a score for each document and rank the documents accordingly. Thus, the score for document, $d_{i,j}$, in class, c_i is given by $\varphi(d_{i,j}) = -s(d_{\psi(i,j)})$.

Documents within the class are then ranked according to their scores, the highest score being ranked first, as before. For notational convenience, we assume the documents to be ordered such that document $d_{i,j}$ has rank j in class c_i.

---

[3] http://www.useit.com/alertbox/reading_pattern.html

3.3   In-Class Rank(ICR)

When a user selects a class, c_i, the *in-class rank* measures the number of class labels and documents that the user examines, when the target document, $d_k = d_{i,j}$ is in class, c_i. The in-class rank is

$$r(d_{i,j}) = i + j, \tag{1}$$

since the user must look at the first i-ranked class descriptions and then the first j-ranked documents within the known class.

However, a classification-based IR system introduces a small overhead. If the target document is ranked high, then this overhead may be noticeable.

3.4   Scrolled-Classification Rank(SCR)

As we shall see shortly, it is often useful to talk about the *scrolled classification rank*, denoted by $s(d_{i,j})$, which we define as the total number of classes and documents a user must examine to find document $d_{i,j}$ by sequentially scrolling through each class and its associated documents in rank order.

In this case, the user will look at i classes, and all of the documents in the previous i-1 classes together with the first j documents of the last class. Thus, the scrolled classification rank of document, $d_{i,j}$ is given by

$$s(d_{i,j}) = i + \sum_{k=1}^{i-1} |c_k| + j. \tag{2}$$

3.5   Out-Class/Scroll-Class Rank(OSCR) and Out-Class/Revert Rank(ORR)

If the target document is not within the selected class, then the user must perform additional work. Upon failing to find the target document in the chosen class, the user may choose to

(i) *scroll* through the classes and the documents in each class in rank order, or
(ii) *revert* to the standard IR display and sequentially scroll through the ranked result set.

The *out-class/scroll-class rank* and *out-class/revert rank* are, respectively, the number of documents the user must then examine in order to find the target in case (i) and (ii) above. We now formalise these notions.

The *out-class/scroll-class rank*, $p(d_{i,j})$, is the total number of class labels and documents that a user must examine in order to find the document for case (i), where the users chooses to scroll through the classes and documents in rank order, after not finding the target in the selected class. Let c_e denote the class the user erroneously selects. Then the out-class/scroll-class rank is given by

$$p(d_{i,j}) = \begin{cases} (e + |c_e|) + s(d_{i,j}) & \text{if } e > i, \\ e + s(d_{i,j}) & \text{if } e < i. \end{cases} \tag{3}$$

The *out-class/revert rank*, $q(d_{i,j})$, is the total number of class labels and documents that a user must examine in order to find the document for case (ii), where the user reverts to the standard result set, i.e. no classification is used in the second phase of the search. The out-class/revert rank is given by

$$q(d_{i,j}) = (e + |c_e|) + s(d_{\psi(i,j)}) = (e + |c_e|) + s(d_k) = (e + |c_e|) + k, \quad (4)$$

where $\psi(i, j) = k$. Note that the out-class/revert rank is a hybrid search strategy that begins with a classification-based strategy and reverts to a ranked-list strategy if the document is not present in the first class selected. This hybrid strategy is different from the presentation in cluster-based search engines, where the user is presented with the ranked listing in a main window and the clusters in another.

3.6   Classification

We have, until now, ignored how classification is performed. In the experiments of Section 5 we assume two cases.

In the first case we assume we have an oracle based on 16 top level categories of the Open Directory that correctly classifies the documents. Of course, in practice, this is not possible. However, analysis of this case provides us with valuable information regarding the best-case performance of the system. Any other system in which classification errors occur will perform worse. In the second case, we assume classification is performed based on a k-nearest neighbour (KNN) classifier [6]. That is, given a document, d_j, we find its k most similar documents in a database of pre-classified documents.

4   Experimental Methodology

Target Testing. The experimental methodology *simulates* a user performing a known-item search, also referred to as target testing.

In our context we make use of target testing to evaluate the performance of a classification-based retrieval system. The motivation is that target testing allows us to evaluate the system automatically without users, and is a precursor to user testing. Additionally, target testing allows us to evaluate the system on numerous queries at a minimal cost in comparison to user testing. However target testing has some shortcoming in that the queries generated for target testing do not necessarily simulate "real" user queries. Moreover, good performance of the system for target testing does not guarantee similar performance when testing the system with "real" users.

Automatic Query Generation. For a given document repository, in our case extracted from the Open Directory, we randomly select target documents. For each target document, a user query is automatically generated by selecting a

Table 1. Summary of the operating conditions and the number of classes and documents examined in each case

simulated user/target	correctly classified	misclassified
knows class	ICR (C1)	OSCR (C4a) or ORR (C4b)
does not know class	SCR (C2a) or SR (C2b)	SCR (C5a) or SR (C5b)
thinks knows class	OSCR (C3a) or ORR (C3b)	OSCR (C6a) or ORR (C6b)

number of words from the target document. This can be performed in a variety of ways (cf. [7], [8]). However, the exact procedure is not important. We only require that queries can be generated such that the target document appears within a designated range of scroll rank. In this way we can simulate a range of good (high ranking) to poor (low ranking) queries.

User/Machine Models. Table 1 summarises the models and corresponding user strategies we described in section 3. The three user models are (i) the user knows class (case 1 and 4); (ii) the user does not know class (case 2 and 5) and, (iii) the user think he knows (case 3 and 6). Note for each user model, there are two cases associated with it because there are two machine models (correct/incorrect classification of the target document). In the Table 1, we assume that the user employs the search strategies we introduced in Section 3.

5 Experiment

The dataset used in our experiments is derived from the Open Directory Project. We have chosen the following 12 top level classes to construct our testset: Arts, Business, Computers, Games, Health, Kids and Teens, Society, Science, Shopping, Home, Sports and Recreation.

We crawled and downloaded all the documents from these 12 top-level classes during September 2006. After removing the noisy data, we divided the remaining 792,030 documents into training set (500,430 documents) and test set (291,600 documents). The training set was used for classification with the k-nearest neighbour classifier.

We randomly selected 600 target documents from the test set, and for each target document we generated 10 queries. The queries were designed so that the scroll rank of the target document output by the standard IR system fell into intervals counting 5 ranks from 1 to 50.Thus, for each target document, we used a set of 10 queries that ranged from "very good" (scroll rank between 1-5) to "very poor" (scroll rank 36-50). The experimental results were averaged over all 600 target documents.

The underlying IR system is based on the open-source search software, Lucene[4]. For stemming we make use of the open-source stemmer, Snowball [5]. The default document ranking algorithm from Lucene was used.

---

[4] http://lucene.apache.org
[5] http://snowball.tartarus.org

5.1 Experimental Results

We performed three sets of experiments. In the first set we used the domz directory as an oracle to classify the result set, the second set used a k-nearest neighbour classifier trained on a subset of the Open Directory to classify the result set, and the last set used our classifier in a more realistic scenario.

Classification Based on an Oracle. Each of the 600 documents has been manually classified into one of 12 classes. Thus, dmoz provides us with an oracle with which to classify all documents in the original result set. This allows us to first examine the best-case performance of our classification-based IR system, i.e. when there are no machine classification errors and the simulated user knows the correct class (case C1 in Table 1).

We can also introduce and control error rates for both the user and the machine classifier. Note that, from Table 1, user errors and machine classification errors both result in the same search length (cases C3, C4, and C6). Moreover, the two cases where the user is aware that they do not know the class (cases C2 and C5) are unaffected by the machine misclassification. Thus, when we report error rates, we do not distinguish between human and machine error rates. Rather, the error rate represents the combination of the two.

Figure 2a summarises the results for these cases. It shows the *cumulative* probability of finding the target document as a function of the rank of the target document. We note that for the standard IR system, the rank corresponds to the scroll rank (SR). For the error-free case, the rank corresponds to the in-class rank (ICR). For non-zero error rates, the rank corresponds to the ranks summarised in Table 1.

For the standard IR system, the scroll rank (SR) is a straight line, since the scroll rank is evenly distributed within the 10 intervals described above. We see that for the error-free case (ICR), the classification-based IR system performs significantly better than the standard IR system (SR). In particular, we observe that approximately 60% of all target documents can be found with a rank of 10 or less. That is, for an ideal user and no machine misclassification (case C1), the

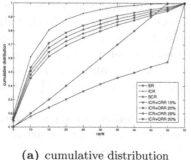 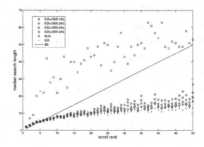

(a) cumulative distribution   (b) Median search length of query results

Fig. 2. Results using dmoz oracle classifier

user must look at no more than 10 classes and documents in order to locate the target document.

The oracle also allows us to control the misclassification rate. And the rates we describe can best be thought of as the combined user and machine error rates. We introduced an error rate of $x\%$ as follows: for $x\%$ of the 600 queries, the user randomly selects a class that does *not* contain the target document, and then uses the out-class/revert (ORR) ranking strategy to locate the document. For the remaining $(100 - x)\%$ of queries, the user chooses the correct class and the target document is found using the in-class ranking (ICR) strategy. Thus, the curves for non-zero error rates represent a combination of two strategies, ICR and ORR.

For an overall error rate of 15%, we observe a decline in performance, as expected. However, at this error rate, the classification-based IR system still performs significantly better than the standard system. For example, over 50% of all target documents are found with a rank of 10 or less. As the overall error rate increases, the performance degrades. However, this degradation is rather smooth and even with an error rate of 30%, the performance remains significantly better than that of the standard IR system.

Finally, for completeness, Figure 2a also shows the cumulative distribution for the scrolled classification rank (SCR). This curve is significantly worse than the scroll rank of the standard IR system. Figure 2a shows that in cases C2 and C5, when the user does not know the class, he is best advised to abandon the classification-based IR system and immediately return to the standard system, i.e. follow the second strategy of scroll-rank (SR) in Table 1. Moreover, in cases C3, C4 and C6, where we have either a user or machine error, then if the user does not find the target in the class he knows or thinks is the correct class, the advice is the same, i.e., revert to the standard system following the second strategy of out-class/revert (ORR) in Table 1. That is, a hybrid-based search strategy performs better than either a category-based or ranked-listing alone.

It is important to recognize that the cumulative distribution does not present the full story. Figure 2b plots the median search length as a function of the scroll rank. Note that here we use the median search length distribution since the search length is highly biased by a small number of outliers, while the median is more robust to this bias. It is clear that for target documents with a low scroll rank (less than 5), the median search length using a classification-based system is slightly longer, on average, due to the overhead of inspecting the class description or when an error occurs. Thus, for very good queries, a classification based system actually increases the search length slightly. Conversely, for poorer queries,the median rank of the target document in the classification-based system is always shorter, on average. Interestingly, both the standard IR system and the classification-based system have regions of superior performance. Only when the initial query is poorer, i.e. the scroll rank is below a certain threshold, does the classification-based system offer superior performance.

Figure 2b also shows, as expected, that this threshold increases as the misclassification rate increases and it is more evident for poor queries. Thus, for example, for a misclassification rate of 25%, the scroll rank must be greater than 7 before a classification-based system is superior.

It is also worth noting that in Figure 2b the quality of the initial queries is uniformly distributed, by design. Thus, 20% of queries have an initial scroll rank between 1-5, another 20% between 6-10, and so on. In practice, the distribution of queries is a function of (i) the user, (ii) the distribution of documents in the database, and (iii) the document scoring function [8]. Thus the benefits of a classification-based system will depend strongly on the distribution of the queries. It is interesting to note that a number of studies such as [9, 10] have reported poor correlations between user judgments of document rankings and those produced by search engines, suggesting that classification-based systems may be useful in practice.

K-Nearest Neighbour Classification. The experiments of the previous section show that very good performance can be expected from a classification-based system, especially for poorer queries. The definition of "poorer queries", i.e. queries for which the scroll rank is larger than a given threshold, varies with the misclassification rate. Simulated misclassification rates of 15-30% suggest that (i) performance degrades gracefully as the error rate increases, and (ii) that useful performance improvements can still be obtained with relatively large error rates.

To investigate what misclassification rate we could expect from a classifier, we implemented a simple non-disjoint k-nearest neighbour (KNN) classifier. The repository of documents remains the same, permitting us to measure the misclassification rate at 16%. Figures 3a and 3b show the cumulative distribution and median search length, respectively, in this case. Clearly, even at this error rate, significant improvements can be obtained, depending on the query distribution.

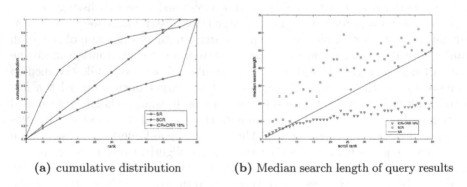

(a) cumulative distribution   (b) Median search length of query results

Fig. 3. Results for the KNN classifier with non-disjoint classes

K-Nearest Neighbour Classification in a More Realistic Scenario. To investigate what misclassification rate expected from a classifier in a realistic scenario, we implemented a k-nearest neighbor classifier over a real search engine. Due to the absence of an oracle for retrieved results, we adopt the classifier's

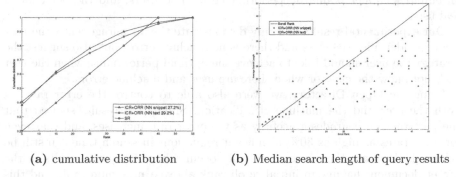

(a) cumulative distribution   (b) Median search length of query results

Fig. 4. Results for the KNN classifier in real case

accuracy on the target document as a measure of the classifier's performance. It may not fully reflect the performance of our classifier, but we can use this measure to approximate our system's performance.

Compared to the previous results shown in Figure 3a, Figure 4a shows the misclassification rate of the k-nearest neighbor is about 30%, which is worse than the previous results. However, we still can see that these results are consistent with the previous conclusions. Moreover, for this more realistic case the classifier trained from dmoz snippets reduces the misclassification rate to about 28%, which is slight better than that trained on the full text of web pages.

Figure 4b shows that for poor queries, the combined class rank will achieve a better performance than scroll rank. However the trend in the curve is not as clear out as the previous curve in Figure 3b. It can also be seen that the curve has high variance. However, our general conclusion that the hybrid-based search strategy performs better than a category-based or ranked-list alone, is still valid.

6   Concluding Remarks

In this paper we have examined how a hybrid model of an IR system might benefit a user. Our study was based on several novel ideas/assumptions.

In order to investigate the best-case performance, we constructed a system using a subset of the Open Directory. All documents in this subset have been manually classified and therefore provide "ground truth" for comparison. The advantage of our approach to rank the class is two-fold. It is simple, however, more importantly, this ranking closely approximates the scroll rank, thus allowing comparison between the class rank and scroll rank. In addition, we identified three classes of simulated users and rational user search strategies. By basing our evaluation on known-item search, we are able to simulate a very large number of user searches and therefore provide statistically significant experimental results. We acknowledge that real users may perform differently and future work is needed to determine the correlation between our simulations, which provide an empirical upper bound on performance for real users, and the behavior of real users.

Our experimental results not only demonstrate the advantage when the user correctly identifies the class and there is no machine error, but also suggest the strategy the user should take to achieve the optimal performance when the user does not know the class or when there are user and machine errors.

Using the Open Directory, we were also able to control the error rates of both the user and the machine classification. Simulation results showed that the performance degrades gracefully as the error rate increases, and that even for error rates as high as 30%, significant reductions in search time can still be achieved. However, these reductions only occur when the query results in the target document having an initial scroll rank above a minimum rank, and this minimum rank increases with the error rate.

This may imply that a classification-based system may be more beneficial for informational queries [11], where the user will probably inspect several search results, rather than for navigational queries [11], which are similar to known-item queries that target a single web page. Such a system could also be useful for novice users who are more likely to generate poor queries.

References

[1] Hearst, M.A., Pedersen, J.O.: Reexamining the cluster hypothesis: Scatter/gather on retrieval results. In: Proceedings of the 19th Annual International ACM SIGIR Conference on Research and Development in Information Retrieval, pp. 76–84.

[2] Chen, H., Dumais, S.: Bring order to the web: Automatically categorizing search results. In: CHI 2000: Proceedings of the SIGCHI conference on Human factors in computing systems, pp. 145–152. ACM Press, New York (2000)

[3] Zeng, H.J., He, Q.C., Chen, Z., Ma, W.Y., Ma, J.W.: Learning to cluster web search results. In: SIGIR 2004: Proceedings of the 27th annual international ACM SIGIR conference on Research and development in information retrieval, pp. 210–217. ACM Press, New York (2004)

[4] Kummamuru, K., Lotlikar, R., Roy, S., Singal, K., Krishnapuram, R.: A hierarchical monothetic document clustering algorithm for summarization and browsing search results. In: Proceedings of the 13th International Conference on World Wide Web, pp. 658–665 (2004)

[5] Osinski, S., Weiss, D.: Carrot 2: Design of a flexible and efficient web information retrieval framework. In: Proceedings of the third International Atlantic Web Intelligence Conference, Berlin. LNCS, pp. 439–444. Springer, Heidelberg (2005)

[6] Duda, R.O., Hart, P.E., Stork, D.G.: Pattern Classification. 2nd edn. Wiley-Interscience, New York (2000)

[7] Azzopardi, L., Rijke, M.D.: Automatic construction of known-item finding test beds. In: Proceedings of the 29th Annual International ACM SIGIR Conference on Research and Development in Information Retrieval, pp. 603–604. ACM Press, New York (2006)

[8] Vinay, V., Cox, I.J., Milic-Frayling, N., Wood, K.: Evaluating relevance feedback algorithms for searching on small displays. In: 27th European Conference on IR Research. ECIR (2005)

[9] Bar-Ilan, J., Keenoy, K., Yaari, E., Levene, M.: User rankings of search engine results. J. American Society for Information Science and Technology 58(9), 1254–1266 (2007)

[10] Su, L.T.: A comprehensive and systematic model of user evaluation of web search engines: Ii. an evaluation by undergraduates. J. American Society for Information Science and Technology 54(13), 1193–1223 (2003)

[11] Broder, A.: A taxonomy of web search. SIGIR Forum 36(2), 3–10 (2002)

Exploiting Morphological Query Structure Using Genetic Optimisation*

Jose R. Pérez-Agüera, Hugo Zaragoza, and Lourdes Araujo

Dpto de Ingeniería del Software e Inteligencia Artificial, UCM
jose.aguera@fdi.ucm.es
Yahoo! Research Barcelona
hugoz@yahoo-inc.com
Dpto. de Lenguajes y Sistemas Informáticos, UNED
lurdes@lsi.uned.es

Abstract. In this paper we deal with two issues. First, we discuss the negative effects of term correlation in query expansion algorithms, and we propose a novel and simple method (query clauses) to represent expanded queries which may alleviate some of these negative effects. Second, we discuss a method to optimise local query expansion methods using genetic algorithms, and we apply this method to improve stemming. We evaluate this method with the novel query representation method and show very significant improvements for the problem of optimising stemming.

1 Introduction

There is an underlying common background in many of the works done in query reformulation, namely the appropriate selection of a subset of search terms among a list of candidate terms. The number of possible subsets grows exponentially with the size of the candidate set. Furthermore, we cannot evaluate a priori the quality of a subset with respect to another one: this depends on the (unknown) relevance of the documents in the collection. For these reasons, standard optimisation techniques cannot be applied to this problem. Instead, we must resort to heuristic optimisation algorithms, such as genetic algorithms, that sample the space of possible subsets and predicts their quality in some unsupervised manner.

Before considering any query reformulation process is very important to take into account that modern Information Retrieval ranking functions apply the *term independence* assumption. This assumption takes on many forms, but loosely it implies that the *effect of each query term on document relevance can be evaluated independently of the other query terms*. This has the effect of rendering all queries *flat*, whithout structure.

However, there are many cases in which queries have some known *linguistic structure*, such as degree of synonymy between terms, term cooccurrence or correlation information with respect to the query or to specific query terms, etc.

---

* Supported by projects TIN2007-67581-C02-01 and TIN2007-68083-C02-01.

E. Kapetanios, V. Sugumaran, M. Spiliopoulou (Eds.): NLDB 2008, LNCS 5039, pp. 124–135, 2008.

This is typical of queries constructed by a query-expansion method, of stemming or normalizing terms, of taking into account multi-terms or phrases, etc. Surprisingly, almost all ranking functions (and experiments) ignore this structural information: after expansion, selection and re-weighting of terms, a flat query (a set of weighted terms) is given to the ranking function which *assumes terms are independent* and scores documents accordingly.

Specifically, we want to investigate two issues: the selection of adequate terms for query reformulation and term independence assumption, to propose a new method that solve the classical problems associated to these issues. The morphological query structure of the query has been chosen to show how our approach is capable to improve state-of-art approaches like Porter's stemmer.

In Section 2 we propose a novel way to represent expanded queries that encodes information about the term correlation using clauses like set of related terms. The proposed representations greatly increase the expressivity power of queries, but at the expense of introducing parameters (weights) which may be hard to set. Section 3 shows one experiment where our clauses representation model is adapted to the problem of morphological query expansion. In section 4 we apply it to the problem of optimising the expansion of a term with respect to its stem. We show that we can significantly improve the performance of Porter stemming by adapting the expansion to every query. Section 5 draw the main conclusions and describe the future lines of work.

2   Ranking Independent Clauses of Dependent Terms

One of the reasons of the high performance of modern ad-hoc retrieval systems is their use of document term frequency. It is well known[12] that i) probability of relevance of a document increases as the term frequency of a query term increases, and ii) this increase is non-linear. For this reason most modern ranking functions use an increasing saturating function to weight document terms that are in the query. An example of this is the term saturating function used as part as BM25[12]:

$$w(d,t) := \frac{tf(d,t)}{tf(d,t) + K1} \qquad (1)$$

where $tf(d,t)$ is the term frequency of term t in document d, and $K1$ is a constant. Similar nonlinear term frequency functions are found in most IR ranking models such modern variants of the vector space model, the language model, divergence from randomness models, etc. Besides, all these ranking functions assume that *the relevance information of different query terms is independent* and therefore the relevance information gained by seeing query terms can be computed separately and added linearly (or log-linearly), for example, in BM25:

$$score(d) := \sum_{t \in q} w(d,t) \cdot idf(t) \qquad (2)$$

This independence assumption is usually reasonable for short queries (i.e. *"Italian restaurant in Cambridge"*), since users use each term to represent a different aspect of the query. However, such assumption breaks down for queries that are sufficiently complex to contain terms with sufficiently close meaning. Consider for example the query *"Italian restaurant cafeteria bistro Barcelona"*. Having seen the term *restaurant* twice in a document, which term is more informative: *Barcelona* or *cafeteria*? Loosely speaking, if a group of terms carries the same meaning, the amount of relevance information gained by their presence should diminish as we see other terms in this group, very much like in equation (1) does for term frequency, and unlike (2).

This situation arises very often in modern IR tasks and systems, in particular in the following areas:

- morphological expansion (e.g. stemming, spelling, abbreviations, capitalization),
- extracting multi-terms from the query
- query term expansion (e.g. user feedback, co-occurrence based expansion),
- lexical semantic expansion (e.g. using WordNet),
- using taxonomies and ontologies to improve search,
- user modeling, personalization,
- query disambiguation (where terms are added to clarify the correct semantic context),
- finding similar documents (where the query is an entire document),
- document classification (where the query is a set of documents),
- structured queries (such as TREC structured topics).

We propose to consider two *levels* of representation: *terms* and *term clauses*. Clauses are *sets of weighted terms* that are intended to represent a particular aspect of the query. The weights w represent their relative importance within the clause (in particular, the strength of the dependence with relevance). Thus, a query can be thought of as a bag of bags of (weighted) terms:

$$c := \left\{ (t_0, w_0), (t_1, w_1), ..., (t_{|c|}, w_{|c|}) \right\}$$
$$q := \left\{ c_1, c_2, ..., c_{|q|} \right\}$$

Boolean retrieval models and the Inquery[3] retrieval model have used query representations even more general than this. Here we restrict ourselves to this representation with two levels to give clear semantics to each level: term and clause. We are going to consider terms *within* a clause as if they were greatly dependent with respect to relevance; in fact we will consider them as if they were virtually the same term. Second, we consider terms *across* clauses as being independent with respect relevance, as is usually done across terms.

Conceptually, what we propose is a projection from the space of terms to the space of clauses. Formally, we represent a document as the vector $d = (tf_1, ... tf_i, ..., tf_V)$ where V if the size of the vocabulary. We represent a query having n clauses as a $n \times V$ matrix of weights: $C = (c_{ij})$ where c_{ij} is the weight of jth term in ith clause. The projected document is then $d|_C := d \times C^\mathsf{T}$.

Consider this example. Imagine that we are given a corpus with four terms

A-D:

Doc	A	B	C	D
d_1	2	1	0	1
d_2	0	1	1	1
d_3	1	2	0	2

Now consider the query with two clauses:

$$q := \{\,\{(A, 1.0),\, (B, 0.7)\},\, \{(C, 1.0)\}\,\}$$

This query can also be represented by the matrix:

$$C := \begin{bmatrix} 1 & .7 & 0 & 0 \\ 0 & 0 & 1 & 0 \end{bmatrix}$$

In the projection $d|_C := d \times C^{\mathsf{T}}$ the original query terms are removed from the collection and replaced with new pseudo-terms representing the clauses; other terms are removed because all terms are included in the clauses. The term frequency of the clause pseudo-term will be equal to the weighted sum of term frequencies of the terms in the clause. In our example:

Transformed Doc	Clause1	Clause2
$d_1|_q$	2.7	0
$d_2|_q$	0.7	1
$d_3|_q$	2.4	0

This projection will be different for every clause, so it cannot be done as a pre-processing step: it must be done online. In practice one can carry out this transformation very efficiently since the projection is performed only on the few terms contained in the query, and document vectors are very sparse. Most important, the information needed for this projection is contained in the postings of the query terms. This means that one can compute the needed term statistics on the fly after postings are returned from the inverted index. This will incur the cost of a few extra flops per document score, but without any extra disk or memory access.

The length of the document is not modified by the projection, nor the average document length. The *clause term frequencies* (ctf) and *clause collection frequencies* (ccf) can be computed as:

$$ctf(d, c) := \sum_{(t, w) \in c} w \cdot tf(d, t)$$

$$ccf(d, c) := \sum_{(t, w) \in c} w \cdot \sum_{d} tf(d, t)$$

$$p_{\mathsf{ML}}(c|d) := \frac{ctf(d, c)}{ctf(d, c) + \sum_{t \notin c} tf(d, t)}$$

$$p_{\mathsf{ML}}(c|Col) := \frac{ccf(d, c)}{ccf(d, c) + \sum_{d, t \notin c} tf(d, t)}$$

The most problematic statistic is the *inverse clause frequency* (*icf*), since this is not clearly defined in the weighted case. One possible choice is the number of postings containing *at least one term in the clause*; we refer this as $icf_{OR}(c)$, and we note that it can be computed directly from the size of the clause result set (documents with non-zero ctf). However, this number may be unfairly large for clauses with lowly weighted common terms. Furthermore, in some settings this number may not even be available (for example if we only score the query term AND set or if we drop from the computation documents unlikely to be highly scored). Another possibility is to use the *expected idf* for a clause term in a document:

$$icf_E(d, c) = \frac{1}{ctf(d, c)} \sum_{(t,w) \in c} w \cdot tf(d, t) \cdot idf(t) \tag{3}$$

In our empirical evaluation we found this is better than using the *min* or the *max clause idf*, and better than using the *mean idf*.

With these statistics at hand we can compute the relevance score of a document with respect to a query with clauses for a number of retrieval systems[2]; we display several in Table 1.

Table 1. Implementing query clauses in several standard ranking models

MODEL	WEIGHTING
BM25	$\frac{ctf}{ctf+K} \cdot icf$
VSM	$\frac{ctf \cdot icf}{\|d_{|q}\|}$
DFR (PL2)	$\frac{1}{ctf+1} \left(ctf \cdot \log_2 \frac{ctf}{\lambda} + (\lambda - ctf) \cdot \log_2 e + 0.5 \cdot \log_2(2\pi \cdot ctf) \right)$
LM (KL)	$p_{\text{smoothed}}(c|q) \log \left(p_{\text{smoothed}}(c|d) \right)$

3   Query Clause Experiments

We have performed experiments to demonstrate the dangers of the term independence assumption for queries with strongly correlated terms, and to test the proposed *query-clauses* ranking idea applied to the stemming problem. Evaluation has been carried out on the Spanish EFE94 corpus which is part of the CLEF collection [10] (approximately 215K documents of 330 average word length and 352K unique index terms) and the 2001 Spanish topic set, with 49 topics of which we only used the title (of 3.3 average word length). All runs employ the standard (equation 1) and the query clause version of BM25 Table 1.

3.1   Stemming

One can view stemming as a form of global query expansion: we expand a term in the query with every term in the dictionary sharing the same stem. However, doing this directly on the query greatly hurts performance (Table 2, rows 1 and 2) . One may think that this loss of performance is due to the noise introduced by the stemming algorithm, but this is not the case: if we replace terms by their stemmed version in the collection and in the query, performance will most often increase and rarely decrease (Table 2 row 3). This is another case of performance being degraded by adding information to the query. In our opinion, this is due to the strong violation of the term independence hypothesis produced by adding so many strongly correlated terms to the query.

A natural way to expand a query by stemming is to construct a set of sets of terms, or a set of *clauses*, where each clause represents all the forms of a stem, possibly weighted (since we may want to weight more strongly the original term typed by the user). The resulting query is a set of sets of clauses which can be ranked with our proposed method. Its performance, using as *idf* the clause's (icf_{OR}) is exactly equivalent to stemming the collection since in both cases term frequencies of stems are collapsed (this is also seen empirically in Table 2, last row)

Table 2. Stemming Performance

Method:	Avg.Prec	Prec10
No Stemming	.37	.47
Stem Expansion (Standard)	.20	.28
Stemming	**.43**	**.52**
Stem Expansion (Clauses)	**.43**	**.52**

3.2 Our Genetic Algorithm

Because we need to perform the selection of a particular set of terms among a huge amount of possible combinations of candidate query terms, the computational complexity of exhaustive search methods is non-viable and we have resorted to a heuristic method such as a genetic algorithm.

Genetic algorithms [7] have been shown to be practical optimization methods in very different areas [9]. Evolutionary algorithms mimic the principles of natural evolution: heredity and survival of the most fit individuals.

A genetic algorithm maintains a population of potential solutions, and is provided with some selection process based on fitness of individuals. The population is renewed by replacing individuals with those obtained by applying "genetic" operators to selected individuals. The usual "genetic" operators are *crossover* and *mutation*. Crossover obtains new individuals by mixing, in some problem dependent way, two individuals, called parents. Mutation gives a new individual by performing some kind of change on an individual. The production of new generations continues until resources are exhausted or until some individual in the population is fit enough. Figure 1 shows the structure of a genetic algorithm. The algorithm works with a collection of individuals $\{x_i, \cdots, x_n\}$, called population. Each individual represents a potential solution to the problem considered, implemented as some data structure, which depends on the problem. The evaluation of each solution gives a measure of its *fitness*. At a new generation step, a new population is formed by selecting the more fit individuals. Some members of the new population suffer transformations as a consequence of applying *genetic operator* to form new solutions. After a number of generations, the program is expected to converge, and it is hoped that then, the best individual represents a solution close to the optimum.

```
evolution program
begin
    generation = 0
    P = initialize_population
    F = evaluation(P)
    while not required_fitness(F) and
            not termination_condition do
    begin
        generation = generation + 1
        I = individuals_selection(P, F) %for genetic operators
        P = new_generation(P, I)
        F = evaluation(P)
    end
end
```

Fig. 1. Structure of a genetic algorithm

Chromosomes of our GA are fix-length binary strings where each position corresponds to a candidate query term. A position with value one indicates that the corresponding term is present in the query. Because some preliminary experiments we have performed have shown that, in most cases, the elimination of the original query terms degrades the retrieval performance, we force to maintain them among the selected terms of every individual. The set of candidate terms is composed of the original query terms, along with related terms provided by the applied thesaurus. Each term of the original query is grouped with the expanded terms related to it, and this set (*term_set*) [11] is submitted as an individual query. The weights assigned to the documents retrieved with each *term_set* are used to sort the total set of retrieved documents.

The applied selection mechanism has been roulette wheel. In roulette wheel selection, the chances of an individual to be chosen for reproduction are proportional to its fitness. We apply the one-point crossover operator and random mutation [6,7,9]. In one-point crossover a single crossover point is chosen on both parents strings. The parts of the parent strings divided by the crossover point are swapped to generate two children, containing a part of each parent. The random mutation operator simply flips the value of a randomly chosen bit (0 goes to 1 and 1 goes to 0). We also apply elitism, the technique of retaining in the population the best individuals found so far. The fitness function used is some measure of the degree of similarity between a document belonging to the system and the submitted query. We will discuss this further in the different experiments.

4   Experimental Results

The system has been implemented in Java, using the JGAP library[1], that provides a generic implementation of a genetic algorithm, on a Pentium 4 processor.

We have carried out experiments to evaluate the fitness functions considered. We have investigated the best range of the GA parameters. Finally, we provide global measures for the whole set of queries that we have considered.

4.1   Selecting the Fitness Function

In the begining, we would like to use Average precision as the fitness function. However, this is not known at query time. Instead, it has been suggested in previous work to use the document scores as the fitness[8]. While this may not be intuitive, it turns out that variations of these scores after expansion are correlated with relevance[1]. One intuitive explanation would be that adding an unrelated term to a query will not bring in new document with high scores, since it is unlikely that it will retrieve new documents; on the other hand adding a term that is strongly related to the query will bring new documents that also contain the rest of the terms of the query and therefore it will obtain high scores.

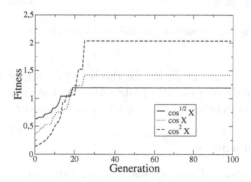

Fig. 2. Fitness functions comparison for the *best_query*, the one for which the greatest precision improvement is achieved

We have considered three alternative fitness functions, $\sqrt{\cos\theta}$, $\cos\theta$ and $\cos^2\theta$. To select the fitness function to be used in the remaining experiments, we have studied the fitness evolution for different queries of our test set. Figure 2 compares the fitness evolution for the query which reaches the greatest improvement

---

[1] http://jgap.sourceforge.net/

(*best_query*). The three functions converge to different numerical values that correspond to the same precision value (.68). We can observe that the square-root cosine function if the first one to converge. A similar behavior is observed in other queries. Accordingly, the square-root cosine has been the fitness function used in the remaining experiments.

4.2   Tuning the GA Parameters

The next step has been tuning the parameters of the GA. Figure 3 shows the fitness evolution using different crossover (a) and mutation (b) rates for the best query. Results show that we can reach a quickly convergence with values of the crossover rate around 25%. Mutation rates values around 1% are enough to produce a quick convergence.

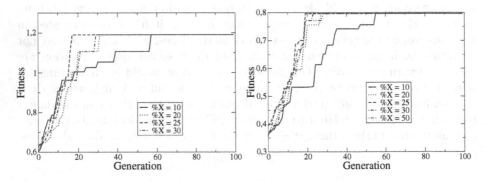

Fig. 3. Studying the best crossover (a) and mutation (b) rates for the best_query

Figure 4 show the fitness evolution for the best (a) and the worst (b) queries, with different population sizes. The plots indicate that small population sizes, such as one of 100 individuals, are enough to reach convergence very quickly.

4.3   Overall Performance

Table 3 (Stem Expansion (Clauses)) shows the results obtained using stemming as query expansion but building a clause for every term. As expected, the results are exactly those obtained in traditionally stemming by collapsing terms to their stems.

In order to show the improvement of our approach, we have compared the genetic algorithm performance with the results of the original user query (*Baseline*) and with the results obtained expanding with the stems provided by the Porter stemming (*Porter Stemming*). We can observe in Table 3 (Genetic Expansion (Clauses)) that the combination of clauses and genetic algorithm achieved an improvement of the performance, greater than the one achieved with other stemming methods traditionally used in the stemming process, such as Porter.

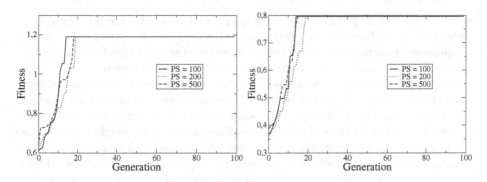

Fig. 4. Studying the population size for the best query (a) and the worst one (b)

Table 3. Stemming Performance

Method:	AvgPrec	Prec10	Rel.Δ
No Stemming	.37	.47	-13.9%
Stemming (Baseline)	**.43**	**.52**	0*
Stem Expansion	.20	.28	-53.5%
Stem Expansion (Clauses)	**.43**	**.52**	0
Genetic Expansion	.39	.49	-9.4%
Genetic Expansion (Clauses)	**.45**	**.53**	+4.4%

5   Related Works

In most query expansion literature terms are selected (globally from the entire corpus or locally from the top retrieved documents), weighted with respect to their potential relevance and then passed on to a standard retrieval system, which is considered a black box. Here we are concerned only with this black box and not with the expansion process; for this reason we will not review the query expansion literature (an up to date overview can be found in [5]). Some work on user and pseudo-feedback has tackled the issue of term re-weighting, from early Rochio algorithms to more modern probabilistic approaches of relevance feedback. While these works discuss the ranking function, to our knowledge they all assume Query Term Independence and concentrate on the re-weighting formula. Again, we are not concerned here on the re-weighting of terms (this is left unspecified in our work), and therefore we do not review this literature further (see for example [4]).

A few papers have dealt with the issue of term *correlation* and its effect on retrieval. In [14] the problem of correlation is discussed in depth. They remark that *term correlation* is only an abstract concept and can be understood in a number of ways. They measure term correlation in terms of *term co-occurrence*. Furthermore they propose to represent documents not in the space of terms but in the space of *minterms* which are sets of highly correlated terms. This has the effect of *decorrelating* the terms in the query with respect to hypothetical *concepts* (formally defined as minterms). Instead of computing all term correlations, [13] proposes to mine association rules to compute the most significant term correlations and the rotates the term vectors to reflect the extracted correlations; this yields a more selective term de-correlation. [11] also proposes mining association rules to find term sets of correlated terms. However, the ranking function adjustment proposed is based on the same idea of this paper: collapsing term frequencies within a clause. In fact, if we disregard relative weights, we use the VSM model, and we construct query clauses using association rules in [11], the ranking function here is exactly the same as in [11]. However our work differs from the previously cited papers in that it is not tied to an extraction method or a ranking model, it does not specify the form of the term correlations and furthermore it assumes that term correlations will be *query-dependant*.

6   Conclusions and Future Work

In this paper we try to show the importance of term dependence issues, how they show up unexpectedly in simple experiments and how they can have a strong adverse effect in performance. Furthermore we propose a method to represent and take into account a simple form of dependence between terms.

On the other hand, we have shown how the clauses can be combined with an evolutionary algorithm to help to reformulate a user query to improve the results of the corresponding search. Our method does not require any user supervision. Specifically, we have obtained the candidate terms to reformulate the query from a *morphological thesaurus*, with provides, after applying stemming, the different forms (plural and grammatical declinations) that a word can adopt. The evolutionary algorithm is in charge of selecting the appropriate combination of terms for the new claused query. To do this, the algorithm uses as fitness function a measure of the proximity between the query terms selected in the considered individual and the top ranked documents retrieved with these terms.

We have investigated different proximity measures as fitness functions without user supervision, such as cosine, square cosine, and square-root cosine. We have also studied the GA parameters, and see that small values such as a population size of 100 individuals, a crossover rate of 25% and a mutation rate of 1%, are enough to reach convergence. Measures on the whole test set of queries have revealed a clear improvement of the performance, both over the baseline, and over other stemming expansion methods.

A study of the queries resulting after the reformulation has shown that in many cases the GA is able to add terms which improve the system performance, and in some cases in which the query expansion worsen the results, the GA is able to recover the original query.

For the future, we plan to investigate the use of other sources of candidate terms to generate the claused queries applying different query expansion approaches like co-occurrence measures or methods based in Information Theory.

References

1. Araujo, L., Pérez-Agüera, J.R.: Improving Query Expansion with Stemming Terms: A New Genetic Algorithm Approach. In: Eighth European Conference on Evolutionary Computation in Combinatorial Optimisation (2008)
2. Baeza-Yates, R.A., Ribeiro-Neto, B.A.: Modern Information Retrieval. ACM Press / Addison-Wesley (1999)
3. Callan, J.P., Croft, W.B., Harding, S.M.: The INQUERY Retrieval System. In: DEXA, pp. 78–83 (1992)
4. Carpineto, C., de Mori, R., Romano, G., Bigi, B.: An Information-theoretic Approach to Automatic Query Expansion. ACM Trans. Inf. Syst. 19(1), 1–27 (2001)
5. Chang, Y., Ounis, I., Kim, M.: Query Reformulation Using Automatically Generated Query Concepts from a Document Space. Inf. Process. Manage. 42(2), 453–468 (2006)
6. Goldberg, D.E.: Genetic Algorithms in Search, Optimization and Machine Learning. Addison Wesley, Reading (1989)
7. Holland, J.J.: Adaptation in Natural and Artificial Systems. University of Michigan Press (1975)
8. Lopez-Pujalte, C., Bote, V.P.G., de Moya Anegón, F.: A Test of Genetic Algorithms in Relevance Feedback. Inf. Process. Manage. 38(6), 793–805 (2002)
9. Michalewicz, Z.: Genetic Algorithms + Data Structures = Evolution programs. 2nd edn. Springer, Heidelberg (1994)
10. Peters, C., Braschler, M.: European Research Letter: Cross-Language System Evaluation: The CLEF Campaigns. JASIST 52(12), 1067–1072 (2001)
11. Pôssas, B., Ziviani, N., Wagner Meira, J., Ribeiro-Neto, B.: Set-based Vector Model: An Efficient Approach for Correlation-based Ranking. ACM Trans. Inf. Syst. 23(4), 397–429 (2005)
12. Robertson, S.E., Walker, S.: Some Simple Effective Approximations to the 2-Poisson Model for Probabilistic Weighted Retrieval. In: ACM SIGIR 1994: Proceedings of the 17th annual international ACM SIGIR, pp. 232–241. Springer, New York (1994)
13. Silva, I.R., Souza, J.N., Santos, K.S.: Dependence Among Terms in Vector Space Model. In: IDEAS 2004: Proceedings of the International Database Engineering and Applications Symposium (IDEAS 2004), Washington, DC, USA, pp. 97–102. IEEE Computer Society, Los Alamitos (2004)
14. Wong, S.K.M., Ziarko, W., Raghavan, V.V., Wong, P.C.N.: On Modeling of Information Retrieval Concepts in Vector Space. ACM Trans. Database Syst. 12(2), 299–321 (1987)

Generation of Query-Biased Concepts Using Content and Structure for Query Reformulation

Youjin Chang, Jun Wang, and Mounia Lalmas

Queen Mary, University of London,
London, E1 4NS, UK
{youjinchang,wangjun,mounia}@dcs.qmul.ac.uk

Abstract. This paper proposes an approach for query reformulation based on the generation of appropriate query-biased concepts. Query-biased concepts are generated from retrieved documents using their content and structure. In this paper, we focus on three aspects of the concept generation; the selection of query-biased concepts from retrieved documents, the effect of the structure, and the number of retrieved documents used for generating the concepts.

Keywords: query reformulation, feature extraction, concept generation, structure, relevance feedback.

1 Motivation

A main issue in information retrieval (IR) is for users to define queries, i.e. the query terms, that properly express their information needs. If we assume that IR engines successfully find all the relevant documents using the terms contained in the initial query, the remaining problem is how to properly formulate the query. In IR, users often need to reformulate their initial queries more than once to obtain better results. There has been wide interest in the selection of terms to reformulate the query [1,4,5]. There are three main approaches: approaches based on relevance feedback information from the user, approaches based on information derived from the set of documents initially retrieved, and approaches based on global information derived from the document collection. The first approach, query reformulation from relevance feedback, has been shown effective if appropriate feedback (i.e. explicit - this document is/is not relevant; or implicit – through click-through data) is given by the user. This paper is concerned with the first type of approach. In this paper, we propose a query reformulation process based on so-called query-biased concepts (QBC). This process is performed as one of the relevance feedback task. We try to enhance the initial query with query-biased concepts generated from the analysis of the content and structural information of documents retrieved by the initial query.

We assume that the retrieved documents have several topics or themes that can be expressed by a set of terms. For example, let us consider an article about 'speech recognition'. The article may discuss the definition of speech recognition, the history of speech recognition, a speech recognition case study, etc. It is necessary to select the themes of the article so that the article can be effectively represented. Furthermore, it

E. Kapetanios, V. Sugumaran, M. Spiliopoulou (Eds.): NLDB 2008, LNCS 5039, pp. 136–141, 2008.

is also necessary to find "overall" concepts by joining those themes that are related. Since there may be similar documents or paragraphs about a 'speech recognition case study' in other documents, we need to integrate those themes across the documents globally. Through these local and global analyses, we aim to construct the concepts that identify the main themes of retrieved documents. The framework for constructing the query-biased concepts is illustrated in Figure 1.

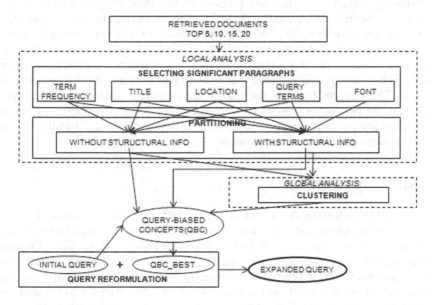

Fig. 1. The procedure of experiments to construct the query-biased concepts

In the local analysis, we select significant characteristics from each (retrieved and relevant) document and name them 'features'. This is done by selecting the significant paragraphs and partitioning those paragraphs. We use the following criteria for scoring each paragraph according to its significance: 1) the ratio of significant terms in a paragraph: the terms that frequently occur in a document are arranged in a significance term list; 2) the location of paragraph; 3) the presence of a title of the document within the paragraph; 4) the presence of query terms within the paragraph; 5) the presence of bold or italic term within the paragraph. The top ranked k paragraphs are chosen as the significant paragraphs. We then partition the selected paragraphs. Through partitioning, the features of each document are generated. It is important to make the selected significant characteristics orthogonal to each other within a document, because orthogonal features are able to represent the main themes of a document separately. In this paper, we extend the framework to deal with structured documents.

Nowadays, with the increased number of documents formatted in the eXtensible Markup Language (XML), it makes sense to investigate whether the structure, as captured by XML, can also be used to generate useful query-biased concepts. We suggest using the structural relations between paragraphs for partitioning. The paragraphs belonging to the same section or subsection can be partitioned. Through this restriction, we can reduce the non-desirable unification caused by common terms and/or specific terms like title terms or query terms within one document. Depending on the level of partitioning, the number of features in a document can be increased or decreased. The results with different levels of partitioning are presented in section 3.

After a local analysis, it is necessary to integrate these features across all the documents to build the concepts. We adopt a single pass method based in early work on clustering analysis [2]. The main purpose of this step, the global analysis, is to prevent the duplication of similar features. The clustering makes it possible to generate the primitive concepts that are approximately orthogonal. Analyzing a set of documents locally and globally has been used in previous studies [1, 4]. The final stage of constructing the query-biased concepts is to combine the generated concepts with the initial query. We compute the similarity of between the initial query and concepts. The concept that has the maximum similarity with the initial query is selected as the best query-biased concepts (QBC_{best}). For a new query, the original query terms are expanded with those associated terms in QBC_{best}. Finally, the new query is resubmitted to the retrieval system.

2   Experimental Set Up

We use the test collection developed at INEX 2005 [3], which consists of a set of XML documents, topics and relevance assessments. The document collection is made up of the full-texts, marked up in XML, of 16,819 articles of the IEEE Computer Society's publications. Generally, one article consists of a front matter (<fm>), body (<bdy>) and back matter (<bm>). The opening and closing tags enclose the main content, which is structured into sections (<sec>), subsections (<ss1>) and sub-subsections (<ss2>, <ss3>). Each of these logical units starts with a title followed by a number of paragraphs (<p>). We use the 23 content-only topics provided by INEX, as we are focusing on document retrieval. The <title> part of the topic is used as an initial query. Although the relevance assessment in INEX is done at element level, we can derive the assessment at document level. Any document that has any relevant content for a query is set as relevant to that query.

All experiments are performed using the HySpirit [7] retrieval system. We carry out a number of query reformulation experiments in the context of a relevance feedback scenario. First, the documents are retrieved using the initial query. To examine the impact of sampling a subset of the top ranked documents, we restrict the set of returned documents to the top 5, 10, 15, and 20 documents, respectively (as this reflects more realistic scenarios). Then we assess the relevance of retrieved documents. Practically, the user's relevance judgment is the most accurate but we use the relevance assessments provided by INEX 2005 to simulate the feedback process. With the selected documents (the retrieved and relevant documents), we construct the query-biased concepts with various approaches. The QBC_CO approach constructs the query-biased concepts with content information only. Neither the font information nor the structural information is used. The QBC_CS approach applies all the techniques with structural information to construct the concepts. Finally, the query-biased concepts from both approaches are combined with the initial query.

We examine four different types of results: one baseline, one pseudo-relevance feedback (PRF), one classis relevance feedback (RF), and various QBC approach (QBC_CO and QBC_CS). For the baseline, we use a traditional $tf*idf$ ranking. For PRF and RF, we use Rocchio's formula [6], where we use the top 5, 10, 15, and 20 retrieved documents of the baseline. For PRF, we use all such documents, whereas for RF, we use those that are relevant. We only use a positive feedback strategy (we only consider relevance), and choose the top 20 terms to expand the query. QBC_CO and QBC_CS are also performed with the top 5, 10, 15, and 20 retrieved documents of the baseline. For QBC_CS, we considered a hierarchical relation between paragraphs such as sections and subsections in the partitioning step. We partition the paragraphs belonging to the same section (QBC_CS_SEC), subsection (QBC_CS_SS1), or sub-subsection (QBC_CS_SS2). We also choose the top 20 terms to form the new query. Finally, we evaluate the results with the full freezing method [8]. There, the rank positions of the top n documents (n = 5, 10, 15, and 20), the ones used to modify the query, are frozen. The remaining documents are re-ranked.

3 Results and Analysis

For space reason, we only compare the results using mean average precision (MAP) over the whole ranking, which we calculate using trec_eval. Table 1 shows all the results. The results of PRF are inferior to the baseline. Since PRF assumes that all top n-ranked documents (n = 5, 10, 15, 20) are relevant, this indicates that we need to use relevance information to find appropriate terms for expanding the query. Although the results of RF are higher than the PRF, they are still lower than the baseline. This indicates that we need better techniques to extract appropriate terms. We can see that QBC_CO and QBC_CS outperform the baseline, PRF, and RF. In the QBC_CO approaches, using both local and global analysis such as summarization, partitioning, and clustering shows the best result. In the QBC_CS approaches, the cases without global analysis (i.e., clustering) show the best result. Here, we only report the best results of QBC_CS where we partition the paragraphs belonging to the same section and do not apply any clustering. We discuss the other results in Table 3.

Table 1. Mean average precision (MAP) of baseline, PRF, RF, and QBC runs. TOP5, TOP10, TOP15, and TOP20 represent the number of retrieved documents

Baseline		PRF	RF	QBC_CO	QBC_CS
0.2073	TOP5	0.1793	0.2056	0.2369	0.2325
	TOP10	0.1793	0.2049	0.2354	0.2211
	TOP15	0.1815	0.2057	0.2380	0.2190
	TOP20	0.1717	0.2045	0.2376	0.2193

It is known that the success of a query reformulation process depends on how the initial query performs. We thus classify the 23 topics into two groups: poor (P), and good (G) performing queries. We investigate whether the QBC approaches are particularly effective in the case of the poorly performing queries. The good/poor decision is based on the MAP achieved by our baseline. If the MAP of the query is above 0.2073, we consider the query to be good. 14 queries with the MAP under 0.2073 are identified to be poor. For simplicity, we only compare the results of RF and QBC approaches with respect to the baseline in Table 2. Due to space limitation, we choose the same results of QBC_CO and QBC_CS approaches with TOP5 in Table 1 again to compare the retrieval performance for the two types of queries.

Table 2. Retrieval performance of QBC_CO and QBC_CS runs in poor(P) and good(G) performing queries. There are 14 poorly performing queries and 9 good performing queries.

	Baseline		RF		QBC_CO		QBC_CS	
	P(14)	G(9)	P(14)	G(9)	P(14)	G(9)	P(14)	G(9)
MAP	0.1061	0.3647	0.1132	0.3493	0.1599	0.3566	0.1399	0.3767
%chg			+6.2	-4.4	+33.6	-2.3	+24.2	+ 3.2
R-precision	0.1581	0.3934	0.1903	0.3981	0.1985	0.3828	0.1694	0.4125
%chg			+16.9	+1.2	+20.4	-2.8	+6.7	+4.6

The MAP of QBC_CO is improved by 33.6% and that of QBC_CS_SEC is also improved by 24.2% in poorly performing queries over the baseline. This indicates that expanding terms by query-biased concepts has a positive effect in the case of poorly performing queries.

Then, we investigate the effect of structural information by comparing the results of various QBC_CS approaches in Table 3. QBC_CS_SEC, QBC_CS_SS1, and QBC_CS_SS2 represent different kinds of partitioning. The cases of QBC_CS_SEC with non-clustering (NONCL) show the best result. Generally, the results of SS1 and SS2 are lower than those of SEC. In QBC_CS approaches, a global analysis does not affect the improved performance for generating query-biased concepts.

Table 3. MAP of QBC_CS runs in different levels of partitioning. CL denotes a clustering and NONCL denotes a non-clustering.

	QBC_CS_SEC		QBC_CS_SS1		QBC_CS_SS2	
	CL	NONCL	CL	NONCL	CL	NONCL
TOP5	0.2035	0.2325	0.2028	0.2290	0.2304	0.2290
TOP10	0.2043	0.2211	0.2036	0.1900	0.1930	0.1900
TOP15	0.2079	0.2190	0.2072	0.2037	0.2090	0.2037
TOP20	0.2079	0.2193	0.2072	0.2041	0.2090	0.2041

Finally, we examine the effect of the number of retrieved documents used for generating concepts. In Table 1, using 15 documents with QBC_CO leads to the best result. In Table 3, using 5 documents with QBC_CS_SEC (NONCL) leads to the best result. This indicates that the number of retrieved documents does not directly affect the performance. This is because we did not use all the retrieved documents but only used those retrieved documents that were relevant. As long as some relevant documents are highly ranked, we are able to generate appropriate concepts for query expansion.

4 Conclusions

In this paper, we proposed an approach for constructing query-biased concepts from retrieved and relevant documents. The generated query-biased concepts were used to expand the queries in a relevance feedback (RF) process. The experimental results showed the improvement of retrieval performance with our various approaches. Particularly, we found an increase of performance when QBC was applied to the poorly performing queries. We also investigated the effect of structural information to construct the query-biased concepts and the number of retrieved documents used for generating the concepts. The use of structural information in a local analysis was effective to select the significant features of documents. But the use of clustering for a global analysis was not beneficial for query reformulation. In QBC_CS approaches, those which generated the query-biased concepts with the content and structural information (without a global analysis) led to the best performance. The retrieval performance of QBC approaches does not seem to rely on the number of retrieved documents. This is because we only used the relevance information of retrieved documents. It is not necessary to have a large number of relevant documents for generating appropriate query-biased concepts. However, it is essential to have them highly ranked for generating appropriate query-biased concepts.

Acknowledgments. This work was financially supported by IT Scholarship Program supervised by IITA (Institute for Information Technology Advancement) & MKE (Ministry of Knowledge Economy), Republic of Korea.

References

1. Chang, Y., Kim, M., Raghavan, V.V.: Construction of query concepts based on feature clustering of documents. Information Retrieval 9(3), 231–248 (2006)
2. Frakes, W.B., Baeza-Yates, R.: Information Retrieval: Data Structures and Algorithms. Prentice Hall, Englewood Cliffs (1992)
3. Malik, S., Lalmas, M., Fuhr, N.: Overview of INEX 2005. In: Fuhr, N., Lalmas, M., Malik, S., Kazai, G. (eds.) INEX 2005. LNCS, vol. 3977, pp. 1–15. Springer, Heidelberg (2006)
4. Nakata, K., Voss, A., Juhnke, M., Kreifelts, T.: Collaborative concept extraction from documents. In: 2nd International Conference on Practical Aspects of Knowledge management, pp. 29–30. Basel (1998)
5. Qiu, Y., Frei, H.P.: Concept based query expansion. In: 16th annual international ACM SIGIR conference on Research and Development in Information Retrieval, pp. 160–170. ACM press, Pittsburgh (1993)
6. Rocchio, J.J.: Relevance Feedback in Information retrieval. In: Salton, G. (ed.) The SMART retrieval system – experiments in automatic document processing, pp. 313–323 (1971)
7. Rölleke, T., Lübeck, R., Kazai, G.: The HySpirit Retrieval Platform. In: ACM SIGIR Demonstration, New Orleans (2001)
8. Ruthven, I., Lalmas, M.: A survey on the use of relevance feedback for information access systems. Knowledge Engineering Review 18(1), 95–145 (2003)

Comparing Several Textual Information Retrieval Systems for the Geographical Information Retrieval Task

José M. Perea-Ortega, Miguel A. García-Cumbreras, Manuel García-Vega, and L.A. Ureña-López

SINAI Research Group, Computer Science Department, University of Jaén, Spain
{jmperea,magc,mgarcia,laurena}@ujaen.es

Abstract. This paper presents a comparison between three different Information Retrieval (IR) systems employed in a particular Geographical Information Retrieval (GIR) system, the GeoUJA IR, a GIR architecture developed by the SINAI research group. It could be interesting and useful for determining which of the most used IR systems works better in GIR task. In the experiments, we have used the Lemur, Terrier and Lucene search engines using mono and bilingual queries. We present baseline cases, without applying any external processes, such as query expansion or filtering. In addition, we have used the default settings of each IR system. Results show that Lemur works better using monolingual queries and Terrier works better using the bilingual ones.

1 Introduction

Geographic Information Retrieval (GIR) is related to a specialized branch of traditional Information Retrieval (IR). GIR concerns the retrieval of information involving some kind of spatial awareness. Existing evaluation campaigns as GeoCLEF[1] whose aim is to provide the necessary framework in which to evaluate GIR systems for search tasks, involving both spatial and multilingual aspects. GeoCLEF is a cross-language geographic retrieval track included in the Cross-Language Evaluation Forum[2] (CLEF) campaign. The selection of a good IR system is essential in this task. The main contribution of this paper is to compare three different textual IR systems for the GIR task.

1.1 The Geographical Information Retrieval Task

We can define GIR as the retrieval of geographically and thematically relevant documents in response to a query of the form <theme, location>, where the spatial relationship may either implicitly imply containment, or explicitly be selected from a set of possible topological, proximity and directional options

---

[1] http://ir.shef.ac.uk/geoclef/
[2] http://www.clef-campaign.org/

E. Kapetanios, V. Sugumaran, M. Spiliopoulou (Eds.): NLDB 2008, LNCS 5039, pp. 142–147, 2008.
© Springer-Verlag Berlin Heidelberg 2008

(e.g. inside, near, north of)[1]. The original queries used in the experiments carried out in this work have not explicitly identified the spatial information, so it has been necessary to extract this *geo-information* (named locations and spatial relationships) from them.

There is a wide variety of approaches to resolve the GIR task ranging from basic IR approaches with no attempts at geographic indexing, to deep natural language processing to extract places and topological information from the texts and queries. Therefore, the main requirement which a GIR system should have with respect to an simple IR system is the recognition and the extraction of entities locations and spatial relations in the document collection and the queries. Another interesting feature would be to have a external geographic knowledge base as a gazetteer.

Section 2 introduces the system description and discusses the information retrieval systems employed in the experiments performed in this work; Section 3 presents the resources used in this empirical comparison and the experimental results; and Section 4 shows the main conclusions of this work.

2   System Overview

The core of a basic GIR system is based on an index-search engine. In general, other preprocessing modules are also used in any GIR system, such as a entity recognizer, a spatial relationship extractor or a geographic *disambiguator*. Our system has several subsystems that we introduce in the next section. The Figure 1 describes the basic architecture of our GIR system. A detailed description of the SINAI[3] GIR system is written in [2].

The modules or subsystems of our system are:

- **Translator subsystem.** We have used the SINTRAM[4] translator[3]. This subsystem translates the queries from several languages into English.
- **Named Entity Recognizer (NER) subsystem.** We have worked with the NER module of the GATE[5] toolkit.
- **Geo-Relation Finder subsystem.** This subsystem is used to find the spatial relations in the geographic queries.
- **Filtering subsystem.** Its main goal is to filter what documents among the recovered ones by the IR subsystem are valid and to make a new ranking with them.
- **Information Retrieval subsystem.** We have worked with several IR systems such as **Lucene**[6], **Terrier**[7] and **Lemur**[8].

---

[3] http://sinai.ujaen.es
[4] SINai TRAnslation Module
[5] http://gate.ac.uk/
[6] http://lucene.apache.org
[7] http://ir.dcs.gla.ac.uk/terrier
[8] http://www.lemurproject.org

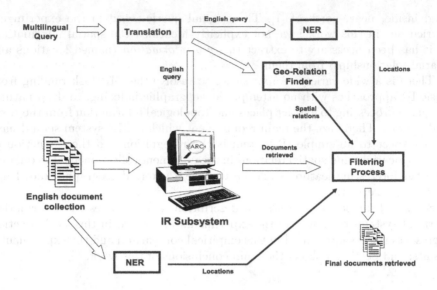

Fig. 1. Basic architecture of the SINAI GIR system

3 Experiments Description and Results

The experiments carried out consist in applying Lucene, Terrier and Lemur separately as textual IR system for the GeoCLEF search task, without considering any additional module. In order to obtain the R-Precision and the Mean Average Precision (MAP) of each experiment, we have used the English relevance assessments defined for GeoCLEF 2007 and the TREC evaluation method.

3.1 Resources

The English collection contains stories covering international and national news events, therefore representing a wide variety of geographical regions and places. It consists of 169,477 documents and was composed of stories from the British newspaper *The Glasgow Herald* (1995) and the American newspaper *The Los Angeles Times* (1994). The document collections were not geographically tagged and contained no semantic location-specific information[4].

A total of 25 topics were generated for GeoCLEF 2007 and they have been used in the experiments. The format of a query consists in three major labels or fields: *title*, *description* and *narrative*. These labels contain the text of the query.

3.2 Lucene Experiments

Lucene comes with several implementations of *analyzers* for input filtering. We have tried with the *Standard* and the *Snowball* analyzers. Another important feature of Lucene is its scoring function. It is a mixture of TF·IDF and boolean retrieval schemas. In our experiments, we have only tested this default scoring function. We have run experiments considering all the text labels from queries and we have worked with queries in German, Portuguese and Spanish. The results using Lucene are shown in the Table 1.

Table 1. Results using Lucene as textual IR system

Query Language	Analyzer	R-Precision	MAP
English	Standard	0.1573	0.1479
	Snowball	0.2210	**0.2207**
German	Standard	0.0669	**0.0543**
	Snowball	0.0527	0.0410
Portuguese	Standard	0.0848	0.0904
	Snowball	0.1068	**0.1095**
Spanish	Standard	0.1415	0.1257
	Snowball	0.1751	**0.1598**

In general, the use of the *Snowball* analyzer improves the results obtained with the *Standard* analyzer. As we expected, the bilingual experiments produce a loss of precision. The translation module works better with the Spanish queries.

3.3 Terrier Experiments

One of the advantages of Terrier is that it comes with many document weighting models, such as Okapi BM25 or simple TF·IDF. For our experiments, we have only used the simple TF·IDF and the Okapi BM25. Same as Lucene experiments, we have also used all the text labels from the topics and multilingual queries. The results using Terrier are shown in the Table 2.

Table 2. Results using Terrier as textual IR system

Query Language	Weighting Function	R-Precision	MAP
English	TF·IDF	0.2567	0.2548
	Okapi BM25	0.2562	**0.2570**
German	TF·IDF	0.0986	0.0888
	Okapi BM25	0.0972	**0.0898**
Portuguese	TF·IDF	0.1794	0.1834
	Okapi BM25	0.1949	**0.1864**
Spanish	TF·IDF	0.2469	0.2466
	Okapi BM25	0.2516	**0.2488**

Analyzing the results, we can see that the use of the Okapi BM25 as weighting model improves the results obtained with TF·IDF. Same as Lucene results, the bilingual experiments return worse MAP values. The translation module works good with the Spanish queries.

3.4 Lemur Experiments

As parameters we have worked with two weighting functions (Okapi or TF·IDF) and the use or not of Pseudo-Relevant Feedback (PRF)[5]. All the text labels from the topics and multilingual queries have been used. The results using Lemur are shown in the Table 3.

Table 3. Results using Lemur as textual IR system

Query Language	Weighting Function	R-Precision	MAP
English	TF·IDF simple	0.1925	0.1789
	Okapi simple	0.2497	0.2430
	TF·IDF feedback	0.1738	0.1675
	Okapi feedback	0.2578	**0.2619**
German	TF·IDF feedback	0.0661	0.0637
	Okapi feedback	0.0666	**0.0652**
Portuguese	TF·IDF feedback	0.1397	0.1274
	Okapi feedback	0.1682	**0.1695**
Spanish	TF·IDF feedback	0.1533	0.1512
	Okapi feedback	0.2200	**0.2361**

The results show that the Okapi weighting function works better than TF·IDF. In addition, the use of PRF combined with Okapi or TF·IDF improves the results obtained with simple Okapi or TF·IDF. In some cases, the improvement with PRF is quite important (around 8% for Okapi using the English queries).

A summary of the best overall results, comparing the three IR systems analyzed in this paper, are shown in the Table 4.

Table 4. Summary of best overall results

Query Language	IR System	Weighting Function	MAP
English	Lucene	TF·IDF + Boolean	0.2207
	Terrier	Okapi BM25	0.2570
	Lemur	Okapi Feedback	**0.2619**
German	Lucene	TF·IDF + Boolean	0.0543
	Terrier	Okapi BM25	**0.0898**
	Lemur	Okapi Feedback	0.0652
Portuguese	Lucene	TF·IDF + Boolean	0.1095
	Terrier	Okapi BM25	**0.1864**
	Lemur	Okapi Feedback	0.1695
Spanish	Lucene	TF·IDF + Boolean	0.1598
	Terrier	Okapi BM25	**0.2488**
	Lemur	Okapi Feedback	0.2361

4   Conclusions

After the analysis of the overall results, and using the default configuration of each IR system, Lemur works better with the English monolingual queries, but the difference is not important. With the multilingual queries, Terrier works better than Lucene and Lemur. Specifically, Terrier obtains around a 38% and 65% of improvement with respect to Lemur and Lucene using German queries. Using Portuguese and Spanish queries, Terrier also improves the results.

Another conclusion is that the simple Okapi weighting function works better than simple TF·IDF. In addition, the use of PRF combined with Okapi or TF·IDF in Lemur improves the results obtained with simple Okapi or TF·IDF.

These conclusions have been obtained using the default configuration of each IR System. As future work, it would be interesting to test several weighting models and the simple TF·IDF, Okapi or PRF schemas in Lucene and Terrier for comparison them with the results obtained using the same weighting models in Lemur.

Acknowledgments

The authors would like to thank CLEF in general, and Carol Peters in particular. This work has been supported by a grant from the Spanish Government, project TIMOM (TIN2006-15265-C06-03).

References

1. Bucher, B., Clough, P., Joho, H., Purves, R., Syed, A.K.: Geographic IR Systems: Requirements and Evaluation. In: Proceedings of the 22nd International Cartographic Conference (2005)
2. Perea-Ortega, J.M., García-Cumbreras, M.A., García-Vega, M., Montejo-Ráez, A.: GEOUJA System. University of Jaén at GEOCLEF 2007. In: Working Notes of the Cross Language Evaluation Forum (CLEF 2007), p. 52 (2007)
3. García-Cumbreras, M.A., Ureña-López, L.A., Martínez-Santiago, F., Perea-Ortega, J.M.: BRUJA System. The University of Jaén at the Spanish task of QA@CLEF 2006. LNCS. Springer, Heidelberg (2007)
4. Mandl, T., Gey, F., Nunzio, G.D., Ferro, N., Larson, R., Sanderson, M., Santos, D., Womser-Hacker, C., Xie, X.: Geoclef 2007: the clef 2007 cross-language geographic information retrieval track overview. In: Proceedings of the Cross Language Evaluation Forum (CLEF 2007) (2007)
5. Buckley, C., Salton, G., Allan, J., Singhal, A.: Automatic query expansion using smart: Trec 3. In: Proceedings of TREC3, pp. 69–80. NIST, Gaithesburg (1995)

Querying and Question Answering

Intensional Question Answering Using ILP: What Does an Answer Mean?

Philipp Cimiano, Helena Hartfiel, and Sebastian Rudolph

Institute AIFB, Universität Karlsruhe (TH)

Abstract. We present an approach for computing intensional answers given a set of extensional answers returned as a result of a user query to an information system. Intensional answers are considered as descriptions of the actual answers in terms of properties they share and which can enhance a user's understanding of the answer itself but also of the underlying knowledge base. In our approach, an intensional answer is represented by a clause and computed based on Inductive Logic Programming (ILP) techniques, in particular bottom-up clause generalization. The approach is evaluated in terms of usefulness and time performance, and its potential for helping to detect flaws in the knowledge base is discussed. While the approach is used in the context of a natural language question answering system in our setting, it clearly has applications beyond.

1 Introduction

Question answering systems for unstructured ([1]) or structured ([2]) information have been a focus of research since a few decades now. They are a crucial component towards providing intuitive access for users to the vast amount of information available world wide offered by information sources as heterogeneous as web sites, databases, RSS feeds, blogs, wikis etc. However, most of the prevalent work has concentrated on providing *extensional* answers to questions. In essence, what we mean with an extensional answer here is a list of those facts which fulfill the query. For example, a question like: *Which states have a capital?*, when asked to a knowledge base about Germany would deliver an (extensional) answer consisting of the 16 federal states ("Bundesländer"): *Baden-Württemberg, Bayern, Rheinland-Pfalz, Saarland, Hessen, Nordrhein-Westfalen, Niedersachsen, Bremen, Hamburg, Sachsen, Brandenburg, Sachsen-Anhalt, Mecklenburg-Vorpommern, Schleswig-Holstein, Berlin,* and *Thüringen*. While this is definitely a correct answer, it is not maximally informative.

A more informative answer would, beyond a mere listing of the matching instances, also *describe* the answers in terms of relationships which can help a user to better understand the answer space. To a question *"Which states have a capital?"* a system could, besides delivering the full extension, also return an answer such as *"All states (have a capital)."*, thus informing a user about characteristics that all the elements in the answer set share. We will refer to such descriptions which go beyond mere enumeration of the answers as *intensional answers* (IAs). The intension is thus a description of the elements in the answer

E. Kapetanios, V. Sugumaran, M. Spiliopoulou (Eds.): NLDB 2008, LNCS 5039, pp. 151–162, 2008.

set in terms of features common to all of them. In the above case, the common property of the answers is that they represent exactly the set of all federal states in the knowledge base.

We could certainly argue that this is over-answering the question. However, in many cases a listing of facts does not help the user to actually understand the answer. First, the extension might be simply too large for a user to make sense of it. In this case, a compact representation of the answer in terms of an intensional description might be useful. In other cases, the user can be dissatisfied with the answer because he simply doesn't know why the elements in the extension actually answer the query. In the above example, the user doesn't learn from the extensional answer that all German federal states have a capital if he doesn't know that the answer consists exactly of all the federal states in Germany. Finally, we will see in the remainder of this paper how intensional answers can enhance the understanding about relations in the knowledge base and can also help in discovering modeling errors.

In this paper, we thus present an approach for computing "intensions" of queries given their extensions and a particular knowledge base. Our research has been performed in the context of the natural language interface ORAKEL (see [2]) and the aim has been to integrate the component for providing intensional answers into the system.We discuss the approach in more detail in Section 2. In Section 3, we describe the empirical evaluation of our approach, which has been carried out based on the dataset previously used in the evaluation of the ORAKEL system (see [2]). We analyze in detail to what extent the intensional answers produced by the system are "useful" and also discuss how the intensional answers can help in detecting errors in the knowledge base. We discuss related work and conclude in Section 4.

2 Approach

2.1 Architecture

The aim of the work described in this paper has been to extend the ORAKEL natural language interface with capabilities for delivering intensional answers. ORAKEL is a natural language interface which translates wh-questions into logical queries and evaluates them with respect to a given knowledge base. ORAKEL implements a compositional semantics approach in order to compute the meaning of a wh-question in terms of a logical query. It is able to deal with arbitrary knowledge representation languages and paradigms (see [2]). However in this work, the implementation is based on F-Logic [3] and Ontobroker [4] is used as the underlying inference engine. For more details about the ORAKEL system, the interested reader is referred to [2]. The workflow of the system is depicted in Figure 1. When the user asks a natural language question referring to the knowledge base, it is processed by the ORAKEL system that computes a logical query which can then be evaluated by the inference engine with respect to the KB and ontology. While the extensional answers are displayed to the user, they are also forwarded to the component which generates intensional answers (named

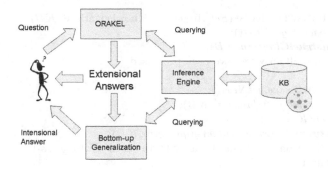

Fig. 1. Workflow of the system

Bottom-up Generalization in the figure). This component provides a hypothesis (in the form of a clause) based on the extensional answers by means of an ILP-algorithm. This hypothesis is the intensional answer and is displayed in addition to the extensional answers. Note that in this paper we are not concerned with the problem of generating an adequate answer in natural language, but rather with the previous step, i.e. the computation of a clause which represents the intensional answer. Obviously, natural language generation techniques can then be applied to generate an answer in natural language. However, this is not the focus of the present paper, but an obvious issue for future work. We assume that even in the case of a useful hypothesis, the user is interested in the direct answers as well, such that we do not forego their presentation. The displaying of the direct answers goes in parallel to the generation of their intensional description, such that the user does not have to wait for them until an intensional answer is possibly generated.

In this article we present an approach for computing *intensions* of queries given their extensions and a particular knowledge base. Our approach builds on Inductive Logic Programming (ILP) [5] as an inductive learning approach to learn descriptions of examples (our answers) in the form of clauses. In particular, it is based on bottom-up clause generalization, which computes a clause subsuming the extension of an answer and thus representing an "intensional" answer in the sense that it describes it in terms of features which all elements in the extension share. In particular, the system iteratively calculates least general generalizations (LGGs) ([5], pp. 36) for the answers by adding one answer at a time and generalizing the clause computed so far. According to our formulation as an ILP problem, the intensional description of the answers is regarded as a hypothesis which covers all the positive answers and is to be found in a search space of program clauses. This search space of program clauses is structured by a relation called θ-subsumption which orders the hypotheses (clauses) in a lattice, thus allowing to effectively navigate the search space (compare [5], pp. 33-37).

We make sure that the resulting clause is consistent by evaluating it with respect to the knowledge base, relying on the inference engine in order to check that it does not cover other examples than the extensional answers. In fact,

INTENSIONALANSWERCLAUSE(*Set Answers*, KnowledgeBase *KB*)
```
 1   a = Answers.getNext();
 2   c = constructClause(a, KB);
 3   while  not all answers have been processed
 4   do
 5       a' = Answers.getNext();
 6       c' = constructClause(a', KB);
 7       c = LGG(c, c');
 8       Answers' = evaluateQuery(query(c), KB);
 9       if c is inconsistent (i.e.Answers ∩ Answers' ⊂ Answers')
10          then
11                  return ∅ ( no consistent clause can be found )
12
13       c'' = reduceClause(c, KB, Answers)
14       if c'' covers all answers
15          then
16                  return c'';
17
18   return reduceClause(c, KB, Answers);
```

Fig. 2. Generalization Algorithm (calculating a reduced clause after each LGG computation)

the resulting clause can be straightforwardly translated into a corresponding query to the knowledge base. Our learning algorithm is thus similar to the FindS algorithm described in [6] in the sense that it aims at finding one single hypothesis which covers all positive examples and ignores negative examples during learning. If the resulting hypothesis which covers all positive examples is consistent in the sense that it does not cover negative examples, then an appropriate clause has been found. Otherwise, there is no clause which covers the positive examples without overgeneralizing. We explain this procedure more formally in the next section.

2.2 Generalization Algorithm

The generalization algorithm is given in Figure 2. The algorithm takes as input the set of (extensional) answers (*Answers*) as well as the knowledge base. The aim is to generate a hypothesis which exactly covers the extensional answers. A hypothesis in our approach is actually a clause of literals that constitutes a non-recursive definite program clause. For the hypothesis generation, the ILP-learner first constructs a specialized clause based on the first positive example. As head of this clause we use an artificial predicate *answer*. This is done by the constructClause procedure, which returns a clause for a certain individual in the knowledge base consisting of all the factual information that can be retrieved for it from the knowledge base. If there is only one answer, then the constructed clause is exactly the intensional answer we are searching for. In such a case, the user thus yields additional information about the entity in terms of the concepts it belongs to as well as other entities it is related to.

In case there are more answers, we loop over these and compute the LGG of the clause c constructed so far and the clause c' constructed on the basis of the next answer a'. The resulting clause is sent as query to the inference engine, which returns the extension of the clause c on the basis of the given knowledge base. If the clause is inconsistent, i.e. it covers more answers than the required ones (this is the case if $Answers \cap Answers' \subset Answers'$) then no consistent clause can be constructed and the algorithm returns the empty clause $\emptyset$. Note that the clause computed as the LGG can grow exponentially in the size of the original clauses. However, the clauses representing the LGG can also be reduced. In case the clause c is consistent, it is reduced as described below and we test if this reduced clause covers exactly the original answers. If this is the case we return the reduced clause c'' and are done. Otherwise, we proceed to consider the next answer and compute the next least general generalization between the last unreduced clause c and the clause c' generated by the next answer a'. This algorithm returns a clause in case it covers exactly the set of extensional answers. The LGG is essentially computed as described in [5], pp. 40 with the only exception that we do not allow functions as terms.

Clause Reduction. The following algorithm performs the reduction of the clause:

REDUCECLAUSE(Clause c, KnowledgeBase $KB, Set\ Answers$)
1   *List literals = orderLiterals(c)*;
2   **for** $i = 1$ to $|literals|$
3   **do**
4         $c' = remove(c, l_i)$;
5         $Answers' = evaluateQuery(query(c'), KB)$;
6   **if** c' is consistent with respect to $Answers$
7   **then**
8                 $c = c'$
9
10 *return c*;

where **evaluateQuery** essentially evaluates a clause formulated as query with respect to the existing knowledge base; **orderLiterals** orders the literals in a clause by an order that fulfills:

- $L_i \leq L_j$ if L_i is an atom of higher arity than L_j
- $L_i \leq L_j$ if L_i and L_j have the same arity and less variables in L_i appear in the remaining literals of the clause.

In fact, we have implemented two different procedures for the elimination of irrelevant literals. These procedures differ with respect to whether the clause is reduced after the computation of each LGG (as sketched in the algorithm depicted in Figure 2) or only at the end. Reduction at the end is unproblematic as the clause covers exactly the positive examples and the reduction does not affect the coverage. In case the reduction is performed after each LGG-computation, this can affect the computation of the further LGGs. Therefore, in the version of our approach in which we compute a reduced clause after each iteration, the reduced clause is only used as output in case it consistently covers all answers. Otherwise, the unreduced clause is used in the next iteration. It is important to mention that the reduction after each iteration is not strictly necessary as it finally yields the same intensional answer. However, our experiments show that by applying the reduction procedure after each iteration we can reduce the number of iterations and speed up our algorithm in most cases.

2.3   A Worked Example

Let's consider our running example, the question: *"Which states have a capital?"*. ORAKEL translates this question into the following logical query: *FORALL X ← EXISTS Y X : state ∧ X[capital → Y]. orderedby X*. The answer of the ORAKEL system can be represented as follows in clause notation:

answer("Saarland")   answer("Mecklenburg-Vorpommern")
answer("Rheinland-Pfalz")   answer("Hamburg (Bundesland)")
answer("Schleswig-Holstein")   answer("Thueringen")
answer("Sachsen-Anhalt")   answer("Sachsen")
answer("Bremen (Bundesland)")   answer("Niedersachsen")
answer("Brandenburg (Bundesland)")   answer("Berlin (Bundesland)")
answer("Baden-Wuerttemberg")   answer("Hessen")
answer("Bayern")   answer("Nordrhein-Westfalen")

Now the goal of our approach is to find a clause describing the *answer*-predicate intensionally. The number of positive examples is 16, while there are no explicit negative examples given. Nevertheless, a clause is inconsistent if, evaluated with respect to the knowledge base by the inference engine, it returns answers which are not in the set of positive examples. This is verified by the condition *Answers ∩ Answers' ⊂ Answers'* in the algorithm from Fig. 2.

In what follows, we illustrate our algorithm using the version which reduces the learned clause after each iteration. The first example considered is *"Saarland"*. The algorithm is initialized with the following clause representing all the facts about *"Saarland"* in the knowledge base:

answer("Saarland") ← state("Saarland"), location("Saarland"),
            inhabitants("Saarland",1062754.0), borders("Saarland","Rheinland-Pfalz"),
            borders("Saarland","France"), borders("Saarland","Luxembourg"),
            borders("Rheinland-Pfalz","Saarland"), location("Saarbruecken","Saarland"),
            capital_of("Saarbruecken","Saarland"), borders("France","Saarland"),
            borders("Luxembourg","Saarland"), flows_through("Saar","Saarland")

The above is the specialized clause produced by constructClause("Saarland") with the artificial predicate *answer* as head. As there are more examples, the next example *"Mecklenburg-Vorpommern"* is considered, which is represented by the following clause:

answer("Mecklenburg-Vorpommern") ← state("Mecklenburg-Vorpommern"),
            location("Mecklenburg-Vorpommern"),
            inhabitants("Mecklenburg-Vorpommern",1735000.0),

borders("Mecklenburg-Vorpommern","Brandenburg"),
borders("Mecklenburg-Vorpommern","Niedersachsen"),
borders("Mecklenburg-Vorpommern","Sachsen-Anhalt"),
borders("Mecklenburg-Vorpommern","Schlesw.-Holstein"),
borders("Brandenburg","Mecklenburg-Vorpommern"),
borders("Niedersachsen","Mecklenburg-Vorpommern"),
borders("Sachsen-Anhalt","Mecklenburg-Vorpommern"),
borders("Schlesw.-Holstein","Mecklenburg-Vorpommern"),
location("Rostock","Mecklenburg-Vorpommern"),
location("Schwerin","Mecklenburg-Vorpommern"),
capital_of("Schwerin","Mecklenburg-Vorpommern")

Now, computing the LGG of these two clauses yields the clause:

answer(X) ← state(X), location(X), inhabitants(X,Y),
        borders(X,Z),..., borders(X,O),   *(12 redundant border(X,_)-predicates)*
        borders(Z,X),..., borders(O,X),   *(12 redundant border(_,X)-predicates)*
        location(P,X), location(Q,X), capital_of(Q,X)

This clause can then be reduced to the following by removing redundant literals:[1]

answer(X) ← state(X), location(X), inhabitants(X,Y), borders(X,Z),
        borders(Z,X), location(P,X), capital_of(Q,X)

Finally, irrelevant literals can be removed, resulting in a clause which does not increase the number of negative examples, but possibly increases the number of positives ones, thus resulting in the clause $answer(X) \leftarrow state(X)$ for our example. This clause then covers exactly the original answers and no others. While this is not part of our current implementation of the system, an appropriate natural language answer could be generated from this clause, thus having *"All states (have a capital)"* as final intensional answer.

Thus, reducing the clause after each step has the effect that a correct clause can be derived in one step for our example. In case the clause is not reduced after each step (by removing irrelevant literals), the algorithm needs to make 7 iterations, considering the examples: Mecklenburg-Vorpommern, Rheinland-Pfalz, Hamburg, Schleswig-Holstein, Thüringen, Sachsen-Anhalt, and Bremen. Due to space limitations, we do not describe this procedure in more detail.

3   Evaluation

The empirical evaluation of the system has been carried out with respect to a dataset used previously for the evaluation of the ORAKEL natural language interface (see [2]). We analyze in detail to what extent the intensional answers produced by the system are "useful" and also discuss how the intensional answers can help in detecting errors in the knowledge base. Thus, we first describe the used dataset in more detail, then we discuss the usefulness of the generated intensional answers. After that, we show how intensional answers can be used to detect flaws (incomplete statements) in the knowledge base and also present some observations with respect to the time performance of our algorithm.

---

[1] Note that this is another kind of reduction than the one presented in Section 2.2. As opposed to the "empirical" reduction described there, the removal of redundant literals yields a clause which is logically (and not just extensionally) equivalent to the original one.

3.1 Dataset

The dataset previously used for the evaluation of the ORAKEL natural language interface consists of 463 wh-questions about German geography. These questions correspond to real queries by users involved in the experimental evaluation of the ORAKEL system (see [2]). The currently available system is able to generate an F-Logic query for 245 of these, thus amounting to a recall of 53%. For 205 of these queries, OntoBroker delivers a non-empty set as answer. Obviously, it makes sense to evaluate our approach to the generation of intensional answers only on the set of queries which result in a non-empty extension, i.e. the 205 queries mentioned before. Of these, the ILP-based approach is able to generate a clause for 169 of the queries. Roughly, we are thus able to generate intensional answers for about 83% of the relevant questions in our dataset. Here, only those questions are regarded as relevant which are translated by ORAKEL into an appropriate F-Logic query and for which OntoBroker delivers a non-empty extension. In general, there are two cases in which our ILP-based approach is not able to find a clause (in 36 out of 205 cases):

- The concept underlying the answers is simply not learnable with one single clause (about 67% of the cases).
- The answer to the query is the result of counting (about 33% of the cases), e.g. *"How many cities are in Baden-Württemberg?"*.

Given this dataset, we first proceeded to analyze the usefulness of the intensional answers. For this purpose, we examined all the answers and determined whether the user learns something new from the answer or it merely constitutes a rephrasing of the question itself.

3.2 Analysis of Intensional Answers

For the 169 queries for which our approach is able to generate an intensional answer, it turned out that in 140 cases the intensional answer could be actually regarded as *useful*, i.e. as not merely rephrasing the question and thus giving new information to the user. For example, to the question: *"Which rivers do you know?"*, the intensional answer *"All rivers"* does not give any additional knowledge, but merely rephrases the question. Of the above mentioned 140 useful answers, 97 are intensional answers describing additional properties of a single answer. Thus, for only 43 of the 140 useful answers the bottom-up generalization algorithm was actually used.

In order to further analyze the properties of intensional answers, we determined a set of 54 questions which, given our observations above, could be suitable for presenting intensional answers. For the new 54 queries, we received as results 49 useful intensional answers and 1 answer considered as not useful. For the other 4 questions, no consistent clause could be found. Together with the above mentioned 140 queries the overall number of useful intensional answers for the actual knowledge base is 189. The queries which provided useful answers can be classified into the following three types:

- Questions asking for all instances with a special property which turn out to coincide exactly with the extension of some atomic concept in the ontology (about 28% of the cases). The user learns from such an answer that all the instances of the class in question share the property he is querying for. An example is the question: *"Which states have a capital?"*, which produces the answer $answer(x) \leftarrow state(x)$ or in natural language simply *"All states"*.
- Questions asking for the relation with a specific entity which is paraphrased (possibly because the user doesn't know or remember it) (16% of the cases) Examples of such questions are: *"Which cities are in a state that borders Austria?"* or *"Which capitals are passed by a highway which passes Flensburg?"*. In the first case, the state in question which borders Austria is *"Bayern"* and the intensional answer is correctly: *"All the cities which are located in Bayern."* In the second case, the highway which passes Flensburg is the A7, so the intensional answer is *"All the capitals located at the A7"*.
- Questions about properties of entities which lead to the description of additional properties shared by the answer set as intensional answer (about 5% of the cases). For example, the question *"Which states do three rivers flow through?"* produces the intensional answer *"All the states which border Bayern and which border Rheinland-Pfalz"*, i.e. *Hessen and Baden-Württemberg.* Another example is the question: *"Which rivers flow through more than 5 cities?"*, yielding the intensional answer: *"All the rivers which flow through Duisburg"*. A further interesting example is the question *"Which cities are bigger than München?"* with the intensional answer: *"All the cities located at highway A24 and which a river flows through"*. These are the cities of Berlin and Hamburg.
- Questions which yield a single answer as a result and for which the intensional answer describes additional properties. An example would be: *"What is the capital of Baden Württemberg?"*, where the extensional answer would be *"Stuttgart"* and the intensional answer (paraphrased in natural language) *"Stuttgart is located at the highways A8, A81 and A85 and A831. Stuttgart has 565.000 inhabitants and is the capital of Baden Württemberg. The Neckar flows through Stuttgart."*

3.3   Debugging the Knowledge Base

In addition to providing further information about the answer space, a key benefit of intensional answers is that they allow to detect errors in the knowledge base. We found that in some cases the intensional answer produced was not the expected one, which hints at the fact that something is wrong with the knowledge base in question. We give a few examples to illustrate this idea:

- For example, for the question *"Which cities have more than 9 inhabitants?"*, we would expect the intensional answer *"All cities"*. However, our approach yields the answer $answer(x) \leftarrow \exists y\ city(x) \wedge inhabitants(x, y)$, i.e. *"all the*

Table 1. Performance measurements for our testsuite

Reduction	Time (in sec.)			# Iterations		
	$\emptyset$	min	max	$\emptyset$	min	max
after each LGG computation	0.528	0.08	13.61	0.621	0	5
at the end	0.652	0.07	33.027	1.702	0	73

cities for which the number of inhabitants is specified". This hints at the
fact that the number of inhabitants is not defined for all declared cities
in the knowledge base. A closer inspection of the knowledge base revealed
that some Swiss cities are mentioned in the knowledge base because some
German rivers have their origin in Switzerland, but no number of inhabitants
is specified for these cities.

- The question *"Which cities are located in a state?"* yields no intensional
  answer at all, while we would expect the answer "all cities". However, as
  mentioned above, there are cities in Switzerland for which a corresponding
  state is missing.
- The query *"Which river has an origin?"* yields as intension: $answer(x) \leftarrow$
  $\exists y\ river(x) \land length(x, y)$ (*"All rivers which have a length"*) instead of *"All
  rivers"*. This is due to the fact that there is one instance of river for which
  neither a length nor an origin is specified. This instance is the river *Isar*.

This shows that intensional answers can be useful to detect where the information in the knowledge base is not complete.

3.4   Performance Analysis

Finally, we have also carried out a performance analysis of the component for in-
ducing intensional answers. Table 1 shows the average time and iterations needed
to compute an intensional answer. In particular, we tested two configurations of
our system: one corresponding to a version in which the clause is reduced after
each LGG computation and one in which the clause is only reduced at the end. A
first analysis of the time performance reveals that our approach can be used efficiently, taking on average about half a second to compute an intensional answer.
As we see extensional and intensional answers as equally important and assume
that both are shown to the user, we can even assume that first the extensional
answer is computed and shown already to the user while the intensional answer
is still computed. While the results differ considerably with respect to the maximum for the different configurations of the system, it is interesting to see that
they hardly differ in the average case. In fact, when the reduction is performed
after each LGG computation, the average time is only very slightly under the
average time for the version in which the clause is reduced only at the end.

4   Related Work and Conclusion

We have presented an approach for computing intensional answers to a query formulated in natural language with respect to a given knowledge base. The

approach relies on Inductive Logic Programming techniques, in particular bottom-up clause generalization, to compute a clause subsuming the extensional answer as returned by the inference engine, which evaluates queries to the knowledge base. This clause represents the intensional answer to the question and could be transformed back into natural language. The latter step is not part of our present contribution and thus remains for future work. The idea of delivering intensional answers to queries is relatively straightforward and has been researched in the context of knowledge discovery from databases (see e.g. [7], [8], [9]), but also logic programs (see e.g. [10]).

Intensional query answering (IQA) emerged as a subject in the research of cooperative response generation, which has been a topic in many works related to natural language interfaces (compare [11] and [12]). The approach closest to ours is the one of Benamara [12] who also presents an approach to induce intensional answers given the extension of the answer. Instead of using ILP, Benamara relies on a rather ad-hoc approach in which parts of the answer are replaced by some generalized description on the basis of the given knowledge base. This generalization is guided by a *"variable depth intensional calculus"* which relies on a metric measuring the distance between two concepts as the number of arcs and the inverse proportion of shared properties. On the one hand, the mechanism for generating intensional answers seems quite adhoc in comparison to ours, but, on the other hand, the WEBCOOP system of Benamara is able to generate natural language answers using a template-based approach.

While the idea of delivering intensional answers is certainly appealing and intuitive, and we have further shown that they can be computed in a reasonable amount of time, we have also argued that their benefit is not always obvious. While our approach is able to generate intensional answers for about 83% of the questions which can be translated by ORAKEL to a logical query and actually have an extension according to the inference engine, we have shown that in some cases the answer is actually θ-equivalent and thus also logically and extensionally equivalent to the question. In these cases the intensional answer is not delivering additional information to the user. However, this case could be easily ruled out by checking θ-equivalence between the query and the clause returned as intensional answer. In these cases, the intensional answer can then be omitted. In the remaining cases, the intensional answers seem to indeed be delivering additional and useful information about the knowledge base to a user (in some cases in a quite oracle-like fashion arguably). The most obvious avenues for future work are the following. First, the bottom-up rule induction approach should explore "larger environments" of the (extensional) answers for clause construction by not only considering properties of the entity in the answer set but further properties of the entities to which it is related. Second, instead of only computing one single clause, we should also consider computing (disjunctive) sets of conjunctive descriptions, thus hopefully increasing the coverage of the approach. However, it could turn out that the disjunctions of intensional descriptions are much harder to grasp by users. Finally, the task of generating natural language descriptions of the intensional answers seems the most pressing issue. On a more general note, we hope that our paper contributes to stimulating discussion on a research issue which has not been on the foremost research frontier lately, but seems crucial towards achieving more natural interactions with information systems.

Acknowledgements. This research has been funded by the Multipla and ReaSem projects, both sponsored by the Deutsche Forschungsgemeinschaft (DFG).

References

1. Strzalkowski, T., Harabagiu, S.: Advances in Open Domain Question Answering. Text, Speech and Language Technology, vol. 32. Springer, Heidelberg (2006)
2. Cimiano, P., Haase, P., Heizmann, J., Mantel, M., Studer, R.: Towards portable natural language interfaces to knowledge bases: The case of the ORAKEL system. Data and Knowledge Engineering (DKE) 62(2), 325–354 (2007)
3. Kifer, M., Lausen, G., Wu, J.: Logical foundations of object-oriented and frame-based languages. Journal of the ACM 42, 741–843 (1995)
4. Decker, S., Erdmann, M., Fensel, D., Studer, R.: Ontobroker: Ontology Based Access to Distributed and Semi-Structured Information. In: Database Semantics: Semantic Issues in Multimedia Systems, pp. 351–369. Kluwer, Dordrecht (1999)
5. Lavrac, N., Dzeroski, S.: Inductive Logic Programming: Techniques and Applications. Ellis Horwood (1994)
6. Mitchell, T.: Machine Learning. McGraw-Hill, New York (1997)
7. Motro, A.: Intensional answers to database queries. IEEE Transactions on Knowledge and Data Engineering 6(3), 444–454 (1994)
8. Yoon, S.C., Song, I.Y., Park, E.K.: Intelligent query answering in deductive and object-oriented databases. In: Proceedings of the third international conference on Information and knowledge management (CIKM), pp. 244–251 (1994)
9. Flach, P.A.: From extensional to intensional knowledge: Inductive logic programming techniques and their application to deductive databases. In: Kifer, M., Voronkov, A., Freitag, B., Decker, H. (eds.) Transactions and Change in Logic Databases. LNCS, vol. 1472, pp. 356–387. Springer, Heidelberg (1998)
10. Giacomo, G.D.: Intensional query answering by partial evaluation. Journal of Intelligent Information Systems 7(3), 205–233 (1996)
11. Gaasterland, T., Godfrey, P., Minker, J.: An overview of cooperative answering. Journal of Intelligent Information Systems 1(2), 123–157 (1992)
12. Benamara, F.: Generating intensional answers in intelligent question answering systems. In: Belz, A., Evans, R., Piwek, P. (eds.) INLG 2004. LNCS (LNAI), vol. 3123, pp. 11–20. Springer, Heidelberg (2004)

Augmenting Data Retrieval with Information Retrieval Techniques by Using Word Similarity

Nathaniel Gustafson and Yiu-Kai Ng

Computer Science Department, Brigham Young University, Provo, Utah 84602, USA

Abstract. Data retrieval (DR) and information retrieval (IR) have traditionally occupied two distinct niches in the world of information systems. DR systems effectively store and query structured data, but lack the flexibility of IR, i.e., the ability to retrieve results which only partially match a given query. IR, on the other hand, is quite useful for retrieving partial matches, but lacks the completed query specification on semantically unambiguous data of DR systems. Due to these drawbacks, we propose an approach to combine the two systems using predefined *word similarities* to determine the correlation between a keyword query (commonly used in IR) and data records stored in the inner framework of a standard RDBMS. Our integrated approach is flexible, context-free, and can be used on a wide variety of RDBs. Experimental results show that RDBMSs using our word-similarity matching approach achieve high mean average precision in retrieving relevant answers, besides exact matches, to a keyword query, which is a significant enhancement of query processing in RDBMSs.

1 Introduction

Information retrieval (IR) and data retrieval (DR) are often viewed as two mutually exclusive means to perform different tasks—IR being used for finding relevant documents among a collection of unstructured/semi-structured documents, and DR being used for finding exact matches using stringent queries on structured data, often in a Relational Database Management System (RDBMS). Both IR and DR have specific advantages. IR is used for assessing human interests, i.e., IR selects and ranks documents based on the likelihood of relevance to the user's needs. DR is different—answers to users' queries are exact matches which do not impose any ranking; however, DR has many advantages—modern RDBMSs are efficient in processing users' queries, which are optimized and evaluated using the optimal disk storage structure and memory allocation.

When using an RDBMS, documents, each of which can be represented as a sequence of words in a data record, can be (indirectly) retrieved by using Boolean queries, which is simple; however, as mentioned earlier, the retrieved results only include exact matches and lack relevance ranking. In order to further enhance existing IR/DR approaches, we propose an integrated system which can be fully implemented within standard RDBMSs and can perform IR queries by using the *word similarity* between a set of document keywords, which represents the content of the document (as shown in a data record), and query keywords to determine their degree of relevance. Hence, at query execution time, keywords in a query Q are evaluated against (similar) words in a data record D. To determine the degree of relevance between Q and D, each keyword

E. Kapetanios, V. Sugumaran, M. Spiliopoulou (Eds.): NLDB 2008, LNCS 5039, pp. 163–174, 2008.

in Q is assigned a weight based on its similarity to the keywords in D, which has been verified to yield accurate query results (see detail in Section 4). The uniqueness and novelty of our integrated approach is on the benefit of being *context free*, i.e., independent of data sets; it can be used on small or large databases without any "training," and its applicability on a wide variety of (unstructured) text data.

We present our integrated system as follows. In Section 2, we discuss related work in query processing in IR and/or DR systems. In Section 3, we introduce our integrated approach using the word-similarity measure. In Section 4, we verify the accuracy of our integrated system using the experimental results. In Section 5, we give a conclusion.

2   Related Work

A number of approaches have been taken to combine the principles of both IR and DR systems—[4] present an integrating method for searching XML documents using a DR system, while additionally providing relevance rankings on retrieved results. The design of this approach, however, is limited to semi-structured XML data. [9] expand SQL queries with terms extracted from Web search results. Since this approach relies on an external source for expansion of individual queries, it is neither self-contained nor context-free. [5] create top-k selection queries in relational databases. Unfortunately, [5] pay little attention to how similarity between two items is calculated.

[12] consider user feedback to expand queries at the expense of additional (user) time and input. Other Works have also been done regarding query expansion by word-similarity matching, but rarely in the context of combining IR and DR as our proposed integrated system. [3] use context-free word similarity to expand query terms in an explicitly IR system, with good results. These works have already shown that word-similarity matching within IR systems leads to relatively accurate retrieved results. [10] provide a number of steps in retrieving information using SQL statements, which include a *query* relation that contains query terms and an *index* relation that indicates which terms appear in which document. Query terms in [10], however, must be matched exactly with the terms stored in the database as oppose to our similarity (partial) matching approach between query and document terms. [14] also introduce an IR ranking which ranks higher documents that contain all the words in the query with moderately weight. We have gone further than that by considering document terms which have high similarity with some, not necessary all, query terms.

[7] use IR indexing methods and the probabilistic database logic to represent the ordered set of tuples as answers to a user's query. [7], like us, consider stemmed document terms; however, their term-similarity measures are calculated using TF-IDF, whereas we use word-correlation factors, which are more sophisticated. [6] develop an information-theoretic query expansion approach based on the Rocchio's framework for ranking relevant documents in a collection using a term-scoring function, which is different from ours, since our query evaluation approach is not attached to any particular framework but still achieves high efficiency in query evaluation. [8], which suggest implementing query expansion using semantic term matching to improve the retrieval accuracy, only consider the k most similar terms to perform the search, whereas we consider the similarity between every keyword in a document and each keyword in a user's query.

3   Our Word-Similarity Matching Approach

We present an approach that searches for documents relevant to a given query in the standard IR context by using *word-similarity matching* (WSM), which is effective in determining document relevance. Our approach, which is fully implemented within an RDBMS and thus takes advantages of the powerful and efficient query evaluation systems embedded into RDBMSs, stores similarity values between words within a table in an RDBMS, which is then used for matching words. When a Web user submits a keyword query requesting data records (i.e., their corresponding documents) from the database, both data records which contain keywords exactly matching the query keywords, as well as those data records which contain words most similar to the query keywords, are retrieved. Our WSM approach can be used with any RDBMS, and is designed to search data records stored in tables—specifically, it searches non-stop, stemmed words[1] extracted from documents and stored in data records. Our word-matching approach works with keyword queries, which are currently the most-commonly-used queries on the Internet. Using a front-end interface, the user submits a keyword query, which is automatically transformed into an SQL query.

3.1   Word-Similarity Table

The values in the original word-similarity table T used in our WSM approach provide a means to determine the degree of similarity between any two words. The set of all non-stop, stemmed words in T was gathered from a collection of roughly 880,000 documents covering numerous topics published on Wikipedia (en.wikipedia.org/wiki/Wikipedia:Database_download) and appears in a number of online English dictionaries: 12dicts-4.0, Ispell, RandomDict, and Bigdict. Due to the volume and variety of this Wikipedia document set S, the words extracted from S cover significant portions of the English language. Prior to detecting any similar documents, we pre-computed T as the *metric cluster matrix* (referred to as MCM) and its *correlation factors* (i.e., the degrees of similarity between any two words) as defined in [2]. MCM (i) is based on the *frequency* of co-occurrences and *proximity* of each word's occurrences in S and (ii) provides fewer false positives and false negatives[2] than other comparable cluster matrices in [2]. MCM is represented as an $n \times n$ structure of floats in the range [0, 1], where n is the total number of unique words extracted from S, and each value in the matrix represents the similarity between the two corresponding words. When using 57,908 words as in the word-similarity matrix, MCM is large, exceeding 6GB of data. However, MCM can be reduced in size, consequently decreasing the processing time and memory requirement. We have constructed the 13%-reduced word-similarity matrix for our WSM approach by selecting the top 13% (i.e., 7,300) most frequently-occurring words[3] in the original MCM, along with their corresponding similarity values.

---

[1] A stop word is any word that is used very frequently (e.g., "him," "with," "a," etc.) and is typically not useful for analyzing informational content, whereas a stemmed word is the lexical root of other words (e.g., "driven" and "drove" can be reduced to the stemmed word "drive").

[2] *False positives (False negatives, respectively)* refer to sentences (documents) that are *different* (the *same*, respectively) but are treated as the *same (different*, respectively).

[3] Word frequency is determined by the word occurrences in the set of Wikipedia documents.

A potential issue with the reduction of the MCM is the consequent loss of words which could be considered in answering users' queries. However, the *ratio* of the occurrences of the top 13% most frequently-occurring words to all word occurrences in the Wikipedia document set is approximately 90%. We will discuss how the remaining 87% of words are also included, but only as exact matches (i.e., not having similarity values stored for these words). This hybrid approach preserves the lower memory requirement of the 13%-reduced matrix, while allowing matching on less-frequent words.

The Reduced Word-Similarity Table. We first represent the 13%-reduced MCM in an RDBMS by converting it to a simple 3-column table, where each tuple $<w_1, w_2, Sim\text{-}Val>$ contains two stemmed, non-stop words w_1 and w_2, and their respective similarity value $Sim\text{-}Val$. We refer to this table as the 13%-reduced word-similarity table, or 13%-WST. Exact-match tuples (e.g., $<$"rubric", "rubric", 1$>$) are also added for each of the 87% least common words excluded from the original 13%-reduced matrix. While the exclusion of $Sim\text{-}Val$ values for 87% of the words used in the original matrix reduces the total number of $Sim\text{-}Val$ values from 3.3 billion to about 50 million— more than a 98% reduction—the resulting 13%-WST is still quite large, but can be further reduced to decrease run-time and memory requirements for processing keyword queries and allow feasible implementation in real-world RDBMSs.

The basis for this further reduction lies in the distribution of $Sim\text{-}Val$ values in the 13%-WST. As shown in Figure 1(a), besides the exact-match word pairs with similarity values of 1, all word pairs have a $Sim\text{-}Val$ on the order of 1×10^{-4} to 1×10^{-9}, with lower values (e.g., around 1×10^{-8}) being exponentially more common than higher values (e.g., around 1×10^{-4} or 1×10^{-5}), i.e., there are relatively few highly similar word pairs, and exponentially more word pairs with little or no similarity. Thus, to perform further reduction, we select and retain word pairs with high similarity, which are more useful in determining the relevance of (the keywords in) a data record R to (the keywords in) a user query Q, and exclude those with lower similarity, as they are less useful (in determining the degree of relevance between R and Q), and are far more numerous. By eliminating low-similarity word pairs, two further-reduced, high-similarity tables are constructed for use in our WSM approach: (i) a smaller table containing only high-similarity word pairs, used for efficiently narrowing down the set of documents for analysis, and (ii) a larger and more detailed table, used in measuring query-data record similarity. Both of these reduced tables are built by collecting all $<w_1, w_2, Sim\text{-}Val>$ tuples from the 13%-WST with $Sim\text{-}Val$ values above a given threshold as follows:

INSERT INTO RTable (SELECT * from WSTable WHERE Sim-Val > threshold);

where RTable is a reduced, high-similarity table, WSTable is the previously discussed 13%-WST, and threshold is a threshold value.

Determining Word-Similarity Threshold Values. We realize that in reducing the 13%-WST, a *lower* threshold value creates a larger reduced table, which results in more computation and a slower implementation, but may also result in more accuracy of data-record retrieval. A *higher* threshold value, on the other hand, creates a smaller reduced table, which is more suited for quickly identifying only those few words in a record most similar to a given keyword in a query. Steps 1 and 2, as presented in

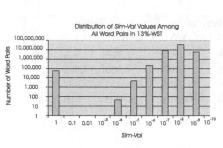
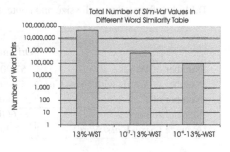

(a) Frequencies of $Sim\text{-}Vals$ from 10^{-4} to 10^{-8} increase exponentially

(b) Relative sizes of 13%-WST, 10^{-7}-13%-WST, and 10^{-6}-13%-WST

Fig. 1. $Sim\text{-}Val$ values in the 13%-WST

Section 3.2, function most efficiently by using reduced tables of two different sizes. We chose two threshold values for these reduced tables to optimize execution time without significantly affecting the accuracy of data-record retrieval. Ideal threshold values have been determined experimentally to be 3×10^{-6} and 5×10^{-7}, which yield two tables containing 0.18% and 1.3% of the word-similarity values contained in the 13%-WST, referred as the 10^{-6}-13%-WST and 10^{-7}-13%-WST, respectively. The size of each similarity matrix and table is displayed in Figure 1(b) and Table 1. Before we can apply WSM to search documents using the 10^{-6}- and 10^{-7}-13%-WST, data records and their corresponding schemes in an RDBMS must first be extracted from the source data.

Document Representation. We use "document" to represent a Web page, which can be an academic paper, a sports news article, etc. We assume that each document is represented by the (non-stop, stemmed) keywords in its "abstract," which can be a title, a brief description, or a short summary of the content of the respective document. Before we search the abstract of a document, it must first be formatted as a data record with non-repeated, non-stop, stemmed keywords in the $KeywordAbstract$ table in where each tuple <Keyword, Doc-ID> contains a keyword and a reference to the (abstract of the) corresponding document record in which it is contained. This essentially creates a *word index* for searching the occurrences of keywords among all the abstracts quickly.

3.2 Calculating Query-Record Similarity

In this section, we describe the steps used to calculate query-data record similarity, which is significantly different from existing methods. In [11] an IR-style, keyword-search method is introduced that can be implemented on RDBMSs with indexing on text, which like ours, allows partial ranking (i.e., not all the terms in a query Q are required to appear in the tuples to be treated as answers to Q). Since [11] present the top-k matches in order to reduce computational times, relevant documents that are positioned lower in the ranking might be excluded, which is unlike our similarity ranking approach that retrieves all documents of various degrees of relevance with little

Table 1. Information on the size and contents of each word-similarity matrix and table

	Matrices		Tables		
	Original	13%-WST	13%-WST	10^{-7}	10^{-6}
Number of Words	57,908	7,300	57,908	57,908	57,908
Number of Words with Sim-$Vals$	57,908	7,300	7,300	7,300	7,300
Number of Sim-Val	3.35GB	53.34M	53.34M	699,994	95,536
Size in Memory	6GB	100MB	1.9GB	25MB	3.4MB
Average (Sim-$Vals$/Word)	57,908	7,300	7,300	89	6.2

computational time. In [13] phrases are used for retrieving (relevant) documents, which like ours allows inexact matching. However, in [13] since the sense of the words in a query is determined by using WordNet, which includes synonyms, hyponyms, and compound words for possible additions to the query, the approach is different from ours.

In performing similarity matching between a user's query Q and a set of data records, we introduce three relatively simple and intuitive steps. In order to shorten running time, the first step quickly filters out data records with little or no similarity to Q. Since keyword queries typically contain more than one keyword, in the second step we calculate the similarity between each query keyword to the keywords in each of the data records retained after the first step. The third and final step "combines" these query keyword-data record similarities to form a single query-data record similarity value, which is used to rank and retrieve relevant data records.

Subset Selection (Step 1). Users typically looks for answers which are at least fairly relevant to a query and disregards those with less relevance. It has been shown that users are typically only interested in the top 10 retrieved results [16]. Thus, our *subset-selection* step can quickly eliminate most of the irrelevant or relatively dissimilar records.

This subset-selection step is performed by choosing records which contain at least a keyword k_1 such that there exists at least a query keyword k_2 for which k_1 and k_2 have a similarity value above a certain threshold value. This can easily be performed by using the 10^{-6}-13%-WST, which has a relatively high threshold value and includes only about the top 0.18% most similar word pairs in the 13%-WST. The 10^{-6}-13%-WST is optimal, since it excludes very few relevant data records, while selecting a relatively small, mostly relevant portion of data records to be ranked and retrieved, which allows for accuracy and short running time in answering a keyword query.

Often the length of a keyword query is very short, only 2.35 words on average [16]. Certainly queries with 4-5 words, or even more, are also made; however, this subset-selection step requires only one query keyword to have the highest similarity, that is at least 3×10^{-6}, to any keywords in a data record in question and is only a pre-processing step which quickly reduces the total number of data records to be considered.

Calculating Keyword-Document Similarity (Step 2). In computing the similarity between a query keyword q_i ($i \geq 1$) and a data record D (represented by the keywords in the abstract of D), we consider the fact that q_i is likely to be similar to D if D contains more than one keyword which is highly similar to q_i. Hence, we calculate the similarity of q_i and D by summing the similarities between q_i and each keyword k_j in D as

$$SumSim(q_i, D) = \sum_{j=1}^{n} Sim(q_i, k_j) \,, \tag{1}$$

where n is the number of keywords in D and $Sim(q_i, k_j)$ is the similarity between q_i and k_j, determined by the 10^{-7}-13%-WST. We use the 10^{-7}-13%-WST, as opposed to the 10^{-6}-13%-WST used in step 1, in order to obtain a more accurate calculation for similarity between q_i and D.

Statistically, longer abstracts will have slightly higher similarities to q_i than shorter abstracts. However, due to the distribution of Sim-Val values, any data record which contains only one or two words which are highly similar to q_i will still be significantly more similar to q_i than data records containing hundreds or thousands of dissimilar words. Also, in order to not weigh any single query keyword too heavily, *repeats* of the same keywords in an abstract are not counted, e.g., "document queries" and "documenting document queries" are treated as identical.

Calculating Query-Document Similarity (Step 3). After the similarity values between each q_i (in a query Q) and a data record D have been calculated, these values can be used to calculate a single similarity value for Q and D. We consider two methods which perform such a calculation. The first method, called $AddSim(Q, D)$, involves summing up $SumSim(q_i, D)$ values ($1 \leq i \leq |Q|$) and is effective at detecting a data record D which contain keywords with high degrees of similarity to at least some query keyword q_i. The second method, $MultSim(Q, D)$, calculates query-data record similarity by multiplying together all the $SumSim(q_i, D)$ values, where $q_i \in Q, 1 \leq i \leq |Q|$.

I. Calculating Similarity by Addition. Given a query Q and a data record D, the similarity of Q and D can be computed by summing the degrees of similarity of each query keyword q_i ($1 \leq i \leq m = |Q|$) to D. Thus, the similarity of Q and D is computed as

$$AddSim(Q, D) = \sum_{i=1}^{m} SumSim(q_i, D) \,. \tag{2}$$

Note that in the distribution of Sim-Val values, exact word matches (having a similarity value of 1) are weighted much more heavily than all other word pairs (which have similarities ranging from 1×10^{-4} to 5×10^{-7} in the 10^{-7}-13%-WST). Thus, $AddSim(Q, D)$ will always equal N plus some small fraction, where N (≥ 0) is the number of exact word matches between Q and D. This effectively means that $AddSim(Q, D)$ will rank any data record D containing N query keywords higher than others containing N-1 query keywords, regardless of other similar content.

$AddSim$ has been shown to effectively identify data records with high similarity to at least some query keywords; however, in some cases, this results in a *false positive*, i.e., data records dissimilar to Q are treated as similar. For example, consider the query "network security." If there exist two data records D_1 and D_2, such that D_1 = "Connecting to Nodes on Distributed Networks," and D_2 = "Authentication over Networks," $AddSim$ would rate D_1 higher than D_2 due to multiple keywords in D_1 having high similarity to "network" (i.e., "network," "connecting," and "nodes"), even though none of them bear much similarity to "security." This is demonstrated in Figure 2.

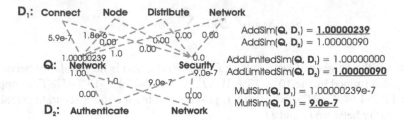

Fig. 2. Computing query-data record similarity using various WSM methods and the query, "network security." Sim-Vals are labeled on edges.

We consider data records that are heavily weighted to only some query keywords by limiting the similarity of each keyword in a query to a data record to 1, which is the similarity value for *identical* words, as shown in Equation 3. This approach has the advantage of assigning a higher ranking to data records with more "variety", i.e., records that contain words relevant to the "majority of" query keywords, as opposed to records containing multiple words relevant to only a single (or a few) query keyword(s).

$$AddLimitedSim(Q, D) = \sum_{i=1}^{m} Min(SumSim(q_i, D), 1) .$$ (3)

AddLimitedSim would rank D_2 as more relevant to Q than D_1, i.e., both data records would have a similarity of 1 to "network", but D_2 would have a higher similarity to "security" than D_1 due to the keyword "Authentication," as shown in Figure 2.

While *AddSim* does have a disadvantage in some cases to *AddLimitedSim*, which prefers high similarity from more keywords, *AddLimitedSim* has shortcomings as well, i.e., in the limited granularity of the calculated query-data record similarity values. Since there is a limit of 1 imposed upon the similarity between each query keyword and D, and all exact word matches yield a similarity of 1, then any data records which contain all query keywords will have the same ranking value (i.e., an integer $k \geq 1$, for a query with k keywords), with no differentiation. Conversely, *AddSim* does not have this problem, i.e., it attributes higher similarity values for data records which contain multiple keywords similar to a single keyword, even above and beyond an exact match.

The question remains, then - Is there some measure that can be used that avoids both the problem of weighing a single query keyword too heavily (as in the *AddSim* method), and the problem of reduced granularity (as in the *AddLimitedSim* method)? We create such a measure by introducing an alternative method, i.e., multiplication.

II. Calculating Similarity by Multiplication. The set of $SumSim(q_i, D)$ values for a given data record D and each query keyword q_i in Q can be viewed as separate *evidences*, or *probabilities*, that D is similar to Q. Using this basis, we apply the Dempster-Shafer's theory of evidence [15], which combines the probabilities of separate events for the calculation of a single evidence (i.e., the *similarity* between keywords in Q and D). Using this rule, the evidence for a single event, i.e., the similarity of Q and D, is calculated by multiplying together its constituent evidences, combining probabilities essentially in an AND operation [15]. However, we do not entirely exclude relevant documents which do not have any similarity to some given query keyword, especially if they have high similarity to other query keywords. Thus, we assign a minimal bound to query keyword-data record similarity values, i.e., 1×10^{-7}, which is less than the minimum value in the 10^{-7}-13%-WST. Hence, we define $MultSim(Q, D)$ as

$$MultSim(Q, D) = \prod_{i=1}^{m} Max(SumSim(q_i, D), 10^{-7}). \qquad (4)$$

Note that in $MultSim$, $SumSim$ is not limited to 1 as it is in $AddLimitedSim$, which is advantageous in that doing so yields relatively *higher* similarity values for data records which have some similarity to a greater number of keywords. Using *multiplication*, overestimating similarities for data records weighted to only a few query keywords is not an issue, since the similarity of a query Q to a data record D suffers severely only if the degree of similarity of D to *any* keyword q_i in Q is low.

Some sample query-data record similarity calculations are shown in Figure 2. Notice the larger gap in the ranking of $MultSim$ in comparison to the other two methods.

4 Experimental Results

We analyze the experimental results of our WSM approach over different sets of documents selected from the ACM SIGMOD Anthology (www.informatik.uni-trier.de/~ley/db/indices/query.html), ACM for short, and Craigslist (http://www.craigslist.org). The document sets, which include classified ads for *books* and *furniture* from Craigslist and bibliographic data for published *articles* from ACM, were parsed into tables in a MySQL server. These data sets contain text representing semantic information, which is the only requirement for our WSM approach. Furthermore, Craigslist regional sites at 355 different cities and countries worldwide were used to extract 10,124 classified advertisements for the *books* database and 21,368 for the *furniture* database. For the third database, 14,394 different bibliographic entries were extracted from 9 different conferences and journals in ACM. The DBs we have created by extracting more than 40,000 records from the data available are sufficiently large and varied in content to be taken as representative of the more complete DBs available online. In our empirical study, we compared *AddSim*, *AddLimited-Sim*, and *MultSim* on a number of queries and data records, against a simple Boolean keyword search, which is common in RDBMSs.

4.1 Structured Query Construction

MySQL queries were formed by progressively nesting the steps presented in Section 3.2 as subqueries, i.e., the subset of data records selected in Step 1 is processed and used as a subquery to calculate query keyword-data record similarity in Step 2, which is then processed as another subquery to calculate query-data record similarity in Step 3.

Consider the sample query Q, "foreign languages", which are reduced to their stems, "foreign" and "languag." These query keywords can be used to process the *subset selection* (i.e., Step 1), which selects all data records that contain at least one word with the similarity to any one of the two keywords in Q that exceeds 3×10^{-6} as follows:

SELECT DISTINCT *id* FROM *keywordabstract k, similarity3en6 c*
    WHERE (*c*.word = "foreign" OR *c*.word = "languag")
    AND *c*.simword = *k*.keyword ORDER BY *id*;   (* Subquery for Step 1 *)

where *id* is the ID number for a given data record, *similarity3en6* is the 10^{-6}-13%-*WST*, which contains only word pairs with a similarity above 3×10^{-6}, and *c*.word and *c*.simword are the words in each word pair in 10^{-6}-13%-*WST*. This subset selection subquery can then be used to compute *query keyword-data record similarity* (Step 2) as

SELECT hitwords.id, SUM(*sim*) as *s* FROM
    (SELECT hits.id, *a*.keyword FROM abstractkeyword *a*, (<Step 1 Subquery>) as hits
    WHERE hits.id = *a*.id ORDER BY *id*) as hitwords, similarity5en7 *c*
WHERE "foreign" = *c*.word AND hitwords.keyword = *c*.simword
GROUP BY id;   (* Subquery for Step 2 *)

where *sim* is the similarity between two words in the *similarity5en7* (a.k.a. 10^{-7}-13%-*WST*) table, and *abstractkeyword* is identical to the *keywordabstract* table, but indexed on *id* instead of *keyword*. This subquery will calculate the similarity between any given query keyword (in this case, "foreign," and this subquery can similarly be created for "languag") and each data record in the subset selected in Step 1.

   The manner in which the subquery for Step 2 is used for processing query-data record similarity varies slightly between the *AddSim, AddLimitedSim* and *MultSim* methods—we will provide here an example for *AddSim*, which will sum the similarity values for "foreign" and "languag" on each data record:

SELECT IDsums.*id*, sum(*s*) as sums FROM
    (<Step 2 Subquery using "foreign"> UNION <Step 2 Subquery using "languag">)
    as IDsums GROUP BY ID ORDER BY sums;   (* Query for Step 3 *)

where *s* is query keyword-data record similarity computed in Step 2. For *AddLimited Sim* and *MultSim*, any values of *s* in Step 2 outside of the predefined range, such as $s > 1$ (and $s < 10^{-7}$, respectively) would be changed accordingly (i.e., to 1 or 10^{-7}), and for *MultSim*, the *product* of *s* would be computed, as opposed to the *sum*. This query for Step 3, which provides an ordered list of *query-data record similarity* values (i.e., sums or products) and the corresponding *id* values for the data records which they pertain to, can be used as a subquery and joined to a table containing more detailed information on the data set and returned to the user with the assigned relevance ranking.

4.2 Analysis of the Retrieved Results

Considerations were taken to determine what measure could effectively illustrate the relative usefulness of each of the three similarity methods. Considering that users are typically most interested in only the few top-ranked query results [16], we desire a measure that conveys information regarding the top results across many queries. We adopt the *Mean Average Precision*, $MAP(r) = \frac{1}{Q} \times \sum_{p=1}^{Q} \frac{r}{t(p)}$ in [1], which is the ratio of relevant data records retrieved to all those retrieved at each retrieved data record, where r is the number of relevant records to retrieve, Q is the total number of queries, and $t(p)$ is the number of records retrieved at the rth relevant record on the pth query.

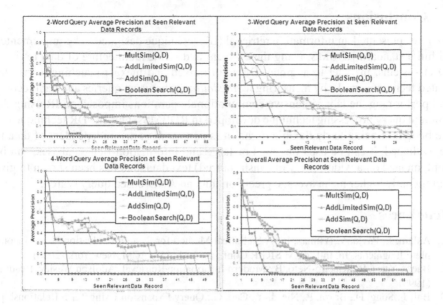

Fig. 3. Charts plotting MAP for various 2-, 3- and 4-word queries and the overall performance

MAP is effective on a set of relatively few (retrieved) data records, as opposed to *recall* [2], which requires a thorough analysis of the entire data set that is not available for our empirical study. Also, MAP provides information primarily concerning the results retrieved (as opposed to excluded data records), which is precisely what we need.

In Figure 3, we plot the average MAP for the top-ranked retrieved data records, using each of the three methods for calculating query-data record similarity in our WSM approach, as well as the simple Boolean AND-based keyword search. In this manner, it can easily be seen how effective each method is within the first 5, 20, or 100 results, etc. Queries which returned many exact matches or with few or no results were less useful, since they differed little from those retrieved by a standard DB query. Thus, we created and selected queries which gave fair number of results, an appreciable number of them being inexact matches. These results were manually verified for relevance.

Across the 2-, 3-, and 4-word queries that we created, each of the three methods implemented by our WSM approach outperforms the Boolean AND-based keyword search, which is expected. Also, for any query for which a Boolean keyword search returns m results, each of the three WSM methods returns the same m results first (though in order of ranking, whereas the Boolean search imposes no ranking), followed by additional, similar matches. For higher n (n being the number of ranked data records read), each method outperforms the keyword search KS (especially for 2-keyword queries), due to the additional relevant data records retrieved that do not contain all query keywords, which are not retrieved by KS. This shows that not only are additional relevant records retrieved by inexact word-matching, but that the ranking imposed by both $AddLimitedSim$ and $MultSim$ (i) is consistently more accurate than the random ranking imposed by the Boolean keyword search at low n (i.e., for records containing all query keywords) and (ii) outperform $AddSim$, especially for *shorter* queries.

5 Conclusions

We have presented an information retrieval (IR) technique designed to be implemented within a standard RDBMS, using the efficient and stable architecture of data retrieval (DR) systems, in order to augment DR with ranked and inexact matching. Our integrated technique works by measuring *word similarity* between queries and data records and performing the labor-intensive computational operations imposed on IR using the efficient query processing technique developed in RDBMSs. Our IR+DR technique has the benefit of being fairly flexible in that it is *context free*, i.e., independent of data to be evaluated; it can be used on small or large and information-rich databases, can be implemented on a wide variety of (unstructured) text data, and has been shown to give significantly higher mean average precision than an AND-based query search.

References

1. Aslam, J., Pavlu, V., Yilmaz, E.: A Statistical Method for System Evaluation Using Incomplete Judgments. In: Intl. ACM SIGIR Conf., pp. 541–548. ACM, New York (2006)
2. Baeza-Yates, R., Ribeiro-Neto, B.: Modern Information Retrieval. Addison Wesley, Reading (1999)
3. Bai, J., Song, D., Bruza, P., Nie, J.-Y., Cao, G.: Query Expansion Using Term Relationships in Language Models for Information Retrieval. In: ACM CIKM, pp. 688–695. ACM, New York (2005)
4. Bremer, J.M., Gertz, M.: Integrating Document and Data Retrieval Based on XML. VLDB J. 1(15), 53–83 (2006)
5. Bruno, N., Chaudhuri, S., Gravano, L.: Top-K Selection Queries over Relational Databases: Mapping Strategies and Performance Evaluation. ACM TODS 2(27), 153–187 (2002)
6. Carpineto, C., de Mori, R., Romano, G., Bigi, B.: An Information-Theoretic Approach to Automatic Query Expansion. ACM Transactions on Information Systems 1(19), 1–27 (2001)
7. Cohen, W.W.: Data Integration Using Similarity Joins and a Word-Based Information Representation Language. ACM Transaction on Information Systems 3(18), 288–321 (2000)
8. Fang, H., Zhai, C.: Semantic Term Matching in Axiomatic Approaches to Information Retrieval. In: ACM SIGIR Conf., pp. 115–122. ACM, New York (2006)
9. Goldman, R., Widom, J.: WSQ/DSQ: A Practical Approach for Combined Querying of Databases and the Web. In: ACM SIGMOD Conf., pp. 285–296. ACM, New York (2000)
10. Grossman, D., Frieder, P.: Information Retrieval: Algorithms and Heuristics. In: Information Retrieval Functionality Using the Relational Model, pp. 168–176. Kluwer Academic, Dordrecht (1998)
11. Hristidis, V., Gravano, L., Papakonstantinou, Y.: Efficient IR-Style Keyword Search over Relational Databases. In: Intl. Conf. on Very Large Data Bases, pp. 850–861. ACM, New York (2003)
12. Kelly, D., Dollu, V., Fu, X.: The Loquacious User: A Document-Independent Source of Terms for Query Expansion. In: ACM SIGIR Conf., pp. 457–464. ACM, New York (2005)
13. Liu, S., Liu, F., Yu, C., Meng, W.: An Effective Approach to Document Retrieval via Utilizing WordNet and Recognizing Phrases. In: ACM SIGIR Conf., pp. 266–272. ACM, New York (2004)
14. Liu, F., Yu, C., Meng, W., Chowdhury, A.: Effective Keyword Search in Relational Databases. In: ACM SIGMOD Conf., pp. 563–574. ACM, New York (2006)
15. Sentz, K., Ferson, S.: Combination of Evidence in Dempster-Shafer Theory. SANDIA SAND2002-0835 (April 2002)
16. Silverstein, C., Marais, H., Henzinger, M., Moricz, M.: Analysis of a Very Large Web Search Engine Query Log. ACM SIGIR Forum 1(33), 6–12 (1999)

Combining Data Integration and IE Techniques to Support Partially Structured Data

Dean Williams and Alexandra Poulovassilis

School of Computer Science and Information Systems, Birkbeck, University of London
Malet Street, London WC1E 7HX, UK
{dean,ap}@dcs.bbk.ac.uk

Abstract. A class of applications exists where the information to be stored is *partially structured:* that is, it consists partly of some structured data sources each conforming to a schema and partly of information left as free text. While investigating the requirements for querying partially structured data, we have encountered several limitations in the currently available approaches and we describe here three new techniques which combine aspects of Information Extraction with data integration in order to better exploit the data in these applications.

Keywords: Information Extraction, Data Integration, Partially Structured Data.

1 Introduction

Despite the phenomenal growth in the use of databases in the last 30 years, 80% of the information stored by companies is believed to be unstructured text [1]. Beyond the workplace, the explosion of predominantly textual information made available on the web has led to the vision of a "machine tractable" Semantic Web, with database-like functionality replacing today's book-like web [2]. In particular, a class of applications exists where the information to be stored consists partly of some structured data conforming to a schema and partly of information left as free text. This kind of data is termed *partially structured data* in [3]. Examples of applications that generate and query partially structured data include: UK Road Traffic Accident reports, where standard format data is combined with free text accounts in a formalised subset of English; crime investigation operational intelligence gathering, where textual observations are associated with structured data about people and places; and Bioinformatics, where structured databases such as SWISS-PROT [4] include comment fields containing related unstructured information.

From our investigation of these applications, we have identified two main reasons for information being stored as text in addition to the information that is stored as structured data: i) It may not be possible in advance to know all of the queries that will be required in the future, and the text captured represents an attempt to provide all the information that could possibly be relevant for future query requirements. ii) Data is captured as text due to the limitations of supporting dynamically evolving schemas in conventional databases — simply changing the size of a column in an existing table can be a major task in production relational database systems.

E. Kapetanios, V. Sugumaran, M. Spiliopoulou (Eds.): NLDB 2008, LNCS 5039, pp. 175–186, 2008.
© Springer-Verlag Berlin Heidelberg 2008

We believe that combining techniques from data integration and Information Extraction (IE) offers potential for addressing the information management needs of these applications. We have previously described ESTEST [5], a prototype system which makes use of the virtual global schema arising from a data integration process over the available structured data sources in order to assist in configuring a subsequent IE process over the available text, and to automatically integrate the extracted information into the global schema, thus supporting new queries encompassing the new data. Experiments with ESTEST in the Road Traffic Accident Informatics domain have shown that this approach is well-suited to meeting the information needs of the users of such data, being able to make use of both the structured and the textual part of their data in order to support new queries as they arise over time.

However, our experiences of using ESTEST in this domain, and our subsequent investigation of the requirements for supporting partially structured data in the Crime Informatics domain, made us aware of several limitations that have led to further research, which we report in this paper. We begin the paper with a brief overview of the initial ESTEST system. We then describe, in Section 3, three limitations with current approaches, including our own initial ESTEST system, each of which can benefit from appropriate combination of techniques from IE and data integration. Our approach to addressing each of these limitations is described in Sections 3.1-3.3, illustrated with an example drawn from the Crime domain. We give our concluding remarks and identify directions of further work in Section 4.

2 Overview of ESTEST

In general, IE is used as a step in a sequence, normally to produce a dataset for further analysis. Our goal, in contrast, is to make the information extracted from text available for subsequent query processing, in conjunction with the structured data, through an integrated virtual schema. Also, over time requirements for new queries may arise and new structured data resources may become available. Therefore, for this class of application, it must be possible to handle incremental growth of the integrated schema, and to repeatedly apply the IE and data integration functionality.

Our ESTEST system supports such an evolutionary approach, allowing the user to iterate through a series of steps as new information sources and new query requirements arise. The steps are illustrated in Figure 1, and comprise: i) integration of the structured data sources via a virtual global schema, ii) semi-automatic configuration of an IE process, iii) running the IE process over the available text, iv) integrating the resulting extracted information with the pre-existing structured information, under the virtual global schema, vi) supporting queries posed on the global schema, that will encompass the extended structured data, and finally vii) optionally enhancing the schema by allowing new data sources to be included before starting a new iteration of steps ii)-vi).

Each step of the ESTEST process may need to be repeated following amendment of the system configuration by the user. The overall process may also need to be restarted from any point. ESTEST makes use of the facilities of the AutoMed heterogeneous data integration toolkit [6] for data integration aspects (steps i, vi and vii), and of the GATE IE System [7] for the IE aspects (step iii). We refer the reader to [5] for further details of the design and implementation of the initial version of ESTEST.

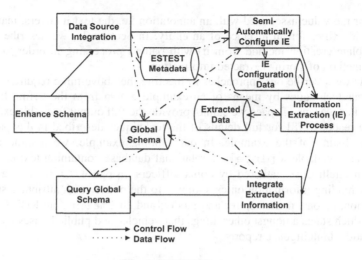

Fig. 1. ESTEST Steps

3 Supporting Partially Structured Data

Since the results reported in [5], we have analysed several real crime databases made available by police forces in the UK. A common feature of the Crime Informatics domain its use of text: for example, applications such as operational intelligence gathering make use of text reports containing observations of police officers on patrol; scene-of-crime applications include textual descriptions of the conditions found at the scene; and serious crime investigations make use of witness statements. In these applications the queries required will only become known over time, for example when looking for patterns between current and earlier incidents.

Our experiences of using ESTEST in practice in the Road Traffic Accident (RTA) domain, and our subsequent investigation of the requirements for supporting partially structured data in the Crime Informatics domain, made us aware of limitations in three areas that are not inadequately supported either by the solutions used by practitioners to date nor by the initial version of our ESTEST system:

i) For domains such as Crime Informatics, where the number and variety of structured data sources is richer than those of the RTA domain, schema matching and the creation of a single global schema covering even just the structured data sources remains a problem. We therefore investigated the possibility of using IE on textual schema metadata to assist in this task, as described in Section 3.1.

ii) The problem of deciding if a reference to an entity extracted from free text refers to a new instance or to one already known in the structured part of the data, became apparent as an issue in both the RTA and Crime Informatics domains. We have investigated combining co-reference resolution techniques from IE with database duplication detection techniques, as described in Section 3.2.

iii) IE systems do not provide general-purpose facilities for storing the annotations found, and developing methods of relating annotations to the instance of the schema element they refer to, and not just the type of the element, may be necessary. In particular the value associated with an annotation i.e. the text it covers, may not be a suitable identifier for an instance of an entity. In Section 3.3 we describe extending our template creation method from ii) with pattern processing in order to provide a general method of storing extracted information.

Below we describe our approaches to meeting the above three requirements, illustrating our techniques by means of an example drawn from the Crime Informatics domain. For reasons of space we cannot provide the full output from this example, but refer the reader to [8] for full details of the techniques described in this paper and a complete listing of the example. In our running example, we assume three data sources are available: OpIntel, a relational database containing textual reports of operation intelligence gathered by police officers on patrol; CarsDB, a relational database holding information on cars known to the police, with attributes such as the registration, colour, manufacturer and model; and CrimeOnt, an RDF / RDFS ontology which states amongst other things that vehicles and public houses are attributes of operational intelligence reports.

3.1 Using Information Extraction for Schema Matching

Schema matching is a long-standing problem in data integration research. The central issue is, given a number of schemas to be integrated, find semantic relationships between the elements of these schemas. A comprehensive survey of approaches to automatic schema matching is [9] where the linguistic approaches are divided into name-based and description matching techniques. The survey also considers the use of synonyms, hyponyms, user-provided name matches and other similarity measures such as *edit distance*. The possibility of using natural language understanding technology to exploit the schema descriptions is mentioned but this is the only category where no prior work is explicitly cited and we are similarly unaware of any previous system which makes use of description metadata for schema matching.

A feature of the class of applications that we target is that new structured data sources may become available over time, and they can be integrated into the virtual global schema and used to assist in IE. Therefore, we have developed an *IE-based schema matcher* which uses IE to process the available schema names and any textual metadata information, and uses the extracted word forms in order to identify correspondences between schema elements across different data sources. Our IE-based schema matcher undertakes the following steps within the overall ESTEST process:

1) For all the available data sources, the available schema metadata is extracted, including the textual names and description metadata for all the elements of the schema. In ESTEST, wrappers are able to extract metadata for data sources organised according to the model they represent including relevant textual description metadata e.g. the relational wrapper retrieves JDBC remarks data, which allows for a free text description of a column or a table, while the ontology wrapper retrieves XML comments.

2) We have developed a SchemaNameTokeniser component to process these schema names and extract word forms from them, to be associated with their schema elements in the virtual global schema. Our schemaNameTokeniser is an extensible component which detects the use of common naming conventions in schema element names and descriptions. It is able to transform names into word forms, making use of abbreviations. For example, "AccountNum", "accNum", and "ACCOUNT-NUMBER"

can all be transformed into the word form "account number". Whereas the GATE `English-Tokeniser` splits sentences into annotations representing words and punctuation, our `SchemaNameTokeniser` produces annotations representing the words extracted from the schema names. Our component makes use of regular expressions to identify naming conventions; these cover the commonly used conventions and can be easily extended.

3) A GATE pipeline is constructed to process the textual description metadata. This pipeline performs named entity recognition on the schema element descriptions, identifying references to schema elements by matching the word forms extracted from the schema names. The pipeline is created automatically by constructing an instance of our `SchemaGazetteer` which treats each element in the schema as a named entity source using the word forms extracted from schema element names and links annotations found in the text back to the associated schema element.

4) Where a match is found between a schema element acting as a named entity source and a description of a schema element from a different data source, then a possible correspondence between these schema elements is inferred.

We now show how this process takes place in the example from the crime domain. The match that should be found is between the `car` table in the `CarsDB` data source and the `vehicle` RDFS class in the `CrimeOnt` data source. The Postgres DDL for the `CarsDB` database includes the following comment on the `car` table: `comment on table car is 'VEHICLE SEEN DURING OPERATIONAL INTELLI-GENCE GATHERING'`.

The first step is for ESTEST to process the schema element names using the `Sche-maNameTokeniser` component. This happens and the word forms "car" and "vehicle" are extracted from the respective schemas and are stored. Next, the GATE pipeline for processing description metadata is created. Now each description is processed, and no matches are found from the other descriptions. However, when the description of the car table is processed, a match is found with the vehicle class and the schema id and schema element id are displayed to identify the elements involved in the match:

```
Document to be processed by IE : 'VEHICLE SEEN DURING
OPERATIONAL INTELLIGENCE GATHERING'
Match between the textual metadata of schema element
84/62, and the schema element 85/104
```

We note that in ESTEST schema elements are identified by a combination of their schema and element number, for example schema element 84/62 refers to element number 62 in schema number 84. Once processing is complete, this remains the only match identified and it is used by ESTEST in creating the global schema.

3.2 Combining Duplicate Detection and Coreference Resolution

In both database and natural language processing there is often a requirement to detect duplicate references to the same real-world entity. We have developed a *duplicate detection component* for partially structured data which, as far as we are aware, combines for the first time techniques from IE and data integration in order to achieve better results than would be obtainable using each independently.

Deciding if two instances in a database in fact refer to the same real-world entity is a long-standing problem in database research. Attempting to solve the problem across a range of application domains has led to a variety of terms being used in the literature for different flavours of the same fundamental task. The statistics community has undertaken over five decades of research into *record-linkage*, particularly in the context of census results. [10] gives an overview the current state of record linkage research and of related topics such as error rate estimation and string comparison metrics. In database integration, the term *merge / purge* [11] is used to describe approaches such as sorting the data and traversing it considering a 'window' of records for possible merging. *Data cleansing* [12] concentrates on finding anomalies in large datasets, and cleansing the dataset by using these anomalies to decide on merging duplicates and on deriving missing values. *Duplicate elimination* [13] refers to variations on merge-sort algorithms and hash functions for detecting duplicates. We are not aware of any attempt to make use of NLP techniques in finding duplicates in databases, beyond character-based similarity metrics such as edit distance.

In IE the term *coreference annotation* [14] has come to be used to describe the task of identifying where two noun phrases refer to the same real-world entity and involves finding chains of references to the same entity throughout the processed text. There are a number of types of coreference, including pronominal coreference, proper names coreference, through to more complicated linguistic references such as demonstrative coreference. [15] shows that a small number of types of coreference account for most of the occurrences in real text: they find that proper names account for 28% of all instances, pronouns 21% but demonstrative phrases only 2%. It is expected therefore that reasonable coreference annotation results can be achieved by effectively handling these main categories of coreference.

The GATE system provides support for these two main categories of coreference, by providing an `OrthoMatcher` component which performs proper names coreference annotation and a JAPE[1] grammar which when executed performs pronominal coreference annotation. The `OrthoMatcher` is executed in a GATE pipeline following the named entity recognition step. No new named entities will be found as a result of finding matches, but types may be assigned to previously unclassified proper names which have previously typed matches. The input to the OrthoMatcher component is a list of sets of aliases for named entities. Also input are a list of exceptions that might otherwise be matched incorrectly. As well as these string comparison rules that apply to any annotation type, there are some specific rules for the core MUC IE types i.e. person, organisation, location and date.

We argue that coreference annotation in NLP is essentially the same task as duplicate detection in databases. The difference is not in the task to be performed but rather

---

[1] GATE's *Java Annotation Patterns Engine* (JAPE) provides finite state transduction over annotations based on regular expressions. The left-hand-side of JAPE rules consist of an annotation pattern that may contain regular expression operators. The right-hand-side consists of annotation manipulation statements, typically by creating an annotation over the matched text which describes its type. JAPE rules are grouped into a *grammar* for processing as a discrete step in an IE *pipeline*, and for each grammar one of a number of alternative control styles is specified to determine the order rules should fire, and from where in the text processing should resume.

in the structure of the data to be processed, free text in the case of NLP and structured data for databases. We have therefore developed a new duplicate detection component for our ESTEST system, comprising three parts: i) a coreference module, ii) a template constructor module, and iii) a duplicate detection module. The coreference module may find new structured data which in turn will be useful for the duplicate detection module, and vice versa. These modules use state-of-the-art techniques, but as there is much active research in both areas they are designed to be extensible and modular. Our research contribution here is not in coreference resolution research nor in database duplication detection as such, but in combining the two approaches.

Our duplicate detection component works as follows:

1) The standard GATE OrthoMatcher component is added to a pipeline. The configuration for this component is automatically created from data in the virtual global schema. When building an IE application using GATE alone, this component would have to be configured by hand for the domain of interest, and the default configuration file provided contains only a handful of examples to show the format of the entries. In contrast, we use the abbreviations and alternative word forms previously collected during the integration of the data sources to automatically create the configuration for the `OrthoMatcher` component.

2) The standard JAPE grammar for pronominal coreference is then executed over the text.

3) Template instances are automatically constructed for the annotations that match schema elements.

4) Our duplicate detection component then extracts the coreference chains from the annotations and uses these to merge templates. Each co-reference chain is examined in turn and its attributes examined and compared pair-wise to decide if they should be merged. This process removes duplicates from within the free text.

5) The resulting set of templates are now compared to the instances already known from the structured data. A decision is made whether to store each template found in the free text as a new instance, or whether to merge it with an existing instance.

6) Any annotations which refer to schema elements but which are not part of any template are stored as before.

The same process is used for both the decision on whether to merge templates found in the free text, and whether to store templates as a new instance or to merge with an existing instance. The available evidence is compared using the size of the extent of each attribute as a straightforward method of weighting the evidence of different attributes in a template. For example, in a template with attributes name and gender, name would be weighted more highly as it is more discriminating as a search argument. In contrast, if the amount of conflicting evidence exceeded a confidence threshold, then the separate instances would be left unmerged and the coreference chain ignored. If there is some contradictory evidence, falling between these two confidence thresholds, then the coreference chain is highlighted for the user to decide whether to merge, together with the conflicting attributes and the confidence level based on the weighting of these attributes.

To decide the confidence level the attributes of each pair of possible matches in the chain are compared. Where there is no value for an attribute for one or both of the concepts then no positive or negative evidence is assumed. If both concepts have the same value for the attribute then that is considered as positive evidence weighted according to the selectivity of the attribute. If they have different values then similarly weighted negative evidence is added to the confidence total.

We made use of the crime example to experiment with our approach. The first step described above is to make use of GATE's OrthoMatcher component by automatically creating an input file based on the word forms associated with the schema elements in the virtual global schema. As mentioned, this component is used to provide alternative names for the same instance of an entity. In the crime example, the following matches are found: {OP, INTEL, OP INTEL}, {ID, REPORT, REPORT ID}, and {CAR, VEHICLE}. Using this component, it became clear that with such general matches the co-reference chains were not producing any value, and in fact every annotation of the same type matched.

It may be that there are uses for this approach in particular domains, and for annotations referring to people there is value in exploiting the extra evidence available in natural language. However we decided instead to look for coreference matches by constructing templates and attempting to merge these where there are no conflicting attributes. It would also be possible to restrict merges to templates found in close proximity in the text.

Annotations found in the text are placed in *templates* which the duplicate detection module creates automatically from the attribute edges in the virtual global schema. To place appropriate weight on the attributes the size of the extents is used. For example, there are 142 car registrations known to CarsDB, but only 11 models, so registrations are more useful evidence. Next, the annotations of interest to be considered in creating templates are extracted. An operational intelligence report processed in the example is "GEORGE BUSH HAS A NEW YELLOW CAR REGISTRATION LO78 HYS. IT IS A FORD MONDEO". From this text, two templates are found, one for each reference to a car in the text:

```
Template: 1 -- <<car>>, Instance ID: estestInstance1
        Attribute: <<car_reg>>, Instance ID: LO78 HYS
        Attribute: <<colour>>, Instance ID: YELLOW
    Template: 2 -- <<car>>, Instance ID: estestInstance2
        Attribute: <<manufacturer>>, Instance ID: FORD
        Attribute: <<model>>, Instance ID: MONDEO
```

These contain no conflicting attributes and so it is assumed that they refer to the same entity (this assumption replaces the alternative co-reference chain approach), and the templates are merged. The resulting merged template contains all the available attributes of the car, and uses the car registration number as the identifier for the car (we explain in Section 3.3 how the identifier determined):

```
Template: 3 -- <<car>>, Instance ID: LO78 HYS
        Attribute: <<car_reg>>, Instance ID: LO78 HYS
        Attribute: <<colour>>, Instance ID: YELLOW
        Attribute: <<manufacturer>>, Instance ID: FORD
        Attribute: <<model>>, Instance ID: MONDEO
```

The merged template is now compared to the instances already contained in the structured data and the size of the attributes' extents is used to weight the attributes.

For example, having the same registration mark is more credible evidence that the car in the text is a reference to one already known than would be the fact that they were both the same colour. The system finds the correct match with 'LO78 HYS'. The details from the template are then stored, including the new fact that this car is yellow. This is the only fact that actually needs to be stored in the ESTEST data store as the others already exist in the CarsDB data source. However there is no disadvantage in duplicating known facts as distinct result sets are returned from queries and in this way there is a useful record of the totality of facts found in the text. Being able to combine the structured and text information has proved advantageous: without doing so queries on the structured database would not include the fact that this car is yellow; just relying on the text would not reveal the manufacturer or model.

3.3 Automatic Extraction of Values from Text

A central task in IE is named entity recognition which at its simplest involves identifying proper names in text by matching against lists of known entity values, although patterns can also be defined to recognise entities. In GATE lookup named entity recognition is performed by gazetteer components, while pattern matching named entity recognition is performed by JAPE However, the facilities offered by IE systems, such as GATE, are restrictive when it comes to automatic further processing of extracted annotations in order to store them: the string the annotation covers will usually not be the string that should be used as the entity identifier; in many cases, the annotation will contain that value as a substring, and the substring may also be associated with another annotation type. This restriction arises from IE systems typically being used in isolation, without consideration for how their results can be automatically processed beyond displaying the annotated text, other than by code specifically written for each application.

We have therefore developed a *pattern processor*: patterns are linked to schema element and these patterns are used to automatically create JAPE rules. These rules have been extended in that the annotations produced contain features which identify the schema element that the annotation relates to schema elements also act as sources of named entities for lookup named entity recognition and the patterns are similarly used to automatically configure a gazetteer, with word forms provided from metadata relating to the schema element, or from their extent, and resulting matches linked back to the schema element. An annotation post-processor then automatically identifies annotations of interest from either lookup or pattern-matching named entity recognition, and stores these results automatically in the ESTEST repository, extending the extents of the corresponding elements of the virtual global schema.

These new facilities have been provided for as follows:

1) It has been necessary to develop a new control style to automatically process annotations. Our pattern processor needs to be able to find all matching annotation types in order to find values to use as identifiers. JAPE's all style does find the complete set of matches but when there are rules with optional parts, then these result in the rule firing twice. To overcome this limitation, we have developed a new style: the JAPE grammars use the all style, however our pattern processor removes annotations covering substrings of the text covered by other annotations of the same type;

for example, in the example above "car" would be deleted leaving the longest match "red car".

2) In order to correctly assign identifiers to instances of schema elements found in the text, the pattern relating to the schema element can optionally specify the name of another schema element to use as their identifier; for example, it is possible to store the car registration number as the identifier of a car by specifying the name of the car_reg schema element in the pattern definition. The JAPE rule generated from this pattern will now include an idAnnotationType feature indicating that the pattern processor should find the value of the car_reg annotation within the string covered by the car annotation.

3) The idAnnotationType feature above enables schema elements that are sources for named entity recognition to be used as identifiers; for example, if a car with a registration mark that is already known was mentioned in the text, the rule would fire. But in order to be able to identify cars not already known it is necessary to be able to specify a text pattern to match, and to associate this pattern with a schema element. For this purpose, an additional new pattern type, value_def, allows a sequence of characters, numerals and punctuation to be specified e.g. to define car registration marks. Now when a pattern such as "AA11 AAA" is found in the text, this will create an annotation linked to the car_reg schema element. Combining this with the car rule will result in new cars being found by the pattern matching named entity recognition and stored identified by their registration marks..

4   Conclusions and Future Work

In this paper we have described three novel techniques for better supporting the requirements of partially structured data, motivated in particular by the requirements arising in the RTA and crime informatics domains. We stress though that our techniques are more generally applicable to a variety of other domains where PSD arises (some examples, discussed in more detail in [8], include Bioinformatics, homelessness resource information systems, scuba-diving accident reporting, and investment banking). The three novel contributions of this paper are as follows:

1) As well as making use of schema elements names in schema matching, we make use of textual metadata, such as JDBC remarks and comments in XML, to suggest correspondences between schema elements in different source schemas. We are aware of no other data integration system that is able to exploit such textual metadata.

2) We perform coreference resolution by merging templates matching the text. We then compare these with the known structured data to decide if they are new instances or if they represent new attributes about existing instances. To our knowledge, this is the first time that approaches for resolving duplicate references in both text and structured data have been combined. While our use of GATE's OrthoMatcher component proved ineffective, we believe that there is potential for making use of pronominal coreference resolution for annotations types that refer to people.

3) We extract values from the text strings covered by annotations, for example storing car registration marks to represent instances of cars. When combined with the template merging extension of 2), this provides a method of automatically storing the results of the IE process. Other than the KIM system [16], which relies on an

ontology of everything, we are aware of no other IE system which provides general facilities for further processing of the annotations produced.

As future work we plan to develop a workbench for end-users and to make use of this to conduct a full evaluation of our approach. In [5] we presented an evaluation of the initial version of ESTEST which showed that in terms of recall and precision it performed as well as a standard IE system, but in addition was able to support queries combining structured and textual information which would not be possible either using a vanilla IE system or a data integration system. However, the development of an end-user workbench is a prerequisite for a full evaluation of the ESTEST approach as discussions with a number of domain experts have shown that a variety of manual workarounds are currently required to fulfill their information needs rather than any system. An appropriate evaluation will therefore require a comparison by an end-user over time of using ESTEST compared with whichever workaround is currently employed when new query or data requirements arise. With our envisaged end-user workbench, it would be possible to develop a real-world application, with the aim of supporting the end-user in undertaking this comparison.

We believe that further possibilities of synergy between data integration and IE are likely to arise as a result of the evaluation phase and our prototype ESTEST system, extended as described in this paper, will be well placed to investigate these possibilities. In particular, our approach to combining text and structured duplication detection techniques can be further developed by experimenting with the effectiveness of using proximity and more elaborate merging algorithms. Also, our end-to-end method of associating patterns with schema elements, using these to automatically perform the later steps of configuring and processing text by IE with the results being stored automatically, provides new opportunities for end-user IE tools without requiring programmers or linguists to configure the systems. Finally, as more comprehensive models for specifying annotations are emerging, it should be possible to produce representations that are capable of describing entities both as elements in a virtual global schema and as they appear within free text.

References

1. Tan, A.H.: Text mining: The state of the art and the challenges. In: Zhong, N., Zhou, L. (eds.) PAKDD 1999. LNCS (LNAI), vol. 1574. Springer, Heidelberg (1999)
2. Berners-Lee, T.: Weaving the Web: The Original Design and Ultimate Destiny of the World Wide Web by Its Inventor. Harper, San Francisco (1999)
3. King, P.J.H., Poulovassilis, A.: Enhancing database technology to better manage and exploit Partially Structured Data. Technical Report, Birkbeck College, University of London (2000)
4. Bairoch, A., Boeckmann, B., Ferro, S., Gasteiger, E.: Swiss-Prot: Juggling between evolution and stability. Briefings in Bioinformatics 5(1), 39–55 (2004)
5. Williams, D., Poulovassilis, A.: Combining Information Extraction and Data Integration in the ESTEST System. In: ICSOFT 2006, CCIS 10, pp. 279–292. Springer, Heidelberg (2008)
6. AutoMed Project, http://www.doc.ic.ac.uk/automed/

 7. Cunningham, H., Maynard, D., Bontcheva, K., Tablan, V.: GATE: A Framework and Graphical Development Environment for Robust NLP Tools and Applications. In: Proc. ACL (2002)
 8. Williams, D.: Combining Data Integration and Information Extraction. PhD Thesis, Birkbeck College, University of London (February 2008)
 9. Rahm, E., Bernstein, P.A.: A Survey of Approaches to Automatic Schema Matching. VLDB Journal (2001)
10. Winkler, W.E.: Overview of Record Linkage and Current Research Directions. US Census Bureau (2006)
11. Hernandez, M.A., Stolfo, S.J.: Real-world Data is Dirty: Data Cleansing and The Merge/Purge Problem. Data Mining and Knowledge Discovery 2(1), 9–37 (1998)
12. Müller, H., Freytag, J.C.: Problems, Methods, and Challenges in Comprehensive Data Cleansing. Humboldt University, Berlin (2003)
13. Bitton, D., DeWitt, D.: Duplicate Record Elimination in Large Data Files. ACM Transactions on Database Systems 8(2), 255–265 (1983)
14. Morton, T.: Coreference for NLP Applications. In: Proc. ACL (1997)
15. Bagga, A.: Evaluation of Coreferences and Coreference Resolution Systems. In: Proc LREC (1998)
16. Popov, B., Kiryakov, A., Ognyanoff, D., Manov, D., Kirilov, A.: KIM - a semantic platform for information extraction and retrieval. Natural Language Engineering 10 (2004)

Towards Building Robust Natural Language Interfaces to Databases

Michael Minock, Peter Olofsson, and Alexander Näslund

Department of Computing Science
Umeå University, Sweden
Phone: +46 90 786 6398; Fax: +46 90 786 6398
{mjm,c02pon,c02nar}@cs.umu.se

Abstract. We seek to give everyday technical teams the capability to build robust natural language interfaces to their databases, for subsequent use by casual users. We present an approach to the problem which integrates and streamlines earlier work based on light annotation and authoring tools. We model queries in a higher-order version of Codd's tuple calculus and we use synchronous grammars extended with lambda functions to represent semantic grammars. The results of configuration can be applied directly to SQL based databases with general n-ary relations. We have fully implemented our approach and we present initial empirical results for the GEOQUERY 250 corpus.

1 Introduction

One factor that has blocked the uptake of natural language interfaces (NLIs) to databases has been the economics of configuring such systems [2,5]. Typically configuration requires high levels of knowledge and long time commitments. The typical work environment has neither of these in great supply and thus when presented with the choice of building a common forms based interface versus an NLI, organizations typically opt for forms. Our work seeks to make NLIs a more attractive option by reducing the time and expertise requirement necessary to build them.

Given the limited linguistic knowledge possessed by most technical teams, modern approaches to configuring NLIs to standard databases come down to one of three approaches:

1. Let authors only lightly *name* database elements (e.g. relations, attributes, join paths, etc.) and reduce query interpretation to graph match[3,14].
2. Offer a GUI based authoring interface where one grunts out a *semantic grammar* that interprets user queries over their database. Such approaches are common in industrial products such as Microsoft's ENGLISHQUERY, Progress Software's EASYASK, Elfsoft's ELF, etc.
3. Use machine learning to induce a semantic grammar from a corpus of natural language query/correct logical interpretation pairs [16,8,6,17,18].

E. Kapetanios, V. Sugumaran, M. Spiliopoulou (Eds.): NLDB 2008, LNCS 5039, pp. 187–198, 2008.

Since the ultimate goal of our project is the delivery of a high impact NLI to database tool, it should not be surprising that we primarily adopt an authoring approach. The contribution of the present paper is to present our formally rooted *name-tailor-define* authoring approach and to show that it may be effectively employed by normal technical personnel. That said, aspects of the first approach are deeply integrated into our work and we have laid the ground for an integration of machine learning techniques to help achieve highly robust NLIs 'in the limit' after experiencing large volumes of user queries.

1.1   Organization of This Paper

Section 2 lays the foundation of concepts necessary to understanding our approach. Section 3 presents our formal approach to analyzing noun phrases, what we consider to be the primary challenge of NLIs to databases. Section 4 presents our GUI-based administration tool in which one populates the system with the formal elements described in sections 2 and 3. Section 5 presents an initial experiment that shows that reasonably skilled subjects can effectively use our authoring tool. Section 6 compares our approach with other approaches to building NLIs to databases. Section 7 concludes.

2   Foundations

2.1   Database Elements, Namings and the Dictionary

While database design may initially involve UML or ER modeling, most ongoing work is at the *representational level* with the underlying relations, views, attributes, tuples and values of the working database. Assume the set of relations **REL** (e.g. CITY), attributes **ATT** (e.g. CITY.NAME) and data values **VAL** (e.g. 'Chicago'). An additional element class that we include here are join conditions **JOIN**. These specify conditions commonly appearing in queries (e.g. CITY.STATE = STATE.NAME). Collectively the set of elements **ELEMENTS** = **REL** ∪ **ATT** ∪ **VAL** ∪ **JOIN**.

In figure 1 we see these elements being named with various phrases. If we consider the set of all alphanumeric strings to be Σ^*, this may be captured by

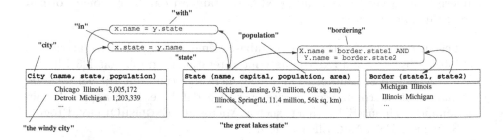

Fig. 1. A fragment of the GEO250 schema with some namings

a naming relation **NAMING** $\subset$ **ELEMENTS** $\times \Sigma^*$ (e.g. (CITY, "city") $\in$ **NAMING**). The population of the naming relation may be accomplished by a mix of automated and manual methods, but a substantial amount of it is simply materialized from the underlying database (e.g. "Chicago" is a name for the database value 'Chicago'.).

2.2 Queries in an Extended Tuple Calculus

Because of the formal difficulties of working directly with SQL expressions, we represent queries as expressions in Codd's Tuple Calculus. The tuple calculus is a widely known syntactic sugar developed over standard first order logic where variables range over tuples within database relations, not over the individual places in n-ary relations. For example the query that expresses the cities in the states with more than 10 million people is $\{x | City(x) \wedge (\exists y)(State(y) \wedge x.state = y.name \wedge y.population > 10,000,000)\}$. Tuple calculus is covered in the majority of introductory database textbooks and may be directly mapped to standard SQL queries or expressions in first-order logic.

To handle various ranking and aggregation capabilities of SQL, we extend the tuple calculus with several higher-order capabilities. For example to support ranking and thus superlatives, we use higher-order predicates.

Example 1. ("cities of over 100,000 people in the largest area mid-western state")

$\{x | City(x) \wedge x.population > 100000 \wedge$
$\quad (\exists y)(State(y) \wedge x.state = y.name \wedge$
$\quad\quad LargestByArea(y, State(y) \wedge y.name \in \{'Indiana', ..., 'Wisconsin'\}))\}$

The two place predicate *LargestByArea* is true for the tuple y that has the greatest area that satisfies the supplied formula, $State(y) \wedge y.name \in \{'Indiana', ..., 'Wisconsin'\}$ in the case here. We assume a set of built-in predicates to express superlative conditions over the various relations and numerical attributes (e.g. There is a higher order predicate *LargestByHeight*$(y, \phi(y))$ which is true for that value of y with the largest height that satisfies $\phi(y)$. Naturally this predicate is only meaningful for mountain tuples.) These expressions have straightforward mappings to SQL for database systems that robustly support sub-selects.

Example 2. (Example 1 translated to SQL for PostgreSQL)

```
SELECT *
FROM City AS x
WHERE x.population > 100000 AND
  EXISTS (SELECT * FROM
          (SELECT * FROM State AS z
           WHERE z.name = 'Indiana' OR ... OR z.name = 'Wisconsin'
           ORDER BY area DESC LIMIT 1) AS y
     where x.state = y.name);
```

2.3   Semantic Grammars in λ-SCFG

Following [17], we model our semantic grammars as *synchronous context-free grammars* [1] augmented with lambda calculus expressions (λ-SCFG). Our λ-SCFG rules define two 'synchronous' trees derived from the same start symbol S. The yield of the first tree is natural language, and the yield of the second tree is a formal language expressing the semantics of the natural language yield of the first tree. The requirement of having variables within the semantic formulas necessitates the use of λ expressions and in turn (slightly) complicates the notion of what a yield is for the second tree – yields are calculated bottom up and involve the well known alpha conversion and beta reduction operations of lambda calculus.

Formally, each λ-SCFG rule has the form:

$$A \rightarrow \langle \alpha, \lambda x_1, ..., \lambda x_k \beta \rangle \tag{1}$$

where A is a single non-terminal symbol and α is a sequence of terminal and non-terminal symbols where the terminals are words or word sequences in natural language. β consists of a sequence of terminals, non-terminals and formal argument terms which compose arguments $x_1, ..., x_k$. We say compose here, because in contrast to [17], we shall at times pass lambda functions in as arguments. In such case we shall use the symbol f_i in the place of x_i. The terminals within β consist of raw material used to build up formal expressions. Each use of a non-terminal symbol in α is paired with the use of the same symbol in β (and vice versa). In cases that there are multiple uses of the same non-terminal in α, one must include index terms to specify which non-terminal in α corresponds to which in β. The paper [17] gives a more complete characterization of how such λ-SCFG rules are used to translate a natural language input to a formal semantic expression.

3   Focusing on Noun Phrases

Through earlier and ongoing work [10,12], we posit that the main analysis task for natural language interfaces to databases is the analysis of noun phrases, often with relative clauses. In fact more experienced users often type just noun phrases to describe the information they would like retrieved. In addition our experience tells us that we can model noun phrases as coordinated pre-modifiers and post-modifiers around head nouns. In the notation of λ-SCFG some of the general rules that guide this process are:

$QUERY \rightarrow \langle \text{"list the"} \cdot NP, answer(\{x|NP(x)\}) \rangle$
$NP \rightarrow \langle PRE \cdot NP, PRE(NP) \rangle$
$NP \rightarrow \langle NP1, NP1 \rangle$
$NP1 \rightarrow \langle NP1 \cdot POST, POST(NP1) \rangle$
$NP1 \rightarrow \langle HEAD, HEAD \rangle$

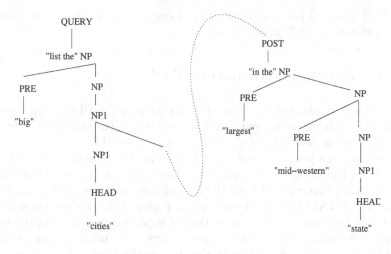

Fig. 2. An example parse of "list the big cities in the largest mid-western state"

The first rule expresses what we call a *sentence pattern*, of which there are many variants. This particular rule enables users to type "list the X" where X is a noun phrase. There are many other sentence patterns that let the user type things like "give me the NP" or "what are the NP?", etc. In total the system has around 75 manually defined sentence patterns and rarely do we find users typing expressions not covered by these basic patterns. Of course as we discover new ones, we, as system designers, simply add them in. The last four rules above are more interesting and enable noun phrases such as those in figure 2.

The authoring process provides the lexical entries that define pre-modifiers, heads and post-modifiers of noun phrases for the given database. Assume the following entries are built over our specific geography database:

$HEAD \rightarrow \langle$ "cities", $\lambda x.City(x)\rangle$
$HEAD \rightarrow \langle$ "state", $\lambda x.State(x)\rangle$
$PRE \rightarrow \langle$ "big", $\lambda f.\lambda x.f(x) \wedge City(x) \wedge x.population > 100,000\rangle$
$POST \rightarrow \langle$ "in the" $\cdot NP$,
 $\lambda f.\lambda x.f(x) \wedge x.City(x) \wedge (\exists y)(State(y) \wedge x.state = y.name \wedge NP(y))\rangle$
$PRE \rightarrow \langle$ "largest",
 $\lambda f.\lambda x.\mathbf{LargestByPop}(x, f(x) \wedge State(x))\rangle$
$PRE \rightarrow \langle$ "largest",
 $\lambda f.\lambda x.\mathbf{LargestByArea}(x, f(x) \wedge State(x))\rangle$
$PRE \rightarrow \langle$ "mid-western", $\lambda f.\lambda x.$
 $f(x) \wedge State(x) \wedge x.name \in \{'Indiana', ..., 'Wisconsin'\}\rangle$

The reader may wish to verify that "list the big cities in the largest mid-western state" parses to the tree in figure 2, which evaluates to the expression *answer*(Q) where Q is the expression in example 1, which in turn is automatically translated to the SQL of example 2.

4   Our GUI-Based Authoring Tool

Our AJAX based administrator tool gives an integrated GUI through which one may import a schema from any ODBC accessible database and commence what we term the *name-tailor-define* cycle of authoring an NLI.

Naming actions provide simple text names for the relations, attributes and join paths of the database schema as well as operations to provide additional names to values materialized from the underlying database. In short one populates the **NAMING** relation of section 2.1. Figure 3 shows the schema browser in our tool where the user has the option of exploring and naming the various database elements. Note the option here of naming the attribute MOUNTAIN.STATE. Note also that in this image the foreign key joins may be clicked on to name join elements. Naming actions trigger the definition of default head, pre-modifier and post-modifier lexical rules. These default rules are formed by coupling logical expressions over the join graph of the database, with associated terms in the **NAMING** relation. The default glosses of these elementary logical expression include function words (e.g. 'of','and',etc.) from the underlying language, English in our case.

To manage comprehensibility and navigation, default rules are gathered into a collection of *entries* that couple a single elementary conceptual expression with a set of m-patterns. Under the hood this expresses m lexical rules, but to the user of the author there is one entry with m patterns. Using a theorem prover

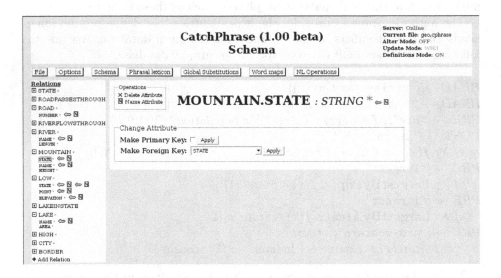

Fig. 3. Naming database elements over the schema

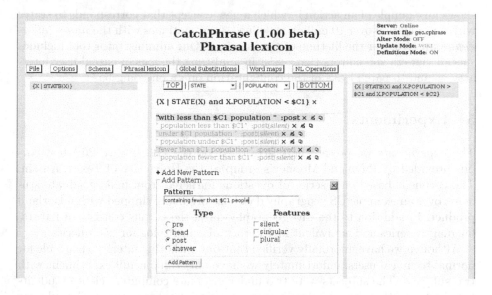

Fig. 4. Tailoring entries

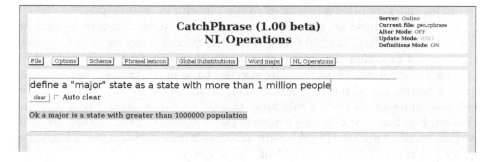

Fig. 5. Defining additional concepts via natural language

these entries are sorted into a subsumption hierarchy for ease of navigation. See [10,12] for a detailed description of this process.

During *Tailoring* one works with patterns that are associated with parameterized concepts expressed in tuple calculus. For example in figure 4 the currently selected entry corresponds to the concept of a state under a certain population. Note that in this case an additional linguistic pattern is being associated with this concept, for a total of seven patterns. Note also that the patterns may be used in the reverse direction to paraphrase queries to clarify to users that the system understood (or misunderstood) their questions. See [11] for an in-depth discussion of this paraphrase generation process.

Once a fair bit of structure has been built up, definition actions may be performed in which the author creates conceptually more specific entries. Figure 5 shows an example of an action in which a new concept that corresponds to states with a population over 10 million is defined. One continues with the *name-tailor-define* process over the lifetime of the interface. Our administrator tool includes special viewers and editors that assist in analyzing the logs of casual user interactions and to patch leaks in the configuration on an on-going basis.

5   Experiments

The experiments we present here are based on the GEOQUERY 250 test corpus provided by Raymond Mooney's group at the University of Texas, Austin. The corpus is based on a series of questions gathered from undergraduate students over an example US geography database originally shipped with a Borland product. In addition to the raw geography data, the corpus consists of natural language queries and equivalent logical formulas in Prolog for 250 queries.

Although we have informally verified that our authoring interface is usable for normal technical users, unfortunately we have only run our full experiment with two subjects. The subjects were two under-graduate computer science students that had recently taken an introductory relational database course. The subjects were instructed to read the user's manual of our administration tool and were trained on its use. The training examples were over several mocked up canonical examples and were in no way related to the GEOQUERY 250 corpus.

After being trained on the system, subjects were presented with a random 'training set' of 100 of the 250 GEOQUERY 250 queries. Subjects were told that they needed to author the system to cover such queries. Of the remaining 150 queries in the corpus, 33 queries were selected to be in our test set. For these 33 queries correct logical queries were manually constructed in our extended tuple calculus representation. As a side note, this process took slightly over 1 hour, thus our first finding is that on a corpus of the complexity of the GEOQUERY 250 corpus, a skilled worker will take approximately 2 minutes per natural language query to write and test the equivalent logical query.

As our subjects worked to cover the 100 queries of the training set, their configurations were saved off at intervals for future analysis. This continued for a two hour time span. Afterwards we automatically tested the resulting sequences of configuration files against the 33 queries of the test set. Exploiting our capability to determine logical query equivalence, queries were marked as correct if their parse to logical form was equivalent to the manually constructed correct result. In the rare case of ambiguity (e.g. "largest state") the answer was marked correct if one of its results was equivalent to the manually constructed correct result. Each configuration file evaluation yielded a measure of *precision*, *recall*[1] and *accuracy* where:

---

[1] We adopt the definition of recall presented in [14]. Unfortunately terms have not been consistently used across the literature on NLI to database evaluation. For example our measure of accuracy corresponds to recall as defined in the UT group's results. We adopt the definition of terms in [14], because they are reminiscent of the trade offs in standard information retrieval between recall and precision. In any case as we compare our results to others, we shall present their results in our terms.

$$\text{precision} = \frac{\text{\# of correct queries}}{\text{\# of parsed queries}} ; \text{recall} = \frac{\text{\# of parsed queries}}{\text{\# of queries}} \tag{2}$$

$$\text{accuracy} = \frac{\text{\# of correct queries}}{\text{\# of queries}} \tag{3}$$

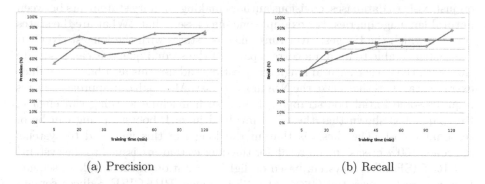

(a) Precision   (b) Recall

Fig. 6. Precision and Recall measures for our initial two trials

Figure 6 shows the resulting precision and recall measures through time.

6   Related Work

Due to the public availability of GEOQUERY 250 corpus, we can compare our initial results to several machine learning approaches [16,8,6,17,18], an approach based on light annotation [14] and an authoring approach over Microsoft's EnglishQuery product (described in [14]).

In comparing our results with machine learning approaches, we focus on results obtained after 120 minutes of effort. Since our informal finding of 2 minutes preparation time for each query in the training set, we thus focus on results with training sets of size 60. The results for such small training sets are not very strong. For example the accuracy of λ-WASP, the latest and currently best performing system developed by the group at the University of Texas, appears to be slightly under 50% with 60 queries in the test set (precision was slightly under 80%, thus in our terminology recall was approximately 60%). In our experiments subjects average slightly under 80% correctness after 120 minutes of authoring with an average precision of 86%. Also of interest is to look at asymptotic results when training samples grow to essentially unbounded size. In these cases, machine learning results are much stronger. The asymptotic precision of λ-WASP appears to be approximately 91.95% with an accuracy of 86.59%, yielding, in our terminology a recall of 94%. Another machine learning experiment over a relaxed-CCG approach obtained similar results [18].

Our comparisons with machine learning approaches highlight a bootstrapping weakness that if overcome would probably make machine learning approaches dominant. The weakness is of course the cost of obtaining the corpus of natural language/logical expression pairs. Mooney talks briefly about an approach to this problem for simulated environments where descriptions in natural language are paired with formal representations of objects and events retained from the simulation [13]. He suggests simulated RoboCup where a commentator describes game events as an interesting test-bed. Our proposed approach, focussed as it is on just NLIs to databases, envisions authors making equality statements between natural language queries. For example one may assert that "What are the states through which the Longest river runs" means "states with the longest river". If the system is able to obtain a correct parse of the second query, it can associate that with the earlier natural language question and use this as a basis to induce extra lexical rules that make the NLI more robust. Although our approach always requires some initial bootstrapping before such machine learning can engage, this paper has shown that the labor involved in such bootstrapping can be of reasonable cost. The thesis is that in the long run this will lead to systems approach 100% precision and recall for the queries that are issued to the system[2]

PRECISE[14] is a system based on light annotation of the database schema that was tested over the GEOQUERY 250 corpus. PRECISE reduces semantic analysis to a graph matching problem after the schema elements have been named. Interestingly the system leverages a domain independent grammar to extract attachment relationships between tokens in the user's requests. The PRECISE work identifies a class of so called *semantically tractable queries*. Although the group did not publish the actual configuration times, they presumably corresponded to the naming phase and thus were rather short durations. We will forgo a discussion of the so called *semantically tractable queries* class and take at face value the claim that they achieved 100% precision and 77.5% recall, yielding a correctness of 77.5%. For such little configuration this is an impressive result and over very simple databases with a stream of very simple queries this may be adequate. However experience tells us that users do actually ask somewhat complex queries at times and they will be frustrated if told that their queries are not semantically tractable and must be rephrased or abandoned.

The PRECISE group reported a side experiment in which a student took over 15 hours to build a GEOQUERY 250 NLI using Microsoft's EnglishQuery tool. The resulting system achieved rather poor results for such an expensive effort – approximately 80% precision and 55% recall, yielding a correctness of approximately 45%. Our limited experience with the Microsoft English query tool was rather frustrating as well, but it would be interesting to repeat this experiment and also see how other commercial systems such as Progress Software's EASYASK and Elfsoft's ELF fare.

It should be noted that historically, transportable systems (e.g.,[7]) arose as a proposed solution to the high configuration costs of NLIs to databases. The idea is to use large scale domain-independent grammars, developed in the linguistics community, to map user requests to intermediate *logical form*. Using *translation*

---

[2] This does not address larger issues such as the limitations of the underlying formal language nor users querying for information outside the scope of the database.

knowledge, the logical form is then translated to a logical query expressed in the vocabulary of the relations of the actual database. Hence building an interface over a new database requires supplying a set of domain-specific lexical entries and the specification of translation knowledge, but does not require new linguistic syntax rules to be defined. A serious problem with the transportable approach however, is that it requires a deep understanding of the particular logical form employed to specify working and robust translation knowledge to the actual database tables in use. Considering the knowledge and tastes of the average technical worker, one wonders how well this can be supported. This said, some very interesting work has recently taken up the transportable approach for authoring NLIs over OWL and F-logic knowledge-bases [4]. Like this work, that work assumes very little computational linguistics knowledge on the part of the person who builds the natural language interface.

The system DUDE[9] presents an easy to use authoring interface to build dialogue systems. The back-end database is essentially a universal relation with relatively simple slot filling queries, but the authoring method is elegantly direct and results are sufficient for many practical applications.

7 Conclusions

This paper has presented a state-of-the-art authoring system for natural language interfaces to relational databases. Internally the system uses semantic grammars encoded in λ-SCFG to map typed user requests to an extended variant of Codd's tuple calculus which in turn is automatically mapped to SQL. The author builds the semantic grammar through a series of naming, tailoring and defining operations within a web-based GUI. The author is shielded from the formal complexity of the underlying grammar and as an added benefit, the given grammar rules may be used in the reverse direction to achieve paraphrases of logical queries (see [11]).

Using the GEOQUERY 250 corpus, our results are compared with contemporary machine learning results and approaches based on light annotation. Our initial experimental evidence shows quick bootstrapping of the initial interface can be achieved. Future work will focus on more complete evaluation and experimentation with more comprehensive grammatical frameworks (e.g. CCG [15]), especially under regimes that enable enhanced robustness [18]. Future work will also explore hybrid approaches using initial authoring to bootstrap NLIs, followed by interactive machine learning that, applied in the limit, will edge such NLIs toward 100% precision and recall.

References

1. Aho, A., Ullman, J.: The Theory of Parsing, Translation and Compiling, vol. 1. Prentice-Hall, Englewood Cliffs (1972)
2. Androutsopoulos, I., Ritchie, G.D.: Database interfaces. In: Dale, R., Moisl, H., Somers, H. (eds.) Handbook of Natural Language Processing, pp. 209–240. Marcel Dekker, New York (2000)

3. Chu, W., Meng, F.: Database query formation from natural language using semantic modeling and statistical keyword meaning disambiguation. Technical Report
   990003, UCLA Computer Science Department (June 1999)
4. Cimiano, P., Haase, P., Heizmann, J.: Porting natural language interfaces between
   domains: an experimental user study with the orakel system. In: Intelligent User
   Interfaces, pp. 180–189. ACM, New York (2007)
5. Copestake, A., Sparck Jones, K.: Natural language interfaces to databases. The
   Natural Language Review 5(4), 225–249 (1990)
6. Ge, R., Mooney, R.: A statistical semantic parser that integrates syntax and semantics. In: Proceedings of the Ninth Conference on Computational Natural Language
   Learning, pp. 9–16 (2005)
7. Grosz, B., Appelt, D., Martin, P., Pereira, F.: Team: An experiment in the design
   of transportable natural-language interfaces. AI 32(2), 173–243 (1987)
8. Kate, R., Mooney, R.: Using string-kernels for learning semantic parsers. In: Proc.
   of COLING/ACL-2006, pp. 913–920 (2006)
9. Lemon, O., Liu, X.: DUDE: a Dialogue and Understanding Development Environment, mapping Business Process Models to Information State Update dialogue
   systems. In: Proceedings of EACL (demonstration systems) (2006)
10. Minock, M.: A phrasal approach to natural language access over relational
    databases. In: Proc. of Applications of Natural Language to Data Bases (NLDB),
    Alicante, Spain, pp. 333–336 (2005)
11. Minock, M.: Modular generation of relational query paraphrases. Research on Lan-
    guage and Computation 4(1), 1–29 (2006)
12. Minock, M.: A STEP towards realizing Codd's vision of rendezvous with the casual
    user. In: 33rd International Conference on Very Large Data Bases (VLDB), Vienna,
    Austria (2007) (Demonstration session)
13. Mooney, R.: Learning language from perceptual context: A challenge problem for
    ai. In: Proceedings of the 2006 AAAI Fellows Symposium (2006)
14. Popescu, A., Etzioni, O., Kautz, H.: Towards a theory of natural language interfaces
    to databases. In: Intelligent User Interfaces (2003)
15. Steedman, M.: The Syntactic Process. The MIT Press, Cambridge (2001)
16. Tang, L., Money, R.: Using multiple clause constructors in inductive logic program-
    ming for semantic parsing. In: Flach, P.A., De Raedt, L. (eds.) ECML 2001. LNCS
    (LNAI), vol. 2167. Springer, Heidelberg (2001)
17. Wong, Y., Mooney, R.: Learning synchronous grammars for semantic parsing with
    lambda calculus. In: Proceedings of the 45th Annual Meeting of the Association
    for Computational Linguistics (ACL 2007), pp. 960–967 (2007)
18. Zettlemoyer, L., Collins, M.: Online learning of relaxed ccg grammars for parsing
    to logical form. In: Proceedings of the 2007 Conference on Empirical Methods
    in Natural Language Processing and Computational Natural Language Learning
    (2007)

Towards a Bootstrapping NLIDB System

Catalina Hallett and David Hardcastle

The Open University, Walton Hall
Milton Keynes, MK6 7AA, UK

Abstract. This paper presents the results of a feasibility study for a
bootstrapping natural language database query interface which uses natural language generation (NLG) technology to address the interpretation
problem faced by existing NLIDB systems. In particular we assess the
feasibility of automatically acquiring the requisite semantic and linguistic resources for the NLG component using the database metadata and
data content, a domain-specific ontology and a corpus of associated text
documents, such as end-user manuals, for example.

1 Introduction

This paper presents the results of a feasibility study for bootstrapping a natural
language database query interface which uses natural language generation (NLG)
technology to address the interpretation problem faced by existing NLIDB systems. The query system presents the user with an interactive natural language
text which can be extended and amended using context sensitive menus driven
by Conceptual Authoring. Using NLG to allow the user to develop the query
ensures accuracy and clarity. When the user submits the query the semantic
representation is transformed into a valid SQL statement. A detailed discussion
of the technology and an evaluation which showed the system to be a reliable
and effective means for domain experts to pose complex queries to a relational
database is presented by Hallett et al. [1].

While this approach delivers clear benefits, they come at a cost; domain expertise is required to construct the semantic resources, linguistic expertise is
required to map this domain knowledge onto the language resources, and knowledge of the database structure is needed to map it onto a valid query structure.
Our proposed solution is for the system to infer the resources and mappings
required from a domain ontology, the database metadata and data content and
a corpus of domain-specific texts. The feasibility study reported in this paper
demonstrated that we can, in principle, infer the required resources from a simple, highly normalised database with well-formed lexical descriptors such as MS
Northwind or Petstore. However, it also highlighted the need to couple metadata
mining with analysis of external sources of information.

1.1 Related Work

Providing user-friendly query interfaces for casual and non-specialist users, which
alleviate the need for programmatic knowledge is a central problem for the data

E. Kapetanios, V. Sugumaran, M. Spiliopoulou (Eds.): NLDB 2008, LNCS 5039, pp. 199–204, 2008.
© Springer-Verlag Berlin Heidelberg 2008

querying community. Whether these interfaces are form-based, visual, or natural language-based, knowledge about the data source structure and content is essential to the construction of intuitive interfaces. Traditionally, natural language interfaces to databases (henceforth, NLIDB) work in two steps:

- query interpretation: a natural language query entered by the user is parsed into a logical representation
- query translation: the logical representation of a query is mapped to a database querying language

It is evident that the query interpretation process requires both extensive linguistic resources for understanding the query, whilst the query translation step requires semantic resources for mapping query terms to database entities. In early NLIDBs, these resources (such as semantic grammars and lexicons) were created through an extensive manual process, resulting in heavily database-dependent systems.

The issue of interface portability was first highlighted in the early 1980's, and the fact that database schemas could be used to acquire domain knowledge has been exploited in systems such as CO-OP [2] and INTELLECT [3]. These systems also demonstrated that a modular architecture might allow query interfaces to be ported without code changes. Although these systems made some use of the database schema to map query terms to database entities, porting the interface to a new database still required extensive reworking of the lexicon, although the introduction of generic linguistic front-ends, in which the query interpretation stage is independent of the underlying database (see [4]), reduced the impact. Current NLIDB systems employ a variety of machine learning techniques in order to infer semantic parsers automatically [5,6] or to improve syntactic parsers with domain specific information [7]. However, these systems require large sets of annotated SQL queries for training purposes. The PRECISE system [8] employs a semantic model in order to correct and enhance the performance of a statistical parser, and so requires far less training data. Customization of the semantic model remains an issue, and the system is restricted to a set of semantically tractable queries, which impairs coverage.

1.2   Query Interfaces Based on Conceptual Authoring

Conceptual Authoring (CA) using NLG [9] has been employed as an alternative to natural language input in order to remove the query interpretation step [1,10]. In querying systems based on CA, the query is constructed incrementally by the user, through successive interactions with a natural language text (termed *feedback text*). Changes to the feedback text directly reflect underlying changes to the semantic content of the query; so whilst the user is always presented with a natural language text, the query is always encoded in a structured internal representation. CA has been used successfully in building query interfaces [1,10] with evaluation showing positive user feedback and a clear preference over using SQL [1].

The feedback text shown in Figure 1 represents a simple SELECT query against the Orders table of the MS Northwind database, expressed by the SQL query in Figure 2. The words in square brackets are anchors and represent sites where the user can reconfigure the query; for example by changing a literal value, setting an aggregate function, removing a selection criterion or adding further criteria or ordering conditions.

List orders which
- were processed by [any employee]
- conisted of [total freight]
- were shipped between [1/4/97] and [3/31/98]
- [further criteria] ordered by [total freight]

Fig. 1. A sample feedback text query

SELECT Orders.EmployeeID, Sum(Orders.Freight) AS Shipping
FROM Orders
WHERE Orders.ShippedDate Between #4/1/1997# And #3/31/1998#
GROUP BY Orders.EmployeeID
ORDER BY Sum(Orders.Freight) DESC;

Fig. 2. The SQL produced by the query represented in Figure 1

2 Feasibility Study

In a previous attempt [11], we investigated the possibility of inferring some basic resources automatically, however this attempt did not reach far enough and it also resulted in relatively clumsy natural language queries. In this section we provide a high level summary of a recent feasibility study undertaken by the authors; a more complete discussion is presented in an auxiliary technical report [12].

The feasibility study focused on a prototype which is a modified version of a previous CA-based query interface [1]. We leave the task of inferring the domain ontology to others [13], and focus on the inferencing of the resources required by the NLG system. The prototype receives as input a model of the database semantics and a domain ontology, and it automatically generates some of the components and resources that in previous Conceptual Authoring querying systems were constructed manually, along with a module which translates the user-composed query into SQL . The resulting query system provides a user interface based on a feedback text (see Section 1.2) which is generated by the query system from a semantic graph. User interaction with the feedback text results in changes to the underlying semantic representation and a new feedback text is generated to update the display. When the query is run the underlying representation is converted to SQL using the model of the database semantics from which the query interface system was inferred.

Table 1. Evaluation results

	Petstore			Northwind		
	Actual	Identified	Accuracy	Actual	Identified	Accuracy
Entities	28	28	100%	49	49	100%
Relations	30	28	93.3%	61	61	100%
Entity lexical desc	28	22	78.5%	49	49	100%
Relation lexical desc	30	20	75%	61	33	54%
Entity part-of-speech	30	30	100%	49	49	100%
subcat frames	30	20	75%	61	33	54%

Although the system is supplied with a domain ontology it still needs to analyse the structure of the database to support the mapping from the semantic model to syntactic and lexical resources. The metadata analysis focused on the following elements as described in the SQL-92 Information Schema: Domain Descriptors (§4.7), Column Descriptors (§4.8), Table Descriptors (§4.9), and Table and Domain Integrity Constraints (§4.10). Since both sample databases are highly normalised a simplistic approach in which tables are identified as kernel entities and their columns are represented as properties was productive, and foreign key definitions sufficed to infer associations between kernel entities. In commercial databases, in the authors' experience, the system would need to locate inner associative entities, and although Query Expression metadata from derived tables and view definitions would aid this process it would be unlikely to prove sufficient. We propose to address this problem by looking at related metadata such as ERDs and ORM mappings.

The system also requires linguistic resources in order to represent the query as text. Some linguistic resources, such as the grammar and lexicon, are reusable, but the mappings from the ontology to the linguistic resources must be inferred. For example, the system needs to choose an appropriate lexicalization for each entity, and an appropriate syntactic frame and lexicalization for each association. It also requires domain-specific semantic and linguistic resources to manage literal values, spatial and temporal modifiers and sub-language jargon. In the case of the prototype, the identification of the lexical descriptors was made easier due to the fact that both databases use clear and reliable column naming conventions, a feature which could not be relied upon in a commercial database. In future versions of the system we intend to use related corpora, such as user manuals or domain-specific technical documentation, to support the inferencing of semantic and linguistic resources and the generation of mappings between the concepts in the domain ontology and the syntactic subcategorisations and lexical anchors required to express them.

We evaluated the prototype using two sample databases, MS Northwind and Petstore. The generated system had a coverage of 71.6% – 179 of 250 questions which could be asked of the database were supported by the interface. We assessed the system's ability to infer resources automatically by comparing the resources constructed by the generator with a manually constructed NLIDB system for each sample database. The results of this comparison are presented in

Table 1, and show that whilst resources that rely on database metadata can be identified quite accurately, linguistic resources which are identified using heuristics over textual metadata are less reliable.

3   Conclusions

Although the prototype made several simplifying assumptions about the nature of the database, it serves as a proof of concept showing that simple inferencing techniques can achieve results, and highlights the areas where more complex techniques are required (in particular, the identification of linguistic descriptions for relations). However, it is not the case that automatic inferencing will always be possible, and the limiting factors set out below may entail supervision, or may even mean that no resources can be inferenced at all. Our future research plans include extending the scope of our metadata analysis to include additional sources of information, as discussed above, to address these limitations.

Structure and Normalisation
Codd [14] proposes an extension to the relational model (RM/T) to align it with predicate logic. In this context he introduces the notion of "property integrity", under which each entity is represented by a single-valued E-relation and its dependent properties (or characteristics) are grouped into subsidiary P-relations. Inferring semantic dependencies from a database normalised to this extent (to 4NF) would be trivial; conversely, a data warehousing database with a flat structure and a large number of columns per relation might prove intractable.

Atomicity
Our task is made much easier if the entities within the database are represented atomically; of course in practice this will seldom be the case as the entities modelled by the database will have feature-based values which will be decomposed into column tuples. Whether or not the system can recognise these tuples will depend on a variety of incidental database-specific factors. There are various heuristics which we can throw at the problem, for example: using a domain ontology; scanning query expressions in derived tables or cached queries for common projections; analysing data content, and so on.

Nonetheless any relational database will almost certainly contain many non-atomic characteristic entities and, unless it is in 4NF, it is highly unlikely that they can all be recovered automatically. This is therefore an area where the system will require supervision if it is to function effectively.

Lexicalisation and Symbolic Values
The process of mapping entities onto concepts in the domain ontology will often involve string matching, either as part of the process of attempting to infer a semantic class or as a fallback strategy. In some instances the column names will be meaningful strings and there will also be string descriptions, in others there may be little lexical information available at all. Similarly, the data values may

be more or less tractable to the system; in particular if the data consists only of symbolic values or field codes then we can infer very little about its meaning.

Metadata Quality

In practice we cannot rely on the quality of metadata in the field. For example, we may encounter databases where foreign key information is not defined in the metadata, it is simply known to developers, where columns are mislabelled, for example due to merging of legacy data sets, where default values, unique constraints and referential constraints are not formally encoded, and so on. In such instances the system will be unable to infer the information required to build the query engine.

References

1. Hallett, C., Scott, D., Power, R.: Composing questions through conceptual authoring. Computational Linguistics (2007)
2. Kaplan, S.J.: Designing a portable natural language database query system. ACM Trans. Database Syst. 9(1), 1–19 (1984)
3. Harris, L.R.: The ROBOT system: Natural language processing applied to data base query. In: ACM 1978: Proceedings of the 1978 annual conference, pp. 165–172. ACM Press, New York (1978)
4. Alshawi, H.: The Core Language Engine. ACL-MIT Press Series in Natural Language Processing. MIT Press, Cambridge (1992)
5. Tang, L.R., Mooney, R.J.: Using multiple clause constructors in inductive logic programming for semantic parsing. In: EMCL 2001: Proceedings of the 12th European Conference on Machine Learning, London, UK, pp. 466–477. Springer, Heidelberg (2001)
6. He, Y., Young, S.: A data-driven spoken language understanding system. In: IEEE Workshop on Automatic Speech Recognition and Understanding (2003)
7. Kate, R.J., Mooney, R.J.: Using string-kernels for learning semantic parsers. In: Proceedings of ACL 2006, pp. 913–920 (2006)
8. Popescu, A.M., Etzioni, O., Kautz, H.: Towards a theory of natural language interfaces to databases. In: IUI 2003: Proceedings of the 8th international conference on Intelligent user interfaces, pp. 149–157. ACM Press, New York (2003)
9. Power, R., Scott, D.: Multilingual authoring using feedback texts. In: Proceedings of COLING-ACL 1998, Montreal, Canada, pp. 1053–1059 (1998)
10. Evans, R., Piwek, P., Cahill, L., Tipper, N.: Natural language processing in CLIME, a multilingual legal advisory system. Natural Language Engineering (2006)
11. Hallett, C.: Generic querying of relational databases using natural language generation techniques. In: Proceedings of the 4th International Natural Language Generation Conference (INLG 2006), Sydney, Australia, pp. 95–102 (2006)
12. Hallett, C., Hardcastle, D.: A feasibility study for a bootstrapping nlg-driven nlidb system. Technical Report TR2008/07, The Open University, Milton Keynes, UK (2008)
13. Shvaiko, P., Euzenat, J.: A survey of schema-based matching approaches, pp. 146–171 (2005)
14. Codd, E.F.: Extending the database relational model to capture more meaning. ACM Trans. Database Syst. 4(4), 397–434 (1979)

Document Processing and Text Mining

Document Processing and Text Mining

Real-Time News Event Extraction
for Global Crisis Monitoring

Hristo Tanev, Jakub Piskorski, and Martin Atkinson

Joint Research Center of the European Commission
Web and Language Technology Group of IPSC
T.P. 267, Via Fermi 1, 21020 Ispra (VA), Italy
{hristo.tanev,jakub.piskorski,martin.atkinson}@jrc.it

Abstract. This paper presents a real-time news event extraction system
developed by the Joint Research Centre of the European Commission.
It is capable of accurately and efficiently extracting violent and disaster
events from online news without using much linguistic sophistication. In
particular, in our linguistically relatively lightweight approach to event
extraction, clustered news have been heavily exploited at various stages
of processing. The paper describes the system's architecture, news geo-
tagging, automatic pattern learning, pattern specification language, information aggregation, the issues of integrating event information in a
global crisis monitoring system and new experimental evaluation.

Keywords: information extraction, event extraction, processing massive datasets, machine learning, finite-state technology.

1 Introduction

In the last decade, we have witnessed an ever-growing trend of utilizing NLP
technologies, which go beyond the simple keyword look-up, for automatic knowledge discovery from massive amount of textual data available on the Internet.

This paper reports on the fully operational event-extraction system developed
at the Joint Research Center of the European Commission for extracting violent
event information and detection of natural and man-made disasters from online news. The news articles are collected through the Internet with the Europe
Media Monitor (EMM) [1], a web based news aggregation system that collects
40000 news articles from 1400 news sources in 35 languages each day. Gathering
information about violent and disaster events is an important task for better understanding conflicts and for developing global monitoring systems for automatic
detection of precursors for threats in the fields of conflict and health.

Formally, the task of event extraction is to automatically identify events in
free text and to derive detailed information about them, ideally identifying *Who
did what to whom, when, with what methods (instruments), where and possibly
why*. Automatically extracting events is a higher-level information extraction
(IE) task which is not trivial due to the complexity of natural language and due
to the fact that in news articles a full event description is usually scattered over

E. Kapetanios, V. Sugumaran, M. Spiliopoulou (Eds.): NLDB 2008, LNCS 5039, pp. 207–218, 2008.

several sentences and documents. Further, event extraction relies on identifying named entities and relations holding among them. Since the latter tasks can be achieved with an accuracy varying from 80 to 90% [2], obtaining precision/recall figures oscillating around 60% for event extraction (usually involving several entities and relations) is considered to be a good result. Although a considerable amount of work on automatic extraction of events has been reported[1] , it still appears to be a lesser studied area in comparison to the somewhat easier tasks of named-entity and relation extraction. Two comprehensive examples of the current functionality and capabilities of event extraction technology dealing with identification of disease outbreaks and conflict incidents are given in [3] and [4] respectively. The most recent trends in this area are reported in [5].

In order to be capable of processing vast amount of textual data in real time (as in the case of EMM) we follow a linguistically lightweight approach and exploit clustered news at various stages of processing ranging from pattern learning to information fusion. Consequently, only a tiny fraction of each text is analyzed. In a nutshell, our system deploys simple 1 and 2-slot extraction patterns on a stream of geo-tagged and clustered news articles for identifying event-relevant entities. These patterns are semi-automatically acquired in a bootstrapping manner via utilization of clustered news data. Next, information about events scattered over different documents is aggregated. Since efficient processing is a prerequisite for being able to extract event information in real time, we have developed our own pattern matching engine. The results of the core event extraction system has been integrated into a real-world global media monitoring system. An evaluation revealed acceptable accuracy and a strong application potential. Although our applications center on the security domain, the techniques deployed in our system can be applied in other domains, e.g., tracking business-related events for risk assessment, humanitarian emergency prediction.

The rest of this paper is organized as follows. First, in Section 2 the architecture of our live event extraction processing chain is described. Section 3 addresses the issues of geo-tagging the news. The pattern acquisition technique and the pattern matching engine are presented in Section 4. Next, Section 5 elaborates on information fusion. Section 6 addresses integration of the event extraction in a global monitoring system. Some evaluation figures are given in Section 7. Finally, we give a summary and future directions in Section 8.

2   Real-Time Event Extraction Process

This section briefly describes the real-time event extraction processing chain, which is depicted in Figure 1. First, before the proper event extraction process can proceed, news articles are gathered by dedicated software for electronic media monitoring, namely the EMM system [1], which regularly checks for updates of news articles across multiple sites. Secondly, the input data is geo-tagged,

---

[1] The research in this area was pushed forward by the Message Understanding Conferences (http://www.itl.nist.gov/iaui/894.02/related/projects/muc) and by the ACE (Automatic Content Extraction) program (http://projects.ldc.upenn.edu/ace).

categorized and grouped into news clusters, ideally including documents on one topic. For each such cluster the system tries to detect and extract only the main event, analyzing all documents in the cluster.

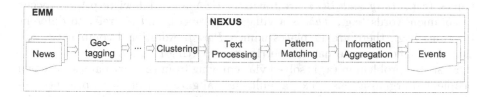

Fig. 1. Real-time event extraction processing chain

Next, each cluster is processed by NEXUS (News cluster Event eXtraction Using language Structures), our core event extraction engine, which for each detected violent event produces a frame, whose main slots are: date and location, number of killed and injured, kidnapped people, actors, and type of event. In an initial step, each document in the cluster is linguistically preprocessed in order to produce a more abstract representation of the texts. This encompasses the following steps: fine-grained tokenization, sentence splitting, named-entity recognition (e.g., people, numbers, locations), simple chunking, labeling of key terms like action words (e.g. *kill, shoot*) and unnamed person groups (e.g. *five civilians*). The aforementioned tasks are accomplished by CORLEONE (Core Linguistic Entity Online Extraction), our in-house core linguistic engine [6], which is an integral part of NEXUS.

Once texts are grouped into clusters and linguistically preprocessed, the pattern engine applies a cascade of extraction grammars on each document within a cluster. Noteworthy, the extraction patterns are matched against the first sentence and the title of each article. By processing only the top sentence and the title, the system is more likely to capture facts about the most important event in the cluster. For creating extraction patterns we apply a blend of machine learning and knowledge-based techniques. Contrary to other approaches, the learning phase exploits clustered news, which intuitively guarantees better precision.

Finally, since information about events is scattered over different articles, the last step consists of cross-document cluster-level information fusion, i.e., we aggregate and validate information extracted locally from each single article in the same cluster via utilization of various information aggregation techniques. In particular, victim counting, semantic role disambiguation and event type classification are performed at this stage of the processing.

The output of NEXUS constitutes input for a global monitoring system. The core event-extraction engine is triggered every 10 minutes on a clustering system that has a 4 hour sliding window in order to keep up-to-date with most recent events. The more thorough description of the geo-tagging, pattern acquisition, our pattern engine, and information fusion follows in the subsequent sections. The data gathering and clustering is addressed in [1].

3   Geo-Tagging and Clustering

Homographs pose a well-known problem in the process of geo-tagging news articles [7]. In particular, words referring to place names may: (a) occur as person names (e.g., *Tony* and *Blair* are towns in USA and Malawi resp.), (b) occur as common words (e.g., *This* is a village in France), and (c) refer to different locations (e.g., there are 12 places named *Paris*). Additional complicacies are caused by inflection and names referring to reporting location.

First, the problem in (a) is solved via removing from the text names of known people and organisations. Next, a multi-lingual gazetteer of place, province, region and country names is used to geo-match a list of candidate locations in the news articles. In order to disambiguate homographs that are common words and place names (see (b) and (c)), the traditional approach is to use language dependent stop word lists. We use a different approach based on two characteristics maintained in our gazetteer. The first characteristic classifies locations based on their perceived size, such that capital cities and major cities have a higher class than small villages. The second characteristic maintains the hierarchical relation of place in its administrative located hierarchy (i.e., town, in province, in region, in country). The disambiguation algorithm lets high class locations pass through as well as locations that have a containment relation with other candidate locations (e.g., *Paris*, *Texas*, *USA*).

In order to handle name inflection, names are maintained with their variants encoded as a regular expression which is interpreted at match detection time. Only matches, for which the language of the name and the article are compatible, are maintained in the candidate location list. Next, a Newswire location filtering is applied, where the word position of the candidate location is used to promote locations occurring after an initial location since the Newswire location generally appears earlier in the article. All the processing up to this point works on a single news article where, depending on the reporting style, a number of candidate locations could be produced. The final location is then selected by scoring all the occurrences of all locations in the articles that compose the cluster.

4   Pattern Learning and Matching

4.1   Pattern Acquisition

While in the past IE systems used patterns created by human experts, state-of-the-art approaches use machine learning (ML) algorithms for their acquisition [8,9]. However, ML approaches are never 100% accurate, therefore we manually filter out implausible patterns and add hand-crafted ones, where it is appropriate.

Our pattern acquisition approach involves multiple consecutive iterations of ML followed by manual validation. Learning patterns for each event-specific semantic role (e.g. *killed* or *kidnapped*) requires a separate cycle of learning iterations. The method uses clusters of news articles produced by EMM [1]. Each cluster includes articles from different sources about the same news story. Therefore, we assume that each entity appears in the same semantic role (actor, victim, injured) in the context of one cluster. The core steps of the pattern acquisition algorithm are depicted in Figure 2.

---

1. Annotate a small corpus with event-specific information, e.g., date, place, actors, affected dead, etc. As an example consider the following two sentences:
   a. <actor>Hezbollah</actor>claimed the responsibility for the kidnapping of the Israeli corporal.

   b. <actor>Al Qaida </actor>claimed the responsibility for the bombing which killed <affected_dead>five people</affected_dead>.

2. Learn automatically single-slot extraction patterns (see [10]), e.g., the pattern [ORGANIZATION] "claimed the responsibility" could be learned from both sentences, where the entity filling the slot [ORGANIZATION] is assigned the role actor(perpetrator)

3. Manually check, modify and filter out low quality patterns. Eventually add new patterns. If the size of the pattern set exceeds certain threshold (the desired coverage is reached)-terminate.

4. Match the patterns against the full corpus or part of it. Next, entities which fill the pattern slots and comply to the semantic constraints of the slot are taken as *anchor entities*. If an anchor entity A (e.g., *five people*) is assigned a role R (e.g., affected_dead) in the news cluster C, we assume with high confidence that in the cluster C entity A appears mostly in the same role R. Consequently, annotate automatically all the occurrences of A in C with the label R, e.g., in our example all the occurrences of *five people* in the cluster from which the second sentence originate will be labeled as affected_dead.

5. Go to step 2.

Fig. 2. Pattern learning algorithm

This algorithm was run separately for *killed, wounded, kidnapped* and the other semantic roles. The algorithm was run for no more than three iterations, since its accuracy constantly degrades. An automatic procedure for syntactic expansion complements the learning. This procedure accepts a manually provided list of words which have identical (or nearly identical) syntactic model of use (e.g. *killed, assassinated, murdered*, etc.) and then it generates new patterns from the old ones by substituting for each other the words in the list. For example, if the patterns X have been killed and X was murdered were learned, the syntactic expansion procedure will generate X have been murdered and X was killed. After single-slot patterns are automatically acquired , we use some of them to manually create 2-slot patterns like X shot down Y. A more comprehensive presentation of the learning algorithm is given in [10].

4.2 Pattern Matching Engine

In order to guarantee that massive amounts of textual data can be digested in real time, we have developed ExPRESS (Extraction Pattern Engine and Specification Suite), a highly efficient extraction pattern engine [11], which is capable of matching thousands of patterns against MB-sized texts within seconds. The specification language for creating extraction patterns in ExPRESS is a blend of two previously introduced IE-oriented grammar formalisms, namely JAPE (Java Annotation Pattern Engine) used in the widely-known GATE platform [12] and XTDL, a significantly more declarative and linguistically elegant formalism used in a lesser known SPRoUT platform [13].

An ExPRESS grammar consists of pattern-action rules. The left-hand side (LHS) of a rule (the recognition part) is a regular expression over flat feature structures (FFS), i.e., non-recursive typed feature structures (TFS)[2] without structure sharing, where features are string-valued and unlike in XTDL types are not ordered in a hierarchy. The right-hand side (RHS) of a rule (action part) constitutes a list of FFS, which will be returned in case LHS pattern is matched.

On the LHS of a rule variables can be tailored to the string-valued attributes in order to facilitate information transport into the RHS, etc. Further, like in XTDL, functional operators (FO) are allowed on the RHSs for forming slot values and for establishing contact with the 'outer world'. The predefined set of FOs can be extended through implementing an appropriate programming interface. FOs can also be deployed as boolean-valued predicates. These two features make ExPRESS more amenable than JAPE since writing 'native code' on the RHS of rules (common practice in JAPE) has been eliminated. Finally, we adapted the JAPEs feature of associating patterns with multiple actions, i.e., producing multiple annotations (eventually nested) for a given text fragment. Noteworthy, grammars can be cascaded. The following pattern for matching events, where one person is killed by another, illustrates the syntax.

```
killing-event :> ((person & [FULL-NAME: #name1]):killed
                   key-phrase & [METHOD: #method, FORM: "passive"]
                   (person & [FULL-NAME: #name2]):killer):event
 -> killed: victim & [NAME: #name1],
    killer: actor & [NAME: #name2],
    event: violence & [TYPE: "killing", METHOD: #method, ACTOR: #name2,
                       VICTIM: #name1, ACTOR_IN_EVENTS: inHowManyEvents(#name2)]
```

The pattern matches a sequence consisting of: a structure of type **person** representing a person or group of persons who is (are) the victim, followed by a key phrase in passive form, which triggers a *killing event*, and another structure of type **person** representing the actor. The symbol **&** links a name of the FFS's type with a list of constraints (in form of attribute-value pairs) which have to be fulfilled. The variables **#name1** and **#name2** establish bindings to the names of both humans involved in the event. Analogously, the variable **#method** establishes binding to the method of killing delivered by the **key-phrase** structure. Further, there are three labels on the LHS (**killed**, **killer**, and **event**) which specify the start/end position of the annotation actions specified on the RHS. The first two actions (triggered by the labels **killed** and **killer**) on the RHS produce FFS of type **victim** and **actor** resp., where the value of the **NAME** slot is created via accessing the variables **#name1** and **#name2**. Finally, the third action produces an

---

[2] TFSs are widely used as a data structure for NLP. Their formalizations include multiple inheritance and subtyping, which allow for terser descriptions.

FFS of type `violence`. The value of the `ACTOR_IN_EVENTS` attribute is computed via a call to a FO `inHowManyEvents()` which contacts some external knowledge base to retrieve the number of events the current actor was involved in the past (such information might be useful in the context of global monitoring systems).

We have compared the run-time behaviour of EXPRESS against the other two pattern engines mentioned earlier. For instance, matching the violent-event extraction grammar (consisting of ca. 3100 extraction patterns) against various news collection of over 200 MBs can be executed from 12 to 25 times faster than the corresponding XTDL grammar, which has been optimized for speed. A more thorough overview of the techniques for compiling and processing grammars as well as the entire EXPRESS engine is given in [11].

5   Information Aggregation

Our event extraction system firstly deploys linear patterns in order to detect entities and their semantic roles in each news cluster. Next, the single pieces of an information are merged into event descriptions via application of information aggregation algorithm. This algorithm assumes that each cluster reports at most one main event of interest. Three main aggregation steps can be distinguished:

Semantic Role Disambiguation: If one and the same entity has two roles assigned, a preference is given to the role assigned by the most reliable group of patterns, e.g., 2-slot patterns like X **shot down** Y are considered the most reliable. Regarding the 1-slot patterns, the ones for detection of *killed*, *wounded*, and *kidnapped* are considered as more reliable than the patterns for extraction of the *actor* (*perpetrator*), the latter one being more generic.

Victim counting: An important part of the information aggregation algorithm is counting the victims involved in an event. It is subdivided into two stages: (a) calculating the estimates for the numbers of killed, wounded and kidnapped locally for each single news article, and (b) using the estimates calculated in (a) computing the corresponding cluster-level estimates.

Article-Level victim number estimation: One article may refer to different noun phrases (NP) from which the number of the victims could be derived, e.g., in: '*Two civilians died in the clashes and one policeman died later in the hospital.*' there are two NPs which refer to victims (*two civilians* and *one policeman*). The numbers to which these NPs refer (one and two) has to be added in order to obtain the total number of the victims. In other cases, an inclusion relation between entities might exist, e.g., '*Two women were killed; four civilians lost their lives in total*' (*two women* are part of *four civilians*). Consequently, the number in the NP, which refers to a more generic concept is chosen. To make such a reasoning feasible, we created a small taxonomy of NPs which refer to people and which frequently occur in the news articles when violent events and disasters are reported. The victim counting algorithm is depicted in Figure 3.

1. Extract from a news article a set of NPs $E = \{E_1, E_2, \ldots, E_n\}$, which refer to individuals or groups with definite number of people and which have certain semantic role.

2. Map each E_i to a concept from the ontology using a list of keyword-taxonomy mappings (we denote it as $conc(E_i)$). As an example consider the text *One journalist was beaten to death and two soldiers died in the clashes in which four civilians lost their lives in total*. Three NPs referring to killed people would be extracted: *one journalist*, *two soldiers* and *four civilians*, and mapped to the taxonomy categories journalist, serviceman, civilian.

3. Delete each E_i from E, if there exist $E_j \in E$ such that $conc(E_i)$ is-a $conc(E_j)$, i.e., $conc(E_i)$ is a direct or indirect successor of $conc(E_j)$ in the taxonomy. In this manner only the most generic NPs are left in E. In our example we may identify one such relation, namely, *journalist is a civilian*, therefore we delete the NP *one journalist* from E.

4. Sum up the numbers reported from the NPs in E. In our example we have to sum up the numbers from the phrases *four civilians* and *two soldiers* (four and two).

Fig. 3. Aricle-level victim counting algorithm

Cluster-level victim number estimation: Finding this number is not always straightforward, since different news sources might report on contradictory information about the number of killed, wounded and kidnapped. We use an ad-hoc technique for computing the most probable estimation for these numbers. It finds the largest group of numbers which are close to each other and subsequently finds the number closest to their average. After this estimation is computed, the system discards from each news cluster all the articles whose reported victim numbers significantly differ from the estimated numbers for the whole cluster.

Event type classification: Whenever possible, we assign a class label to the detected violent event or disaster. Some of the most used event classes are *Terrorist Attack, Bombing, Shooting, Air Attack, ManMadeDisaster, Floods*, etc. The classification is based on a blend of keyword matching, taxonomy and domain specific rules. The classification algorithm is depicted in figure 4.

In order to clarify the step 2 of the algorithm consider a text in which the words: *blast, explosion* and *bombing* appear. The first two will give a score of two for the type *Explosion* and the third one will give a score of 1 for the type *Bombing*. Since *Bombing* is a kind of *Explosion* and the text does not refer to other sub-types of *Explosion* the score for *Bombing* will be boosted by adding the score for *Explosion*. Consequently, *Bombing* would obtain the highest score (3). As an example of a domain-specific rule, consider the following one: if the event description includes named entities, which are assigned the semantic role *kidnapped*, as well as entities which are assigned the semantic role *released*, then the type of the event is *Hostage Release*, rather than *Kidnapping* even if the later type got a higher keyword-based score. If the event refers to kidnapped people and at the same time the news articles contain words like *video* or *videotape*, then the event type is *Hostage Video Release*. The latter rule has a higher priority, therefore it impedes the *Hostage Release* rule to fire erroneously, when release of hostage video is reported. Another rule is that if more than half of the news articles in a cluster contain constructions like *arrested PERSON, nabbed PERSON*, etc., then the type of the event is *Arrest*, even if another event type obtains higher keyword score. This rule comes into play, when the arrest of some criminal or terrorist is reported together with the crimes for which he or she is arrested. The reference to the crime in such cases may erroneously trigger violent event type which refer to the specificity of the crime, for example *Shooting, Robbery, TerroristAttack*, etc. This rule prevents such erroneous classifications and correctly classifies the event as *Arrest*.

---

1. For all types in $T = \{T_1, T_2, \ldots, T_k\}$ compute a rank ($rank(T)$) based on a number of occurrence of keywords related to that type. For each event type a dictionary of specific keywords have been created semi-automatically.

2. Boost the the rank of type T_1 if the following holds: $rank(T_1) > 0$ and there exists T_2 with: (a) $rank(T_2) > 0$, (b) T_1 *is-a* T_2 and (c) there is no other type T_i which does not subsume T_1 and T_i *is-a* T_2. If the condition holds, then set $rank(T_1) = rank(T_1) + rank(T_2)$ (the score of the more generic type is added to the score of the more specific one). The logic behind this step is that when some event takes place, it can be referred to at different levels of abstraction and the most specific reference should be preferred, since it is the most informative.

3. Select the type T_i for which $rank(T_i) > rank(T_j)$ for all other T_j unless a domain-specific event type classification rule can be applied. These rules have higher precedence over the ranking and were introduced based on empirical observations.

Fig. 4. Event type classification

The information aggregation algorithm has some limitations. In particular, it considers only one main event per news cluster, ignoring events of lower importance or incidents subsumed by the main event. In the security related domain it is often necessary to detect links between events, e.g., a kidnapping typically includes capturing a hostage, a statement and a video release by the kidnappers, in which they declare what they want to liberate the abducted person, police action to liberate the hostage, and finally his/her liberation. Currently, these events are detected separately and they are not merged into one complex event.

6 Support to Global Crisis Monitoring

The general end user requirements for Global Monitoring emerge from two different contexts, namely situation rooms and actors in the field. The former generally require large graphical displays showing geographical and related territorial information, alerting functionality like audio alarms, zooming to new events, fast indication of the gravity and typology of the event. Actors in the field also have the similar functional requirements, in addition, they might require the possibility to have universal access without the need to download or install dedicated client software applications. A third but nonetheless important end user is a web service that consumes the event data for further integration.

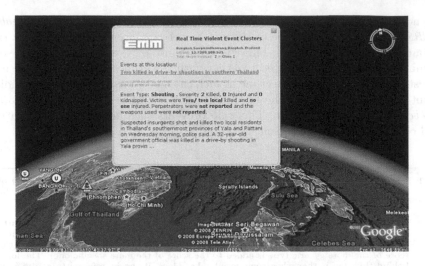

Fig. 5. Event visualization in Google Earth (To see open Google Earth with KML: http://press.jrc.it/geo?type=event&format=kml&language=en)

Table 1. Event Extraction Performance

Detection task	Accuracy (%)	Detection task	Accuracy (%)
Dead counting	70	Geo-tagging (country)	90
Injured counting	57	Geo-tagging (place name)	61
Event classification	80		

There are two emerging standards for integrating and transmitting geo-located data to Web-based clients: Geographical Simple Syndication (GeoSS) (http://www.georss.org) supported by a number of map based visualization clients (e.g., GoogleMaps and GoogleEarth), an a proprietary format Keyhole Markup Language (KML) (http://code.google.com/apis/kml) supported directly by GoogleMaps and GoogleEarth. KML allows style information to be included in the markup of the information transmitted. We support both standards.

A major issue to address is the handling of multiple events occurring at the same location. We decided that we create geo-clusters where the single events are combined into the detail of the cluster as a list ordered by individual event gravity, which is calculated as the sum of victims prioritized by deaths then injuries and finally kidnappings. The magnitude of the geo-cluster can be calculated analogously. The content of the RSS/KML follows as much as possible each standard, e.g., the title is the location, the description has 2 main parts, one is textual – the location hierarchy (location, province, region, country) and the other is an embedded description of all the events in the geo-cluster as a list.

We developed two applications in order to support Global Crisis Monitoring. One is a dedicated Web Page using the Google Maps API, which is accessible at (http://press.jrc.it/geo?type=event&format=html&language=en). The second uses the Google Earth desktop application (see Fig. 5). Both applications support: event zooming, event type and gravity indication, showing related territorial information, whereas audio alarms are only supported by the first one.

7   Evaluation

An evaluation of the event extraction performance has been carried out on 94 English-language news clusters based on news articles downloaded on 1st and 7th of April 2008. 30 violent events and disasters were described in these clusters. Our system detected 23 out of these 30 violent event descriptions (**77% coverage**). In some cases several clusters referred to the same event. We consider that an event is detected by the system, if at least one cluster referring to it was captured. Our system detected 31 news clusters as referring to violent events and disasters. 23 out of these 31 detected events were violent events and disasters (**74% precision**). Table 1 shows the accuracy of the performance of the victim counting, classification and geo-tagging. For each single task, the fraction of the news clusters is given, for which a correct result was returned. The evaluation shows that event classification and country-level geo-tagging have relatively high performance. Victim counting and place-level geo-tagging need further improvement. Currently, coordinate phrases like *two civilians and a soldier* are not captured. Therefore in some cases the number of the victims can be underestimated. An important issue is the situation in which two different incidents are reported as a part of one event. For example, a terrorist attack might include bombing and shooting. In such cases our system will assign one label to the whole event, where two labels are relevant.

8   Conclusions and Future Directions

We have presented real time event extraction from on-line news for global crisis monitoring, which has been fully operational 24/7 since the December 2007.[3] In particular, we introduced NEXUS – which performs cluster-level information fusion in order to merge partial information into fully-fledged event descriptions. The results of the evaluation on violent event extraction show that NEXUS can be used for real time global crisis monitoring. All the text core processing tools, which we use, are based on finite-state technology in order to fulfill the requirements of a real-time text processing scenario.

In order to improve the quality of the extracted event descriptions, several system extensions are envisaged. Firstly, some improvement of the event classification is planned, i.e., we are working towards fine grained classification of natural and man made disasters. We also plan to improve the classification accuracy for the violent events. Secondly, we aim at multilingual event extraction and crisis monitoring (Romance and Slavonic languages). This is feasible, since

---

[3] We are indebted to our colleagues without whom the presented work could not have been be possible. In particular, we thank Erik van der Goot, Ralf Steinberger, Bruno Pouliquen, Clive Best, Jenya Belyaeva and other colleagues.

our algorithms and grammars are mostly language independent. Finally, a long-term goal is to automatically discover structure of events and relations between them, i.e., discovering sub-events or related events of the main one.

References

1. Best, C., van der Goot, E., Blackler, K., Garcia, T., Horby, D.: Europe Media Monitor. Technical Report EUR 22173 EN, European Commission (2005)
2. Nadeau, D., Sekine, S.: A Survey of Named Entity Recognition and Classification. Lingvisticae Investigationes 30(1), 3–26 (2007)
3. Grishman, R., Huttunen, S., Yangarber, R.: Real-time Event Extraction for Infectious Disease Outbreaks. In: Proceedings of Human Language Technology Conference (HLT) 2002, San Diego, USA (2002)
4. King, G., Lowe, W.: An Automated Information Extraction Tool For International Conflict Data with Performance as Good as Human Coders: A Rare Events Evaluation Design. International Organization 57, 617–642 (2003)
5. Ashish, N., Appelt, D., Freitag, D., Zelenko, D.: Proceedings of the Workshop on Event Extraction and Synthesis, held in conjunction with the AAAI 2006 conference, Menlo Park, California, USA (2006)
6. Piskorski, J.: CORLEONE – Core Linguistic Entity Online Extraction. Technical report, Joint Research Center of the European Commission, Ispra, Italy (2008)
7. Pouliquen, B., Kimler, M., Steinberger, R., Ignat, C., Oellinger, T., Blackler, K., Fuart, F., Zaghouani, W., Widiger, A., Forslund, A., Best, C.: Geocoding multilingual texts: Recognition, Disambiguation and Visualisation. In: Proceedings of LREC 2006, Genoa, Italy, pp. 24–26 (2006)
8. Jones, R., McCallum, A., Nigam, K., Riloff, E.: Bootstrapping for Text Learning Tasks. In: Proceedings of IJCAI 1999 Workshop on Text Mining: Foundations, Techniques, and Applications, Stockholm, Sweden (1999)
9. Yangarber, R.: Counter-Training in Discovery of Semantic Patterns. In: Proceedings of the 41st Annual Meeting of the ACL (2003)
10. Tanev, H., Oezden-Wennerberg, P.: Learning to Populate an Ontology of Violent Events. In: Perrotta, D., Piskorski, J., Soulie-Fogelman, F., Steinberger, R. (eds.) Mining Massive Data Sets for Security. IOS Press, Amsterdam (in print, 2008)
11. Piskorski, J.: ExPRESS Extraction Pattern Recognition Engine and Specification Suite. In: Proceedings of the International Workshop Finite-State Methods and Natural language Processing 2007 (FSMNLP 2007), Potsdam, Germany (2007)
12. Cunningham, H., Maynard, D., Tablan, V.: JAPE: a Java Annotation Patterns Engine. 2nd edn. Technical Report, CS–00–10, University of Sheffield, Department of Computer Science (2000)
13. Drożdżyński, W., Krieger, H.U., Piskorski, J., Schäfer, U., Xu, F.: Shallow Processing with Unification and Typed Feature Structures — Foundations and Applications. Künstliche Intelligenz 2004(1), 17–23 (2004)

Topics Identification Based on Event Sequence Using Co-occurrence Words

Kei Wakabayashi and Takao Miura

Dept.of Elect.& Elect. Engr., HOSEI University
3-7-2 KajinoCho, Koganei, Tokyo, 184–8584 Japan

Abstract. In this paper, we propose a sophisticated technique for topic identification of documents based on event sequences using co-occurrence words. Here we consider each document as an event sequence, each event as a verb and some words correlated with the verb. We propose a new method for topic classification of documents by using Markov stochastic model. We show some experimental results to examine the method.

Keywords: Topic Classification, Markov stochastic model.

1 Motivation

Recently there have been a lot of knowledge-based approaches for documents, and much attention have been paid on classification techniques of documents. Generally in the classification we assume each document is translated into a *vector* and compared *cosine-based* similarity with each other based on vector-space model.

However, document contents depend heavily on temporal aspects, such as accidents or affairs in news papers, and we can't identify them directly by vector-space modeling approach. This is because that we construct vectors by using set of words but not by *sequences* of words. It seems desirable to model documents based on sequences to classify them for these contents. In this investigation, we consider a sequence of news articles about a particular affair as one document. Our main purpose is targeted for *topic classification* of these affairs in documents.

Certainly one of the typical approach to tackle with the problem is *Topic Detection and Tracking*iTDTj[1]. In TDT, a topic is characterized as a sequence of *events*. By an event we mean an individual fact that can be identified in a spatio-temporal manner. Event tracking tasks in TDT contain classification of documents describing events. In this investigation, we consider *sequence of event* as one affair, and classify them based on event transition.

Here we propose a new and sophisticated technique for the classification of event sequence in documents by using Markov stochastic process. We have discussed HMM approach over a collection of news articles so far[5]. Generally, in each sentence in Japanese, the last verb can be seen as a predicate of the sentence. Using this property we can believe a fact that in the first paragraph in each news article, the last verb provides with a *summary of the article*. And

E. Kapetanios, V. Sugumaran, M. Spiliopoulou (Eds.): NLDB 2008, LNCS 5039, pp. 219–225, 2008.

we consider the verb as an output symbol in *Hidden Markov Model* (HMM). We have shown HMM plays as an excellent classifier to the documents. On the other hand, sometimes this causes oversimplified situation to describe events. In this investigation, we put our attention on co-occurrence words with verbs and consider them as characteristic features of events to classify news documents more precisely.

In Barzilay[2], they discuss how to model documents using stochastic process approach. They estimate structure of documents by HMM with 2-gram syntax as an output. Compared to the approach, we put our focus on estimating *topic structure* and there is no direct relationship. In another approach [4], they examine what's going on in specific documents (transcription of cooking courses on TV program) based on HMM. However they focus on single topic (cooking single cuisine) and no general discussion on concurrent multiple topics.

In this work, we discuss classification issue of topics based on event sequences in section 2. In section 3 we review Markov stochastic model and we develop the classification technique using the stochastic model in section 4. Section 5 contains experimental results. We conclude our discussion in section 6.

2   Identifying Event Sequences

First of all, let us describe how we classify event sequences (considered as affairs).

In this investigation, we define two topics are similar if two sequences of events are similar. For example, given two different murder cases in Tokyo and Kyoto, we can say they are similar although they occurred in distinct cities, at different times, by distinct offenders. This is because there happen similar event sequences, such as "*someone was killed*", "*suspecious person was wanted*" and "*police arrested offender*".

Let us illustrate our approach in a figure 1. We give a series of news articles describing a case of murder, and they are arranged in a temporal order. In the right side of the figure 1, we show a sequence of events in this affair. By going through the articles, we see an event that some guy was found dead. Then police examines and finds some hints about one suspicious person, and finally the person is arrested. These are all the events that constitute a topic of this *murder*.

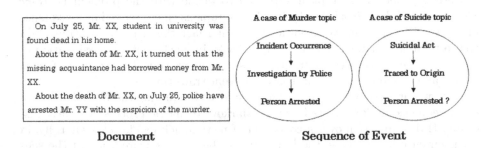

Document　　　　　　　**Sequence of Event**

Fig. 1. Identifying Topics from Documents

Clearly the event sequence depends on the characteristics of *murder* story. For instance, in a case of a suicidal act, we might have a different sequence: some guy was found dead but considered as a suicide, and the person is traced to the origin. In a case of suicide, the suspicious might commit suicide, but it is not common as a suicidal act since some other guy had been found dead. We can say that the two sequences are *not* similar with each other, and that a murder case carries its own pattern of an event sequence (or sequences).

Because every topic carries own pattern of event sequences, we believe we can estimate topics in documents and identify them.

3   Markov Stochastic Model

In this section we introduce Markov stochastic model to identify topics. A *stochastic automaton* is nothing but an automaton with output where both the state transition and the symbol output are defined in a probabilistic manner. This framework is similar to *Hidden Markov Model*, but in this investigation states are not hidden. We consider the states as verb, and the output symbols as co-occurrence words of verb in documents. For instance, looking at a sentence `police started a search`, we extract the verb `start` as a state, and the symbols `police` and `search` as output symbols.

Our stochastic model consists of (Q, Σ, A, B, π) defined below:

(1) $Q = \{q_1, \cdots, q_N\}$ is a finite set of states
(2) $\Sigma = \{o_1, \cdots, o_M\}$ is a family of output symbols where each o_t means a subset of V, i.e., $\Sigma \subseteq 2^V$.
(3) $A = \{a_{ij}, i, j = 1, ..., N\}$ is a state transion matrix of state transition where each a_{ij} means a probability of the transition at q_i to q_j. Note $a_{i1} + ... + a_{iN} = 1.0$.
(4) $B = \{b_i(o_t), i = 1, ..., N, t = 1, ..., M\}$ is a probability of outputs where $b_i(o_t)$ means a probability of an output o_t at a state q_i and $b_i(o_1) + ... + b_i(o_M) = 1.0$. Since o_t consists of co-occurrence words and a verb may appear with *some words* as co-occurrence, we should examine how to define $b_i(o_t)$. That is, when $o_t = \{o_{t,1}, o_{t,2}, \cdots, o_{t,|t|}\}$, we define $b_i(\{o_{t,1}, o_{t,2}, \cdots, o_{t,|t|}\}) = \prod_k b_i(o_{t,k})$ for simplicity.
(5) $\pi = \{\pi_i\}$ is an initial probability where π_i means a probability of initial state q_i.

Let us note that the probability matrix A shows the transition probability within a framework of simple Markov model, witch means a next verb, say `arrest`, arises in a probabilistic manner depending only on the current one, `start` for instance. Similarly the output symbols `police, search` appear depending only on the current verb `start`.

When we obtain models, we should obtain the transition probability matrix A, the output probability B and the initial probability π. In this approach, we assume *training data* for each topic and we take *supervised learning* to calculate the model $\mathcal{M}$. First, we extract sequence of states and output symbols from training data of the topic, then we count their frequencies and consider the relative values as the probabilities.

Given Markov stochastic model $\mathcal{M}$, we can calculate the probability of any sequence. Formally, given a test document consisting of a verb sequence $q = (q_1, q_2, \cdots, q_T)$ with the sequence of output symbols $o = (o_1, o_2, \cdots, o_T)$, then a Markov model $\mathcal{M}$ generates *likelihood*, i.e., the probability $p(q, o|\mathcal{M})$ defined as follows:

$$p(q, o|\mathcal{M}) = \pi_{q_1} b_{q_1}(o_1) \prod_{t=1}^{T-1} a_{q_t q_{t+1}} b_{q_{t+1}}(o_{t+1}).$$

4 Estimating Topics

In this section let us develop our theory to estimate topics in documents using Markov stochastic model. Basically we take a *Principle of Likelihood* (POL). That is, given a test sequence and Markov models $\mathcal{M}_1, ..., \mathcal{M}_{\mathcal{L}}$ according to topics respectively, we calculate the likelihoods and estimate a topic m of the maximum one. Here we describe the procedures for the estimation in detail.

Here we consider a document as a sequence of news articles concerning one topic. Generally it is said that every news article contains the most important and summarized content in its first paragraph. Thus we extract all the first paragraphs and put them together into one document in temporal order. We extract sequences of last verbs and the co-occurrence words (nouns) related to the verb as *event sequences* from documents as shown in figure 2.

But in Japanese, unfortunately, the relevant noun phrases can appear in any position *prior* to the verb. In this work, we examine the co-occurrence words by using specialized EDR Japanase Dictionary[3] where each item describe correlationship among words with POS tags and the semantics collected from many corpus. We have extracted only relationship between nouns and verbs. Figure 2 contains some example of the dictionary content in the right side.

Given a document D, let g be a function to generate $< q_1 q_2 \cdots q_n, o_1 o_2 \cdots o_n >$ to D where q_i, o_i mean an i-th state and a set of the co-occurrence words to q_i respectively. Let us see how we construct such a function g. To each sentence delimited by a punctuation mark such as a period, we obtain a sequence of words

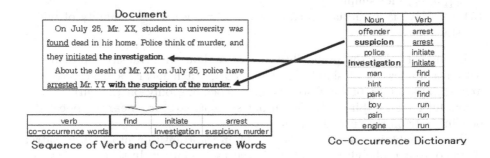

Fig. 2. Extracting Verbs and Co-Occurrence Words

by "Chasen", a morphological analysis. Then we remove all the sentences not in *past tense* in a document and extract the final verb v_t from each sentence remained as a state[1]. We skip sentences that are not in past tense because they don't capture *change of circumstances* but very often they have some forecast or perspective.

After obtaining a verb v_t, we extract all the co-occurrence words o_t with v_t. To do that, we extract the *all the noun phrases* prior to v_t within the sentence containing v_t, and we examine the *last noun* in each noun phrase to the EDR dictionary to see whether the noun can be really the co-occurrence word of v_t or not. If it is, we add all the nouns in the phrase to a set o_t of words as shown in figure 2. We apply g to all the sentences in D and we obtain a function $g(D) =< (q_1, q_2, \cdots, q_T, EOS), (o_1, o_2, \cdots, o_T, NONE) >$ where EOS means "End-Of-Sequence", $NONE$ a null set.

In the final step, we learn model parameters and discuss how to estimate topics. Given a topic c, we construct a Markov stochastic model $\mathcal{M}_c$ by using a set $\mathcal{D}_c$ of training documents of the topic. To each document D in $\mathcal{D}_c$, we obtain $g(D)$ described in the previous section. Then we count the total frequencies of the transitions of the states (the verbs), the co-occurrence words to estimate the probabilities.

To estimate a topic of a test document D, we apply POL to D. First, we obtain $g(D) = < q, o >$. To each topic c, we obtain the likelihood $p(q, o | \mathcal{M}_c)$ as described in the previous section. Then, to estimate the most likely class c_d, we apply Principle Of Likelihood (POL) as $c_d = argmax_c P(q, o | \mathcal{M}_c)$.

5 Experiments

5.1 Preliminaries

As a test corpus for our experiments, we examine all the articles in both Mainichi Newspaper 2001 and 2002 in Japanese and select 3 kinds of news topics, *one-man crime* (crimes by single or a few person such as murder and robbery), *organizational crime* (crimes by organization such as companies) and *corruption scandal*. We have extracted 256 documents in total by hands shown in a table 1. We have divided each of them into two groups, one for training and another for test.

As we said, we apply "Chasen" for morphological analysis to documents in Japanese. Also we take EDR Japanese dictionary of Co-Occurrence words[3].

5.2 Results

Let us show the results. First we show the structures of Markov stochastic models we have constructed and examine whether we can interpret them suitably or not.

We illustrate the topology of a topic *One-man Crime* and *Corruption Scandal* in a figure 3 where a circle means a state (verb) and an directed arrow between

---

[1] In Japanese, we can determine "past tense" easily by examining *auxiliary verbs*.

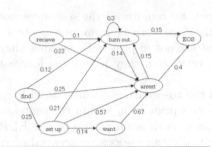

Verb	Co-occurrence Words
receive	emergency, call
find	dead, room, road
set up	Investigation, headquarter, police
want	suspicion, murder, robber, assault
turn out	investigation, confession, suspicion
arrest	suspicion, murder, abandon, police, robber

One–man Crime Model

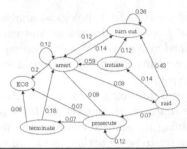

Verb	Co-occurrence Words
initiate	investigation, house, search, suspicion
raid	house, special, investigation, prosecutor
prosecute	incident, prosecutor, suspect, court, crime
terminate	disciplinary, action, incident
turn out	investigation, witness, confession, suspicion
arrest	suspicion, bribery, fraud, breach, obstruct

Corruption Scandal Model

Fig. 3. Model Structure

two states means the transition with the probability. There are output symbols (co-occurrence words) in tables with the probabilities.

In One-man Crime model, we see the high probability from **arrest** to *EOS*, i.e., one-man crimes are solved by the arrests of criminals. It is possible to say that this situations are distinctive of this topic. In Corruption Scandal model, we have characteristic words of co-occurrence to each state compared to other topics. For example, at the state **arrest**, we have co-occurrence words **bribery**, **fraud** and **treach**. In Organizational Crime model, we see the high probability of the path from **arrest** to itself. This is because there are more than one persons arrested very often in this topic.

Let us show all the results of classification in a table 1. Here we get the correctness ratio 64.3% in total.

Then let us discuss how we can think about our experimental results. First of all, we get the better result to One-man Crime case. In these documents, the cases are generally solved with the arrestment since there is only one criminal. This is the main characteristic of one-man crime and we see why we get the good classification ratio.

We have discussed a topic of *Corruption Scandal* so far[5] where we have examined only verbs and we got 45.5% of the classification ratio, while we have improved the classification 63.6% in this work. This comes mainly from a fact

Table 1. Test Corpus

Topic	Training Documents	Test Documents	Correct Answers	Correctness ratio (%)
One-man Crime	91	45	32	71.1
Organizational Crime	35	17	8	47.1
Corruption Scandal	46	22	14	63.6
(total)	256	84	54	64.3

that we have examined co-occurrence words. In fact, *Corruption Scandal* topic has many characteristic words of co-occurrence compared to other topics.

We see the lowest classification ratio of a topic of Organizational Crime. This is because a set of the states and a set of the co-occurrence words are similar to One-man Crime but there exist a wide variety patterns of state transitions. However, we get the classification ratio 47.1%. Although this is not good, the algorithm doesn't depend on co-occurrence words and is completely different from the conventional ones.

6 Conclusion

In this investigation, we have proposed a new approach to classify sequences of news articles in Japanese by using Markov stochastic model. We have also proposed how to examine nouns as well as verbs for the classification. And we have shown how well the approach works to sequences of news articles by experimental results.

References

1. Allan, J., Carbonell, J., Doddington, G., Yamron, J., Yang, Y.: Topic Detection and Tracking Pilot Study: Final Report. In: proc. DARPA Broadcast News Transcription and Understanding Workshop (1998)
2. Barzilay, R., Lee, L.: Catching the Drift: Probabilistic Content Models, with Applications to Generation and Summarization. In: Proceedings of the NAACL/HLT, pp. 113–120 (2004)
3. Japan Electronic Dictionary Research Institute (EDR): Japanese Co-occurrence Dictionary, http://www2.nict.go.jp/r/r312/EDR/J_index.html
4. Shibata, T., Kurohashi, S.: Unsupervised Topic Identification by Integrating Linguistics and Visual Information Based on Hidden Markov Model, COLING (2005)
5. Wakabayashi, K., Miura, T.: Identifying Event Sequences using Hidden Markov Model. In: Kedad, Z., Lammari, N., Métais, E., Meziane, F., Rezgui, Y. (eds.) NLDB 2007. LNCS, vol. 4592, pp. 84–95. Springer, Heidelberg (2007)

Topic Development Based Refinement of Audio-Segmented Television News

Alfredo Favenza[1], Mario Cataldi[1], Maria Luisa Sapino[1], and Alberto Messina[2]

[1] Universita' di Torino, Italy
alfredofa@tiscali.it, {cataldi,mlsapino}@di.unito.it
[2] RAI - Centre for Research and Technological Innovation, Torino, Italy
a.messina@rai.it

Abstract. With the advent of the cable based television model, there is an emerging requirement for random access capabilities, from a variety of media channels, such as smart terminals and Internet. Random access to the information within a newscast program requires appropriate segmentation of the news. We present text analysis based techniques on the transcript of the news, to refine the automatic audio-visual segmentation. We present the effectiveness of applying the text segmentation algorithm CUTS to the news segmentation domain. We propose two extensions to the algorithm, and show their impacts through an initial evaluation.

1 Introduction and Related Work

For television program repurposing applications, it is important to be able to combine and reuse fragments of existing programs. In order to identify and combine fragments of interest, the editing author needs instruments to query, and efficiently retrieve relevant fragments. Text analysis techniques can support and refine the segmentation results returned by the audio/visual processing methods. Video segmentation has been a hot research topic since many years [4]. The common goal of most technical approaches is to define automatic techniques able to detect the *editorial parts* of a content item, i.e. modal-temporal segments of the object representing a semantic consistent part *from the perspective of its author* [8]. Very recent approaches achieve this goal with interesting results [3]. Significant results in the area of Digital Libraries have been achieved by the partners of the Delos European project [1].

In this paper, we address the problem of segmenting the television news in self contained fragments, to identify the points in which there is a significant change in the topic of discussion. Different units will be indexed for efficient query retrieval and reuse of the digital material. We experiment our method on data available in the archives of RAI CRIT (Center for Research and Technological Innovation) in Torino. In particular, we analyze the transcript of the news, with the goal of combining the resulting text-based segments with the ones extracted by algorithms based on the visual features of the news programs.

In the next sections we first describe the text segmentation algorithm, *CUTS (CUrvature based development pattern analysis and segmentation blogs and other Text Streams)* [7], which focusses on the segmentation of texts of different nature (originally, blog files) with the guidance of the information about how the addressed topics evolve along time.

E. Kapetanios, V. Sugumaran, M. Spiliopoulou (Eds.): NLDB 2008, LNCS 5039, pp. 226–232, 2008.
© Springer-Verlag Berlin Heidelberg 2008

Our choice of CUTS, as opposed to text-tiling [5], or to the approaches based on the detection of minima in text similarity curves [6,2] is based on the specific application domain we are dealing with. We apply CUTS on the spoken text, as it is extracted by an Automatic Speech Recognition engine. We then extend the original CUTS method to take into account multi-dimensional curves, and the actual temporal duration of the entries. We discuss our initial experimental results, which show that the extended versions improve on the precision of the segmentation technique within the context of news transcript segmentation.

2 Background: CUTS Algorithm

In [7], authors propose CUTS, curvature based development pattern analysis and segmentation for blogs and other text streams. Given a sequences of ordered blog entries, CUTS captures the topic development patterns in the sequence, and identifies coherent segments as opposed to sequences in which the discussion is smoothly drifting from one topic to another, or a main subject is interrupted. There are three main phases in CUTS algorithm.

(i) First, the sequence of entries is analyzed, and a representative surrogate (a keyword vector) is generated for each entry. The N text entries are represented in the standard TF/IDF form. As usual, the surrogate generation includes a preliminary phase of stop word elimination and stemming. Each vector component, $w_{k,j}$ represents the weight of the $j - th$ term in the vocabulary w.r.t the $k - th$ entry. Weights are at the basis of the entry-similarity evaluation:

$$s_{i,j} = \sum_{k=1}^{n} w_{i,k} . w_{j,k}$$

The topic evolution is actually computed by referring to the *dissimilarity* among the entries. Pairwise entries dissimilarity is stored in a dissimilarity matrix, D, whose elements $D_{i,j} = 1 - s_{i,j}$, represent the degree of dissimilarity between the $i - th$ and the $j - th$ entries.

(ii) The sequence of entries is then mapped onto a curve, which highlights the development patterns in terms of the (dis)similarity between adjacent entries. CUTS maps the initial data in a new, 1-dimensional space by applying multidimensional scaling algorithm [9]. The mapping preserves to the best the distances between the points. A second dimension plays the role of a temporal dimension[1]. In the resulting space, the consecutive entries form a curve (referred to as the CUTS-curve). By analyzing this curve, in [7] the authors identify different development patterns, illustrated in figure 1. *Dominated* segments are characterized by a sort of "stability" of the topic, which in the CUTS curve, corresponds to an almost horizontal segment. In *Drifting* patterns a smooth transition from a topic to another is observed. As the CUTS curve reflects the differences between consecutive entries, this topic development reflects in a slope in a

---

[1] In section 3.2 we will show how the final segmentation can benefit from a different dimensional choice.

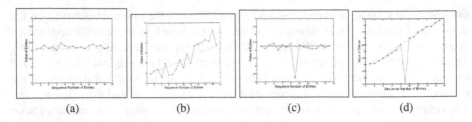

Fig. 1. Example of (a) dominated curve segment (b) drifting curve segment (c) interrupted dominated curve segment and (d) interrupted drifting curve segment

sloping the curve. The slop of the curve measures how fast one topic evolves in the following one. *Interrupted* patterns are detected when a sudden, significant change occurs. An interruption is seen as a new topic, which is addressed for a very short amount of time (w.r.t. the duration of the other topics). Interruptions may occur both within dominated and within drifting segments. In audio-segmented text, due to low quality of segmentation, interrupts are common and need to be recognized as such for effective news topic identification. This is one of the reasons we chose CUTS over other schemes.

(iii) The pattern development curve is then analyzed, and the topic segments, as reflected by the changes in the slopes of the curves, are identified. The algorithm introduces an iterative method to represent the curve with a series of straight line segments, each denoted by a $4 - tuple\ g_i = (k_i, \sigma_i, (x_s, y_s)_i, (x_e, y_e)_i)$, where (i) k_i is the slope of the segment, and reflects the speed of the topic change in the segment; (ii) σ_i measures the concentration of the entries around the dominant development pattern, and reflects how well the segment approximates the original curve. Good approximation reflects in a small value for σ_i, which is the average of the distances from the original points to the line segment; (iii) $(x_s, y_s)_i$, and $(x_e, y_e)_i$ denote the first and the last points of the segment, respectively.

The decision on whether two consecutive topic segments should be combined in the same topic segment is based on the comparison of their corresponding curve parameters, to measure how homogeneous the segments are in terms of their slope and concentration. Two segments are considered homogeneous (and thus combinable) if

$$| k_i - k_{i+1} | < \lambda_{drifting} \text{ and } | \sigma_i - \sigma_{i+1} | < (\sigma_i + \sigma_{i+1})/2$$

The new parameter $\lambda_{drifting}$ is a threshold to determine when two different topic evolution speed can be considered as homogeneous. It is described in terms of the difference of slopes of the curve segments. After two curve segments are combined, the k and σ values of the resulting segment are calculated. Then the combination process is iteratively repeated.

The annotation process associates each base topic segment with a label characterizing its topic development pattern as *dominated*, *drifting*, or *interrupted*.

3   Extending CUTS for News Segmentation

Using a naive application of CUTS to the domain of the news we had misalignment in terms of information loss due to the choice of (i) the dimensionality reduction, scaling the initial multidimensional domain to a monodimensional one, and (ii) the uniform treatment of the duration of the entries.

To address these issues, we propose two extensions to the original CUTS algorithm. The first one (Section 3.1) extends the method to the 3-dimensional case, in which the first two dimensions are returned by the multidimensional scaling method, and the third one is - as in the original method - the positional temporal dimension. The second extension (Section 3.2) takes into account the actual duration of the entries.Finally, we discuss the impact of the combination of the two extensions, returning a 3-dimensional, temporal approach.

3.1 Multi-dimensional CUTS

In our first extension, *the sequence of news entries is represented by curves in a 3-dimensional space*. The x-axis will be associated with the temporal dimension (mapping consecutive entries to consecutive unit time instants). The y and z axes are associated to the dimensions returned by the 2-dimensional scaling. While the main idea of the CUTS algorithm remains unchanged, the extension affects the way distances, breakpoints, drifting, and all the parameters (segments slopes, segment composition conditions) needed to analyze the 3-dimensional curves, are defined.

In the original CUTS method [7], the choice of the break points to identify segments is based on the distance between pairs of points and lines, both of them in the two dimensional space. A similar strategy is adopted in the 3-dimensional space, of course by referring to the appropriate methods to evaluate the distances between a point and a segment in the 3-dimensional space.

Generalization of curve model to n-dimensional space. The drifting condition can be expressed as a condition on the difference between two consecutive entries of $J_{\hat{S}(t)}$, i.e. the finite difference of the Jacobian vector of the multidimensional curve.

$$\parallel \Delta J_{\hat{S}(t)} \parallel^2 \leq \Lambda^2 \tag{1}$$

being Λ the discriminating threshold. Under the hypothesis of unit duration associated to the entries, the intervals of duration of two consecutive entries are $[i, i+1]$, and $[i+1, i+2]$, respectively, and their duration is, by hypothesis, exactly 1. Thus

$$\parallel \Delta J_{\hat{S}(t)} \parallel^2 = \sum_{i=0}^{n} [(d_i(i+2) - d_i(i+1)) - (d_i(i+1) - d_i(i))]^2$$

In general, a possible way to estimate the value for the parameter Λ, when dealing with two dimensions, is to compute the global variation of the similarity along the two corresponding axes. For any dimension d_i, we will have $\lambda_i = \alpha_i \frac{\delta d_i}{\delta t}$, where δd_i is the excursion of d_i, δt is the corresponding excursion of t, and α_i is the weight of the variation along the $i - th$ direction. The above condition (1) is thus rewritten as

$$\parallel \Delta J_{\hat{S}(t)} \parallel^2 \leq \sum_{i=0}^{n} \lambda_i^2$$

Table 1. Automatic monodimensional and two-dimensional temporal CUTS based annotation

Seg	Base S.	Sec.	Ann	Seg	Base S.	Sec.	Ann	Seg	Base S.	Sec.	Ann
1	0-3	0-34	drif	12	60-62	892-921	dom	1	0-3	0-34	drif
2	3-4	34-61	drif	13	62-63	921-937	dom	2	3-4	34-61	drif
3	4-26	61-383	dom	14	63-73	937-1064	dom	3	4-53	61-773	dom
4	26-28	383-415	dom	15	73-86	1064-1269	dom	4	53-59	773-872	inter
5	28-37	415-502	drif	16	86-98	1269-1469	dom	5	59-108	872-1595	dom
6	37-38	502-517	dom	17	98-100	1469-1503	dom	6	108-109	1595-1614	drif
7	38-52	517-755	dom	18	100-103	1503-1535	dom	7	109-110	1614-1632	drif
8	52-53	755-773	dom	19	103-112	1535-1669	dom	8	110-120	1632-1806	dom
9	53-54	773-780	drif	20	112-119	1669-1794	dom				
10	54-59	780-872	drif	21	119-120	1794-1806	dom				
11	59-60	872-892	dom								

Notice that this is a general formulation of the drifting condition, which, in the case $n = 1$, and entries of duration 1, reduces to the condition applied by the original CUTS,

$$[(d_1(i + 2) - d_1(i + 1)) - (d_1(i + 1) - d_1(i))]^2 \leq \alpha \delta d_1$$

that is,

$$\mid k_{i+1} - k_i \mid \leq \lambda_{drifting}$$

3.2 Extensions to the Temporal Dimension

Our input text fragments also contain information about their actual temporal duration within the news programme.

Technically, taking into account the temporal dimension only requires the actual instantiations of the time intervals appearing in the general definition of the function $\parallel \Delta J_{\hat{S}(t)} \parallel^2$ in Section 3.1. In fact, in Section 3.1 we introduced simplified expressions for the function, reflecting the fact that all the entries were associated to a time unit duration. In this case, denoting with $[st_i, et_i]$ and $[st_{i+1}, et_{i+1}]$ the time intervals associated to the $i - th$ and the $(i + 1) - th$ entries respectively[2], the corresponding expression for $\parallel \Delta J_{\hat{S}(t)} \parallel^2$ becomes

$$\parallel \Delta J_{\hat{S}(t)} \parallel^2 = \sum_{i=0}^{n} \left(\frac{d_i(et_{i+1}) - d_i(st_{i+1})}{et_{i+1} - st_{i+1}} - \frac{d_i(et_i) - d_i(st_i)}{et_i - st_i} \right)^2$$

Table 1 reports the results of the segmentation and annotation methods discussed in the previous sections, extended with the temporal dimension.

We notice that in both cases we see that extensions improve the qualities of the results, giving a closer approximation of the human domain expert classification.

4 Analysis of Results

To evaluate the methods, we measure the coherence of the automatically detected topic segment boundaries (S_1) wrt, the segment boundaries in the ground truth (S), i.e., a

---

[2] In this specific application it holds that $et_i = st_{i+1}$.

segmentation given by a human domain expert. We define *precision* as the ratio of entry numbers from the automatically extracted sequence which are common boarder list ($Prec = (entry(S_1) \cap entry(S))/(entry(S_1))$), and recall as the ratio of entries which delimit topic segments in the ground truth, and are also returned by the automatic system ($Rec = (entry(S_1) \cap entry(S))/(entry(S))$. We also notice that the above measure penalizes the cases in which the disagreement between the ground truth and the returned sequence of topic segments is small (for example, the cases in which there is a one entry displacement between the automatically detected topics and the ground truth). To take into account the displacement, we define an embedding procedure which aligns the two topic sequences to be compared, and we compute the alignment cost as follows: $Align\_cost = d + \sum_i((s_i - 1) + (s_i \times da_i))$, where, d is the summation of the differences between the i-th element, $e1_i$ of the shorter sequence and the closer element, $e2_i$ appearing in the other sequence of entries, s_i is the number of segments between between $e2_i$ and $e2_{i-1}$, i.e., the number of segments in the second sequence that all together correspond to a single segment in the first one, and da_i captures the highest disagreement in the annotation between the $i - th$ segment in the first sequence and any of the corresponding s_i segments, computed by measuring the positional distance between the annotations, wrt.the ordering $dominated \prec drifting \prec interrupted$.

Table 2 shows the results on the example presented in the previous sections.

Table 2. Statistical evalutation of results

	2MDS	2MDS + Time	1MDS	1MDS + Time
Precision	0.681818	0.709415	0.271916	0.295726
Recall	0.616666	0.645833	0.205349	0.270398
Alignment cost	28	17	67	56

5 Conclusions and Future Work

We have presented a text analysis based technique on the transcript of the news, to refine the automatic audio-visual segmentation. In particular, we have discussed and evaluated the effectiveness of applying the text segmentation algorithm CUTS [7], and our two extensions of it, to the news segmentation domain. The initial evaluation results are consistent with our expectations. We are now conducting intensive user studies to evaluate the methods on different data and users samples. We are also working on the integration of the text based segmentation results with the segments extracted through audio-visual methods.

References

1. Delos network of excellence on digital libraries. Internet Site,
   http://www.delos.info/
2. Andrews, P.: Semantic topic extraction and segmentation for efficient document visualization. Master's thesis, School of Computer & Communication Sciences, Swiss Federal Institute of Technology, Lausanne (2004)

3. Venkatesh, S., Phung, D.-Q., Duong, T.-V., Bui, H.H.: Topic transition detection using hierarchical hidden markov and semimarkov models. In: Proc. of ACM Multimedia 2005 (2005)
4. Smeaton, A., Lee, H., O'Connor, N.E.: User evaluation of físchlár-news: An automatic broadcast news delivery system. ACM Transactions on Information Systems 24(2), 145–189 (2006)
5. Hearst, M.A.: Texttiling: Segmenting text into multi-paragraph subtopic passages. Computational Linguistics 23(1), 33–64 (1997)
6. Hearst, M.A., Plaunt, C.: Subtopic structuring for full-length document access. In: Proc. of SIGIR (1993)
7. Qi, Y., Candan, K.-S.: Cuts: Curvature-based development pattern analysis and segmentation for blogs and other text streams. In: Proc. of Hypertext 2006 (2006)
8. Snoek, C.G., Worring, M.: Multimodal video indexing: A review of the state-of-the-art. In: Proc. Multimedia Tools and Applications, pp. 5–35 (2005)
9. Torgerson, S.: Multidimensional scaling: I. theory and method. Psychometrika 17(4) (1952)

Text Entailment for Logical Segmentation and Summarization

Doina Tatar, Andreea Diana Mihis, and Dana Lupsa

University "Babes-Bolyai"
Cluj-Napoca
Romania
{dtatar,mihis,dana}@cs.ubbcluj.ro

Abstract. Summarization is the process of condensing a source text into a shorter version preserving its information content ([2]). This paper presents some original methods for text summarization by extraction of a single source document based on a particular intuition which is not explored till now: the logical structure of a text. The summarization relies on an original linear segmentation algorithm which we denote logical segmentation (LTT) because the score of a sentence is the number of sentences of the text which are entailed by it.

The summary is obtained by three methods: selecting the first sentence(s) from a segment, selecting the best scored sentence(s) from a segment and selecting the most informative sentence(s) (relative to the previously selected) from a segment. Moreover, our methods permit dynamically adjusting the derived summary size, independently of the number of segments.

Alternatively, a Dynamic Programming (DP) method, based on the continuity principle and applied to the sentences logically scored as above is proposed. This method proceeds by obtaining the summary firstly and then determining the segments.

Our methods of segmentation are applied and evaluated against the segmentation of the text "I spent the first 19 years" of Morris and Hirst ([17]). The original text is reproduced at [26]. Some statistics about the informativeness of the summaries with different lengths and obtained with the above methods relatively to the original (summarized) text are given. These statistics prove that the segmentation preceding the summarization could improve the quality of obtained summaries.

1 Introduction

Text summarization has become the subject of an intense research in last years and it is still an emerging field. The research is done in the extracts and abstracts areas. The extracts are the summaries created by reusing portion of the input verbatim, while the abstracts are created by regenerating the extracted content ([13]). However, there does not exist a theory of construction of a good summary.

E. Kapetanios, V. Sugumaran, M. Spiliopoulou (Eds.): NLDB 2008, LNCS 5039, pp. 233–244, 2008.

The most important task of summarization is to identify the most informative (salient) parts of a text comparatively with the rest. Usually the salient parts are determined on the following assumptions [16]: they contain words that are used frequently; they contain words that are used in the title and headings; they are located at the beginning or end of sections; they use key phrases which emphasize the importance in text; they are the most highly connected with the other parts of text. If the first four characteristics are easy to achieve and verify, the last one is more difficult to establish. For example, the connectedness may be measured by the number of shared words, synonyms, anaphora [18],[20]. On the other hand, if the last assumption is fulfilled, the cohesion of the resulting summary is expected to be higher than if this one is missing. In this respect, our methods assure a high cohesion for the obtained summary while the connectedness is measured by the number of logic entailments between sentences.

Another idea which our work proposes is that a process of logical segmentation before the summarization could be benefic for the quality of summaries. This feature is a new one comparing with [23].

We propose here a method of text segmentation with a high precision and recall (as compared with the human performance). The method is called logical segmentation because the score of a sentence is the number of sentences of the text which are entailed by it. The scores form a structure which indicates how the most important sentences alternate with ones less important and organizes the text according to its logical content. Due to some similarities with TextTiling algorithm for topic shifts detection of Hearst ([12,11]) we call this method Logical TextTiling (LTT). The drawback of LTT is that the number of the segments is fixed for a given text and it results from its logical structure. In this respect we present in this work a method to dynamically correlate the number of the logical segments obtained by LTT with the required length of the summary. The method is applied to the Morris and Hirst's text in [17] and the results are compared with their structure obtained by the Lexical Chains.

The methods presented in this paper are fully implemented in Java and C++. We used our own systems of Text Entailment verification, LTT segmentation, summarization with Sum_i and AL methods. For DP algorithm we used a part of OpenNLP ([25,24]) tools to identify the sentences and the tokens in the text. OpenNLP defines a set of Java interfaces and implements some basic infrastructure for NLP components. The DP algorithm verifies the continuity principle by verifying that the set of common words between two sentences is not empty.

The paper is structured as follows: in Section 2, some notions about textual entailment and logical segmentation of discourse by Logical TextTilling method are discussed. Summarization by logical segmentation with an arbitrarily length of the summary (algorithm AL) is the topic of Section 3. An original DP method for summarization and segmentation is presented in Section 4. The application of all the above methods to the text from [17] and some statistics of the results are presented in Section 5. We finish the article with conclusions and possible further work directions.

2   Segmentation by Logical TextTiling

2.1   Text Entailment

Text entailment is an autonomous field of Natural Language Processing and it represents the subject of some recent Pascal Challenges ([7]). As we established in an earlier paper ([22]), a text T entails a hypothesis H, denoted by $T \to H$, iff H is less informative than T. A method to prove $T \to H$ is presented in ([22]) and consists in the verification of the relation: $sim(T, H)_T \leq sim(T, H)_H$. Here $sim(T, H)_T$ and $sim(T, H)_H$ are text-to-text similarities introduced in [6]. Another method in [22] calculates the similarity between T and H by *cosine* and verify that $cos(T, H)_T \leq cos(T, H)_H$. Our system of text entailment verification used in this paper relies on this second method for text entailment verification as a directional relation. The reason is that it provides the best results compared with the other methods in [22].

2.2   Logical Segmentation

A segment is a contiguous piece of text (a sequence of sentences) that is linked internally but disconnected from the adjacent text. The local cohesion of a logical segment is assured by the move of relevance information from the less important to the most important and again to the less important. Similarly with the topic segmentation ([10,12,11]), in logic segmentation we concentrate on describing what is a shift in relevance information of the discourse. Simply, a valley (a **local minim**) in the obtained logical structure is a boundary between two logical segments (see Section 5, Fig.1). This is an accordance with the definition of a boundary as a perceptible discontinuity in the text structure ([4]), in our case a perceptible discontinuity in the connectedness between sentences.

The main differences between TextTiling (TT) and our Logical TextTiling (LTT) method are presented in [23]. The most important consist in the way of scoring and grouping the sentences.

As in [17] is stated, cohesion relations are relations among elements in a text (references, conjunctions, lexical cohesion) and coherence relations are relations between clauses and sentences. From this point of view, the logical segmentation is an indicator of the coherence of a text.

The algorithm LTT of segmentation has the following function:

INPUT: A list of sentences $S_1, ..., S_n$ and a list of scores $score(S_1), ... score(S_n)$
OUTPUT: A list of segments $Seg_1, ...Seg_N$ (see Section 5)

The obtained logical segments could be used effectively in summarization. In this respect, the method of summarization falls in the discourse-based category. In contrast with other theories about discourse segmentation, as Rhetorical Structure Theory (RST) of Mann and Thompson (1988), attentional / intentional structure of Grosz and Sidner (1986) or parsed RST tree of Daniel Marcu (1996), our Logical TextTiling method (and also TextTiling method [10]) supposes a linear segmentation (versus hierarchic segmentation) which results in an advantage from a computational viewpoint.

3 Summarization by Segmentation

3.1 Scoring the Segments

Given a set of N segments we need a criterion to select those sentences from a segment which will be introduced in the summary. Thus, after the score of a sentence is calculated, we calculate a score of a segment. The final score, $Score_{final}$, of a sentence is weighted by the score of the segment which contains it. The summary is generated by selecting from each segment a number of sentences proportional with the score of the segment. The method has some advantages when a desired level of granularity of summarization is imposed.

The denotations for this calculus are:

- $Score(S_i) =$ *the number of sentences implied by* S_i
- $Score(Seg_j) = \frac{\sum_{S_i \in Seg_j} Score(S_i)}{|Seg_j|}$
- $Score_{final}(S_i) = Score(S_i) \times Score(Seg_j)$ *where* $S_i \in Seg_j$
- $Score(Text) = \sum_{j=1}^{N} Score(Seg_j)$
- *Weight of a segment* : $C_j = \frac{Score(Seg_j)}{Score(Text)}, 0 < C_j < 1$
- $N=$*the number of segments obtained by LTT algorithm*
- $X=$ *the desired length of the summary*
- $NSenSeg_j =$*the number of sentences selected from the segment* Seg_j

The summarization algorithm *with Arbitrarily Length of the summary* (AL) is the following:

INPUT: The segments $Seg_1, ...Seg_N$, the length of summary X (as parameter), $Score_{final}(S_i)$ for each sentence S_i ;

OUTPUT: A summary SUM of length X, where from each segment Seg_j are selected $NSenSeg_j$ sentences. The method of selecting the sentences is given by definitions $Sum1, Sum2, Sum3$ (section 3.2).

Calculate the weights of segments (C_j), rank them in an decreased order, rank the segments Seg_j accordingly.

Calculate $NSenSeg_j$: while the number of selected sentences does not equal X, calculate $NSenSeg_j = min(| Segment_j |, Integer(X \times C_j))$, if $Integer(X \times C_j)) \geq 1$ or $NSenSeg_j = 1$ if $Integer(X \times C_j) < 1$.

Reorder again the selected sentences as in initial text.

Remarks: A number of segments Seg_j may have $NSenSeg_j > 1$. If $X < N$ then a number of segments Seg_j must have $NSenSeg_j = 0$

3.2 Strategies for Summary Calculus

The method of extracting sentences from the segments is decisive for quality of the summary. The deletion of an arbitrary amount of source material between two sentences which are adjacent in the summary has the potential of losing essential information. We propose and compare some simple strategies for efficiently including sentences in the summary.

Our first strategy is to include in the summary the first sentence from each segment, as this is of special importance for this segment. The corresponding method is:

Definition 1
Given a segmentation of initial text, $T = \{Seg_1, \cdots, Seg_N\}$ the summary is calculated as:

$$Sum_1 = \{S'_1, \cdots, S'_X\}$$

where, for each segment $Seg_i, i = 1, \cdots, N$, first $NSenSeg_i$ sentences are selected.

The second way is that for each segment the sentence(s) which imply a maximal number of other sentences are considered the most important for this segment, and hence they are included in the summary. The corresponding method is:

Definition 2
Given a segmentation of initial text, $T = \{Seg_1, \cdots, Seg_N\}$ the summary is calculated as:

$$Sum_2 = \{S'_1, \cdots, S'_X\}$$

where $S'_k = argmax_j\{score(S_j) \mid S_j \in Seg_i\}$. For each segment Seg_i a number of $NSenSeg_i$ of different sentences S'_k are selected.

The third way of reasoning is that from each segment the most informative sentence(s) (the least similar) relative to the previously selected sentences are picked up. The corresponding method is:

Definition 3
Given a segmentation of initial text, $T = \{Seg_1, \cdots, Seg_N\}$ the summary is calculated as:

$$Sum_3 = \{S'_1, \cdots, S'_X\}$$

where $S'_k = argmin_j\{sim(S_j, Seg_{i-1}) \mid S_j \in Seg_i\}$. Again, for each segment Seg_i a number of $NSenSeg_i$ of different sentences S'_k are selected. The value $sim(S_j, Seg_{i-1})$ represents the similarity between S_j and the last sentence selected in Seg_{i-1}. The similarity between two sentences S, S' is calculated in this work by $cos(S, S')$.

4 Dynamic Programming Algorithm

4.1 Summarization by Dynamic Programming

In order to meet the coherence of the summary our method selects the chain of sentences with the property that two consecutive sentences have at least one common word. This corresponds to the continuity principle in the centering theory which requires that two consecutive units of discourse have at least one entity in common ([18]) . We make the assumption that each sentence is associated with a score that reflects how representative is that sentence. In these experiments the scores are the logical scores as used in the previous sections. The score of a selected summary is the sum of individual scores of contained sentences. The summary will be selected such that its score is maximum.

The idea of the Dynamic Programming Algorithm for summarization is the following: let us consider that δ_i^k is the score of the best summary with length k that begins with the sentence S_i. This score will be calculated as:

$$\delta_i^k = max_{j\ (i<j)} \begin{cases} score(S_i) + (\delta_j^{k-1}) & \text{if } S_i, S_j \text{ have common words} \\ penalty * (score(S_i) + (\delta_j^{k-1})) \text{ otherwise} \end{cases}$$

When running the algorithm we used a penalty with value equal to 1/10.

INPUT:
- the text to be summarized, that is a sequence of sentences $S_1, ..., S_n$;
- $score(S_i)$, the scores associated to each sentence;
- the length X of the summary.
OUTPUT: A summary SUM of length X

A detailed version of this algorithm for summarization occurred in [23].

4.2   Segmentation by Dynamic Programming

A sentence is chosen in the summary if its score is high and satisfy the continuity principle. On the contrary, segment boundaries are sentences that have minimal scores. More exactly, the algorithm considers as boundary sentences of a segment those that have minimal scores and are situated between two consecutive sentences chosen for summary. Consequently, the text will have a number of segments correlated with the length X of the summary. This is not a constraint for the number of segments because the length of a summary is an input parameter of the summarization method. The algorithm of segmentation is:

INPUT:
-the text to be segmented, that is a sequence of sentences $S_1, ..., S_n$ and their scores.
-the set of sentences in a Summary by their positions in text: $P_1, ..., P_X$, where X is the length of the summary.
OUTPUT: $Seg = Seg_1, ..., Seg_{nrseg}$ - the list of segments

```
nrseg = 1
begpoz = 1
for i = 1 to X − 1 do
        endpoz = argmin_{j,j∈{P_i,P_{i+1}−1}} score(S_j)
        *Seg_{nrseg} is from the position begpoz to the position endpoz
        nrseg = nrseg + 1
        begpoz = endpoz + 1
endfor
```

An improvement of the segmentation method is the idea that an important segment in a document can contain a chain of consecutive sentences chosen for the summary. Similar as before, the boundaries of a segment are sentences that have minimal scores and are situated between two consecutive sentences chosen for summary, with the supplementary condition that they are not consecutive in text.

5 Experiments

This section offers a framework for lighting the question about the relationship between segmentation and summarization. As the experiment proves, the segmentation brings benefits to the summarization (see Section 5.3).

We have applied our algorithms of segmentation and summarization to the narrative literary text from [17] reproduced at [26]. Let us remark that the text in [17] and the text considered in our experiment differ by three sentences, eliminated by us because they are too short (five words or less). These are the sentences 24, 38, 39. The remained 41 sentences are renumbered.

The structure of logically scored sentences for the above text is presented in Fig.1.

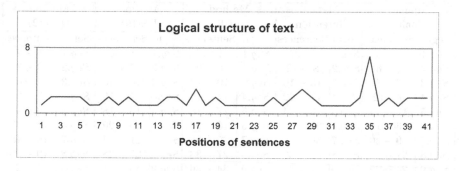

Fig. 1. The logical structure of the text

5.1 Evaluation of Segmentation

There are several ways to evaluate a segmentation algorithm. These include comparing the segments against that of some human judges, comparing the segments against other automated segmentation strategies and, finally, studying how well the results improve a computational task [18]. We will use all these ways of evaluation, including the study how our segmentation method effect the outcome of summarization.

Regarding the comparison with the human judge, the method is evaluated according to the precision and the recall: precision is the percentage of the boundaries identified by the algorithm that are indeed true boundaries, the recall is the percentage of the true boundaries that are identified by the algorithm ([19]).

A problem with the precision and the recall is that they are not sensitive to "near misses" ([11]): if two algorithms fail to match any boundary precisely, but one is close to correct, whereas another is entirely off, both algorithms would receive scores 0 for precision and recall. This is the reason we introduce a modified version for precision and recall. Namely, we count:

- how many of the same (or very close) boundaries with the human judge the method selects out of the total selected (precision);
- how many true (or very close) boundaries are found out of the total possible (recall)

In Table 1 the result of LTT segmentation with 11 segments and the results of DP segmentation with 9 and 12 segments are presented. Also the Lexical Chains structure (12 segments) and the Intentional structure (9 segments) given in [17] are reproduced. Let us remark that the hierarchical structure discovered by the authors in [17] is transformed in a linear structure by considering the smallest intervals.

Table 1. The result of LTT segmentation and the results of DP segmentation with 9 and 12 segments against Manual Lexical Chains and Manual Intentional variants

Manual Chains		Manual Intention		Method LLT		Method DP (9)		Method DP (12)	
Segm	Sentences	Segm	Sentences	Segm	Sentences	Segm	Sentences	Segm	Sentences
$Seg.1$	$1-3$	$Seg.1$	$1-7$	$Seg.1$	$1-7$	$Seg.1$	$1-6$	$Seg.1$	$1-2$
$Seg.2$	$4-6$	$Seg.2$	$8-12$	$Seg.2$	$8-9$	$Seg.2$	$7-9$	$Seg.2$	$3-3$
$Seg.3$	$7-8$	$Seg.3$	$13-16$	$Seg.3$	$10-13$	$Seg.3$	$10-13$	$Seg.3$	$4-6$
$Seg.4$	$9-12$	$Seg.4$	$17-22$	$Seg.4$	$14-16$	$Seg.4$	$14-15$	$Seg.4$	$7-9$
$Seg.5$	$13-15$	$Seg.5$	$23-24$	$Seg.5$	$17-18$	$Seg.5$	$16-18$	$Seg.5$	$10-13$
$Seg.6$	$16-18$	$Seg.6$	$25-29$	$Seg.6$	$19-24$	$Seg.6$	$19-20$	$Seg.6$	$14-15$
$Seg.7$	$19-20$	$Seg.7$	$30-32$	$Seg.7$	$25-26$	$Seg.7$	$21-24$	$Seg.7$	$16-18$
$Seg.8$	$21-23$	$Seg.8$	$33-36$	$Seg.8$	$27-33$	$Seg.8$	$25-38$	$Seg.8$	$19-20$
$Seg.9$	$24-29$	$Seg.9$	$37-41$	$Seg.9$	$34-36$	$Seg.9$	$39-41$	$Seg.9$	$21-24$
$Seg.10$	$30-32$			$Seg.10$	$37-38$			$Seg.10$	$25-30$
$Seg.11$	$33-36$			$Seg.11$	$39-41$			$Seg.11$	$31-38$
$Seg.12$	$37-41$							$Seg.12$	$39-41$

The evaluation of our segmentation methods against Manual segmentations ($Manual = Chain, Intention$) is made with the following measures:

$$Precision_{Method,Manual} = \frac{Number\,of\,correct\,gaps}{Number\,of\,gaps\,of\,Method}$$

$$Recall_{Method,Manual} = \frac{Number\,of\,correct\,gaps}{Number\,of\,gaps\,of\,Manual}$$

Here $Number\,of\,correct\,gaps$ is the numbers of begins of segments found by $Method$ which differ with -1,0,+1 to the begins of segments found by $Manual$. This definition for $Number\,of\,correct\,gaps$ increases the sensitivity of the precision and the recall to "false negatives" and "false positives" when they are near-miss. Moreover, our definition for the precision and the recall is more intuitive than P_k measure ([3]), because they directly reflect the competence of algorithms.

The comparison between the Precision and the Recall of different methods presented in this paper reported to Chain and Intention manual segmentation is given in the Fig.2.

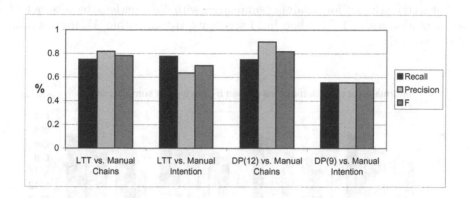

Fig. 2. Precision and Recall for LTT segmentation and DP segmentation

5.2 Evaluation of the Summarization

There is no an unique formal method to evaluate the quality of a summary. In this paper we use as a measure of quality of a summary, the similarity (calculated as *cosine*) between the original (summarized) text ([26]) and the summaries obtained with different methods.

For example, the summaries determined from the segments obtained by LTT method (and then AL algorithm) with 9 segments are formed by the following sentences:

$Sum_1 = \{1, 8, 14, 17, 25, 27, 34, 37, 39\}$.
$Sum_2 = \{2, 8, 14, 17, 25, 28, 34, 35, 39\}$
$Sum_3 = \{1, 7, 9, 16, 18, 33, 35, 36, 41\}$.

The summary obtained with DP for all lengths are:

DP(4)={3,7,35,41}
DP(6)={3,7,10,14,35,41}
DP(8)= {3,7,10,14,17,19,35,41}
DP(9) ={3,7,10,14,17,19,23,35,41}
DP(11)= {2,3,7,10,14,17,19,23,28,35,41}
DP(12) = {2,3,4,7,10,14,17,19,23,28,35,41}
DP(18) ={2,3,4,5,7,8,10,11,12,14,15,16,17,19,23,28,35,41}
DP(20)={2,3,4,5,7,8,10,11,12,14,15,16,17,19,23,25,27,28,35,41}.

The informativeness of the different types of summaries Sum_1, Sum_2, Sum_3 (see section 3.2) of different lengths (including the above 9 length summaries) is presented in Fig.3. The better results for Sum_1 method are explained by the fact that the first sentence in a segment usually contents the decisive information about it. Moreover, these results validate the quality of our segmentation. Also, for Sum_3 method, each sentence in the summary adds new information as much as it is possible, thus the good results obtained with this method.

Let us remark that for obtaining summaries with different lengths, after a first segmentation with LTT method in 11 segments, the algorithm AL from section 3.1 is applied.

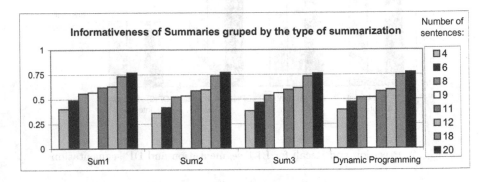

Fig. 3. Evaluation of summaries with different lengths, calculated with Sum_1, Sum_2, Sum_3 and DP methods

5.3   Segmentation Effects on Summarization

Some views from Fig.3. are detailed in Table 2, emphasizing the better results for first segmentation and then summarization ($SEG + SUM$) with Sum_1, Sum_2, Sum_3 methods compared with directly summarization (SUM) with DP method.

In 5 from 6 cases of $SEG + SUM$ methods (for the lengths 9 and 12) the results are better than SUM method. Let us observe that for extreme lengths

Table 2. The informativeness of SUM and $SEG + SUM$ summaries

	SUM	$SEG + SUM$		
$Length$	DP	Sum_1	Sum_2	Sum_3
9	0.5182	**0.5693**	**0.5336**	**0.5631**
12	0.5951	**0.6292**	0.5951	**0.6143**
4	0.3938	**0.4025**	0.3600	0.3819
6	0.4739	**0.4907**	0.4167	0.4665
8	0.5182	**0.5590**	**0.5226**	**0.5379**
11	0.5794	**0.6199**	**0.5853**	**0.5971**
18	0.7469	0.7312	0.7360	0.7312
20	0.7728	0.7683	0.7713	0.7607

(4, 6, 18, 20) the influence of the segmentation is not so benefic. Namely, only for 13 cases from 24 cases $SEG + SUM$ results are better than SUM method.

6   Conclusion and Further Work

As the abstraction is harder to be developed in summarization systems, the effort in the extraction should be done with a view to increasingly coherent summaries. This paper shows that the discourse segmentation by text entailment relation between sentences is a good basis for obtaining highly accurate summaries. Also, scoring the sentences on a logical base can give good results with a method such as Dynamic Programming, when the segmentation is not required before the summarization.

The algorithms described here are fully implemented and use text entailment between sentences without requiring thesaural relations or knowledge bases. The evaluation indices acceptable performance when compared against human judgement of segmentation and when the informativeness of summaries are considered. However, our method for computing the logical score of a sentence has the potential to be improved. We intend to improve our text entailment verification tool and to study how the calculation of logical score of a sentence (considering only neighbors of a target sentence, or considering also the number of sentences which imply the target sentence, etc) effects the segmentation and the summarization.

A drawback of our method is the following: extracted sentences contain anaphora links to the rest of the text. The wild used heuristic, to include together with the extracted sentence the one immediately preceding it, is not applicable for $Sum1$ method, when we select the first sentence of a segment. However, we intend to apply for Sum_2 and Sum_3 this heuristic.

Our method was tested only on narrative texts. We intend to extend the evaluation using other types including the newspaper corpus provided by DUC conferences.

Acknowledgments. This work has been supported by CNCSIS Grant 91-037/19.09.2007 CONTO.

References

1. Barzilay, R., Elhadad, M.: Using lexical chains for Text summarization. In: Mani, J., Maybury, M. (eds.) Advances in Automated Text Summarization, pp. 111–121. MIT Press, Cambridge (1999)
2. Barzilay, R., Lapata, M.: Modelling local coherence: an entity based approach. In: 43rd Annual Meeting of the ACL, pp. 141–148 (2005)
3. Befferman, D., Berger, A.: Statistical models of text segmentation. Machine Learning 34(1-3), 177–210 (1999)
4. Boguraev, B., Neff, M.: Salience-based content characterization of text document. In: Mani, J., Maybury, M. (eds.) Advances in Automated Text Summarization, pp. 99–110. MIT Press, Cambridge (1999)

5. Choi, F.Y.: Advances in domain independent linear text segmentation. In: 6th Applied Natural Language Processing Conference, NAACL , pp. 26-33 (2000)

6. Corley, C., Mihalcea, R.: Measuring the semantic similarity of texts. In: Proceedings of the ACL Workshop on Empirical Modeling of Semantic Equivalence and Entailment, pp. 13–18 (2005)

7. Dagan, I., Glickman, O., Magnini, B.: The PASCAL Recognising Textual Entailment Challenge. In: Quiñonero-Candela, J., Dagan, I., Magnini, B., d'Alché-Buc, F. (eds.) MLCW 2005. LNCS (LNAI), vol. 3944, pp. 177–190. Springer, Heidelberg (2006)

8. Ferret, O., Grau, B.: A topic segmentation based on sematic domains. In: Horn, W. (ed.) ECAI, pp. 426–430. IOS Press, Amsterdam (2000)

9. Grosz, B., Sidner, C.: Attention, intentions and the structure of discourse. Computational Linguistics 12(3), 175–204 (1986)

10. Hearst, M.: TextTiling: A Quantitative Approach to Discourse Segmentation. Technical Report 93/24, University of California, Berkeley (1993)

11. Hearst, M.: TextTiling: Segmentig Text into Multi-Paragraph Subtopic Passages. Computational Linguistic 23, 33–64 (1997)

12. Hearst, M.: Multi-paragraph segmentation of expository text. In: 32nd Annual Meeting of ACL, pp. 9–16. ACL (1994)

13. Hovy, E.: Text summarization. In: Mitkov, R. (ed.) The Oxford Handbook of Computational Linguistics, pp. 583–598. Oxford University Press, Oxford (2003)

14. Kaufmann, S.: Cohesion and collocation: using context vectors in Text Segmentation. In: 37th Annual Meeting of the ACL, pp. 591–599. ACL (1999)

15. Mani, J.: Automatic summarization. John Benjamins Publishing Comp., Amsterdam (2001)

16. Marcu, D.: From discourse structure to text summaries. In: Mani, I., Maybury, M. (eds.) ACL/EACL Workshop on Intelligent Scalable TS, pp. 82–88 (1997)

17. Morris, J., Hirst, G.: Lexical Cohesion Computed by Thesaural Relations as an Indicator of the Structure of Text. Computational Linguistics 17(1), 21–48 (1991)

18. Orasan, C.: Comparative evaluation of modular automatic summarization systems using CAST. PhD Thesis. University of Wolverhampton, UK (2006)

19. Pevzner, L., Hearst, M.: A critique and improvement of an Evaluation Metric for Text segmentation. Computational Linguistics 28(1), 19–36 (2002)

20. Radev, D., Hovy, E., McKeown, K.: Introduction to the Special Issues on Summarization. Computational Linguistics 28, 399–408 (2002)

21. Silber, H.G., McCoy, K.: Efficiently computed lexical chains, as an intermediate representation for automatic text summarization. Computational Linguistics 28, 487–496 (2002)

22. Tatar, D., Serban, G., Mihis, A., Mihalcea, R.: Text Entailment as directional relation. In: Orasan, C., Kuebler, S. (eds.) CALP Workshop at RANLP, pp. 53–58. Incoma Ltd, Bulgaria (2007)

23. Tatar, D., Tamaianu-Morita, E., Mihis, A., Lupsa, D.: Summarization by logic segmentation and text entailment. Research in Computing Science 33, 15–26 (2007)

24. Journal of Machine Learning Research, http://jmlr.csail.mit.edu/papers/volume5/lewis04a/a11-smart-stop-list/english.stop

25. The OpenNLP CCG Library, http://opennlp.sourceforge.net/

26. Babes-Bolyai University, http://www.cs.ubbcluj.ro/~dtatar/nlp/Hirst.txt

Comparing Non-parametric Ensemble Methods for Document Clustering

Edgar Gonzàlez and Jordi Turmo

TALP Research Center
Universitat Politècnica de Catalunya
{egonzalez,turmo}@lsi.upc.edu

Abstract. The biases of individual algorithms for non-parametric document clustering can lead to non-optimal solutions. Ensemble clustering methods may overcome this limitation, but have not been applied to document collections. This paper presents a comparison of strategies for non-parametric document ensemble clustering.

1 Introduction

As the availability of large amounts of textual information is unlimited in practice, supervised processes for mining these data can become highly expensive for human experts. For this reason, unsupervised methods are a central topic of research on tasks related to text mining. One of these tasks is document clustering. Most of the work in this area deals with parametric approaches [1, 2], in which the number of clusters has to be provided *a priori*.

On the contrary, non-parametric document clustering can be defined as the process of grouping similar documents without requiring *a priori* either the number of document categories or a careful initialization of the process from a human user. Some approaches to this task consist in repeatedly applying an iterative clustering algorithm (e.g., k-Means) to obtain a set of clusterings with a different number of clusters and starting conditions each one, and then selecting the best clustering using some model criterion [3]. Some others estimate the number of clusters *a priori* considering mathematical properties of the input documents, and then apply an iterative clustering algorithm [4]. Other approaches are based on the use of a hierarchical clustering algorithm (e.g., Hierarchical Agglomerative Clustering (HAC)) and a criterion function to select the best number of clusters in the dendrogram [5]. Recently, hybrid methods have been experimented, using the output generated from one clustering algorithm to initialize another one [6, 7].

However, each proposed approach implements some algorithm, which has an intrinsic and particular bias; uses a certain document representation; and depends on a different document similarity measure. All these assumptions lead the clustering process to a particular solution that may not be the optimal document clustering. In order to overcome this limitation, ensemble methods can

E. Kapetanios, V. Sugumaran, M. Spiliopoulou (Eds.): NLDB 2008, LNCS 5039, pp. 245–256, 2008.

be used. From a general point of view, given multiple clusterings, these methods aim at finding a combined clustering with better quality [8].

Most work in ensemble document clustering has focused on parametric approaches [9, 10, 11]. However, non-parametric ensemble approaches for generic clustering have appeared recently, such as [12].

We believe that two questions remain hence unanswered in the state of the art with respect to the use of ensemble methods for document clustering:

- **How well do ensemble methods perform for non-parametric document clustering?** Non-parametric methods have not been tested thoroughly on document collections so far.
- **How well do different individual clustering strategies perform in the context of non-parametric ensemble document clustering?** The influence of the strategy used to find individual clusterings to be later combined has often been overlooked. Different strategies need to be compared.

This paper deals with both questions. It evaluates non-parametric clustering algorithms on document collections; and it presents an empirical comparison of the effectiveness of two different strategies for the generation of clustering ensembles: one relying on massive randomization of a single algorithm, and another relying on few but heterogeneous different algorithms.

The rest of the paper is organized as follows: Section 2 settles the problem of non-parametric document ensemble clustering. Sections 3 and 4 describe the two considered generation strategies for the clustering ensembles. Section 5 then gives an overview of the experiments performed and their results. Last, Section 6 draws conclusions of our work.

2   Non-parametric Document Ensemble Clustering

Having $\mathcal{D} = \{d_1 \ldots d_n\}$ a set of documents, a clustering, Π, of this set is a partition of $\mathcal{D}$ into a set, $\{\pi_1 \ldots \pi_k\}$, of k disjoint clusters, π_i. The clustering, Π, can also be viewed as a function mapping documents, d_l, onto labels $\{1 \ldots k\}$ corresponding to clusters $\{\pi_1 \ldots \pi_k\}$, where $\Pi(d_l) = i \leftrightarrow d_l \in \pi_i$.

Bearing this in mind, the aim of clustering combination is to find a clustering, $\bar{\Pi}$, which is the consensus of r clusterings, $\{\Pi_1 \ldots \Pi_r\}$, by means of a consensus function Γ.

Two settings are classically considered for this problem, according to whether the consensus function accesses or not the original representation of the data. It is usual to refer to the case when the original data are not accessed as *cluster ensemble* [9]. This setting allows combination of clusterings obtained using different document representations. We stick to it in this paper, as it is a more general framework than the former and, in addition, it is widely used by the machine learning research community [8, 13, 14].

For our experiments, we have focused on the non-parametric ensemble clustering approach of [12], which includes methods for the determination of the number of clusters. Among the methods proposed in the paper, we have chosen

---

Algorithm 1. Major ensemble strategy

Parameter: $\mathcal{D}$ a document collection
Parameter: r a natural number
Parameter: k_{max} a natural number
Parameter: φ a supervised clustering algorithm

1: **for** $j = 1 \ldots r$ **do**
2:    Select a number of clusters at random
     $k_j \in \{2 \ldots k_{max}\}$
3:    Select k_j documents at random as starting centroids
4:    Apply φ to $\mathcal{D}$ to obtain clustering Π_j
5: **end for**
6: Return ensemble $\{\Pi_1 \ldots \Pi_r\}$

---

the AGGLOMERATIVE algorithm, enhanced with LOCALSEARCH. This combination was found in preliminary experiments to outperform the rest of the proposed approaches on the evaluation data collections[1].

3   Major Ensemble Strategy

There has been recent interest in research on ensemble clusterings from repeated runs of randomly initialized algorithms [8, 13]. In these works, the results obtained were competitive to other proposed approaches for a variety of classical clustering problems in machine learning.

For this reason, as a first strategy we have considered repeatedly applying a single individual clustering method a high number of times, with different starting conditions selected at random. The main properties of this strategy are the following:

- The resulting clusterings share the same data representation.
- The algorithm is unique, hence, the implicit bias introduced by the clustering process is always the same.
- The size of the ensemble can be high.

The procedure is detailed in Algorithm 1. First, a number of clusters k from 2 to k_{max} is selected at random. Then, k documents are selected at random from the collection, and are given as starting centroids to a clustering algorithm, φ. This process is repeated a number of times r, and the r resulting clusterings are combined using the ensemble clustering function.

The parametric clustering algorithm, φ, is a parameter of the method. For our experiments, we have used the EM-based clustering algorithm of [15]. This algorithm has obtained competitive results for text classification, and has already been used for document clustering [7]. Other parameters that need to be chosen are the number of individual clusterings, r, and the maximum number of clusters,

---

[1] Further details about these algorithms can be found on the original paper.

k_{max}. For the considered document collections, the best results among the set of explored parameter values were obtained with $r = 50$ and $k_{max} = 10$.

We will refer to this method as **Major**.

4 Minor Ensemble Strategy

Whereas the **Major** combination strategy we have described in the previous section is based on the repetition of a randomly initialized single clustering algorithm, the second strategy we have considered, **Minor**, is based on the use of a small number of heterogeneous, unsupervised and deterministic clustering algorithms. As in this case there is no randomization, it is crucial to the method that the biases provided by the algorithms be substantially different from each other. For this reason we have selected the following unsupervised clustering methods:

- A classical method, consisting of a hierarchical algorithm and a clustering criterion function.
- A hierarchical-iterative hybrid method. Specifically, the hybrid method of [7], which has been shown to give good performance for unsupervised document clustering of different real-world collections.
- A new version of the previous hybrid method, based on information theory, which we have devised specially for this purpose.

A description of each one of them follows.

4.1 Hierarchical Method

In order to generate a dendrogram, the Agglomerative Information Bottleneck algorithm (aIB) is used. [16] applies the algorithm to a variety of standard supervised clustering problems. Particularly, aIB showed good performance for the task of supervised document clustering.

After the dendrogram is built, the Calinski and Harabasz C score [17] is used to determine the level of the dendrogram at which the best clustering occurs. This score has been compared to other similar criteria to determine the number of clusters in a collection, and shown to be the most efficient one [3]. Its value is the normalized ratio of the inter-cluster distances (between documents of different clusters) against intra-cluster distances (within documents of the same cluster). The level at which this value is highest is selected as the best estimation of the number of clusters.

We will refer to this method as **Hi**.

4.2 Geometric Hybrid Method

The method presented in [7] tries to find a good initial clustering for an iterative refinement algorithm. Iterative refinement algorithms are known to be efficient and give good performance, but nevertheless are sensitive to the choice of the initial model, and require the number of clusters to be provided. In particular, a good estimation of the number of clusters is mandatory for a good initial model, even if this model does not cover all documents in the collection.

An outline of the procedure follows:

1. A hierarchical algorithm is used to find a dendrogram.
2. The inner nodes in the dendrogram are scored according to different heuristics, based in minimizing the distances within documents covered by the node, and maximizing distances to the rest of the documents[2].
3. The nodes the best scored according to the heuristics are chosen as clusters for an initial clustering candidate. A different candidate is built for each heuristic.
4. These candidates are scored using a global quality function, and the best scored candidate is selected.
5. This candidate is used as initial model for an iterative refinement algorithm, to produce a final clustering solution.

In its original implementation, the method is specified using a geometric point of view:

- Documents are represented as $tf \cdot idf$ vectors of words.
- The distance metric is cosine distance.
- The hierarchical algorithm used is HAC with group average distance as distance between clusters, which was pointed as the most suitable distance in HAC context by published evaluations of the algorithm [1].
- The global quality function is Calinski and Harabasz C score.
- The iterative refinement algorithm applied is the EM-based algorithm of [15].

We will refer to this method as **Geo**.

4.3 Information Theoretical Hybrid Method

Even if geometric clustering methods remain the state of the art, there has been a recent interest in applying information theoretical measures to the task of document clustering [16, 18]. Following this general direction of research, and to find a view of the data different from that of **Geo**, we have made a new version of the aforementioned hybrid method using information theoretical concepts:

- Documents are represented as conditional probability distributions of words.
- The distance metric is Jensen-Shannon divergence. There are other measures coming from information theory that could be useful to define a document distance, such as Kullback-Leibler divergence or mutual information. However, on the contrary of Jensen-Shannon divergence, they are not symmetric or require absolute continuity.

---

[2] For simplicity, the details about these heuristics have been elided in this paper.

- The hierarchical algorithm used is aIB.
- The global quality function used is a specially devised Message Length Criterion, described below in Section 4.3.
- The iterative refinement algorithm applied is Divisive Information Theoretical Clustering (DITC) [18]. This algorithm includes devices to deal with sparseness and high dimensionality of data, and was shown to give good performance on document collections.

We will refer to this method as **IT**.

Message Length Criterion. Classical information theoretical selection criteria, such as Minimum Description Length or Minimum Message Length, require a probability distribution, which cannot be directly derived from the dendrogram. However, we have devised a criterion to select the best clustering in the same spirit, based in coding, messages and lengths.

The idea is to use the information in a clustering Π to send a collection of documents $\mathcal{D}$ as a message. We first send the send the centroid of each cluster using a code based on the meta-centroid of the collection (a first message of length $L_C(\Pi)$), and then send the distribution of words in each document using a code based of the centroid of the cluster to which it belongs (a second message of length $L_D(\Pi)$). Using formulae from Information Theory, the total length of this message, $L(\Pi)$, is roughly:

$$L(\Pi) \approx L_C(\Pi) + L_D(\Pi)$$
$$L_C(\Pi) \approx - \sum_{\substack{\pi_i \in \Pi \\ w}} p(w|c_i) \cdot \log p(w|mc)$$
$$L_D(\Pi) \approx - \sum_{\substack{\pi_i \in \Pi \\ d_l \in \pi_i \\ w}} p(w|d_l) \cdot \log p(w|c_i)$$

where w are words, c_i are the cluster centroids and mc is the meta-centroid.

We expect *better* clusterings (i.e. more suited to the data) to allow better compression of the data and hence, shorter messages. Therefore, we select the clustering Π which has the lowest $L(\Pi)$, expecting it to be the *best*.

This formula was the one to give the best results in preliminary experiments, compared to a version of the C score using Jensen-Shannon divergence.

Moreover, this formula was appealing to us because it includes an implicit measure of the goodness of the number of clusters (more clusters imply largest $L_C(\Pi)$ but smallest $L_D(\Pi)$, and vice versa).

5   Experiments

In order to evaluate and compare the performance of the two proposed ensemble strategies, **Major** and **Minor**, between them and to individual clustering approaches (of which **Geo** can be considered a baseline in the state of the art), we have carried out a series of experiments. The following sections explain the experimental framework, and present their results.

5.1 Evaluation Data

Six different real-world English document collections have been used in our experiments:

APW. The Associated Press (year 1999) subset of the AQUAINT collection. Due to memory limitations in our test machines, the collection was reduced to the first 5000 documents.

EFE. A collection of news-wire documents from year 2000 provided by the EFE news agency.

LAT The Los Angeles Times subset of the TREC-5 collection. For the same reason as in APW, again only the first 5000 documents were selected.

REU. A subset of the Reuters-21578 text categorization collection, which includes only the ten most frequent categories. Similarly to previous work, we use the ModApte split [7, 15], but, since our algorithms are unsupervised, we use the test partition directly.

SMT. A collection previously developed and used for the evaluation of the SMART information retrieval system.

SWB. A subset of the Switchboard conversational speech corpus, which contains the 22 topics which were treated in more than fifty conversations. Each side of the conversation was considered a separate document.

Following other research work [2, 7], the documents were pre-processed by discarding stop words and numbers, converting all words to lower case, and removing terms occurring in a single document. Table 1 lists relevant collection characteristics after pre-processing (number of documents, categories and terms).

Table 1. Evaluation data sets

Collection	Docs	Cats	Terms
APW	5000	11	27366
EFE	1979	6	10334
LAT	5000	8	31960
REU	2545	10	6734
SMT	5467	4	11950
SWB	2682	22	11565

5.2 Evaluation Metrics

The quality of the clustering solutions is measured using the metrics of purity, inverse purity and F_1. These metrics have been widely used to evaluate the performance of document clustering algorithms [2], and are based in comparing the clustering to a partition which is considered *true*.

If we have a partition of the documents in $\mathcal{D}$ into a set of disjoint categories considered *true*, these metrics can be defined as:

Pur. Purity evaluates the degree to which each cluster contains documents from a single category. The purity of a cluster is the fraction of the documents in the cluster that belong to its majoritarian category. The overall purity is the average of all cluster purities, weighted by cluster size.

IPur. Inverse purity evaluates the degree to which the documents in a category are grouped in a single cluster. The inverse purity of a category is the fraction of the documents in the category that are assigned to its majoritarian cluster. The overall inverse purity is the average of all category inverse purities, weighted by category size.

F$_1$. F_1 is a global performance score, and is calculated as the harmonic mean of purity and inverse purity.

5.3 Experimental Setup

Each collection was clustered using each of the proposed methods. For the **Geo**, **Hi**, **IT** and **Minor** methods, a single run was performed, as these methods are deterministic.

For the **Major** method, we performed five runs and the results presented are the average of all the runs. As mentioned in Section 3, the results are those obtained with $r = 50$ and $k_{max} = 10$, which were the parameter values to provide the best F_1 scores in average across all collections.

5.4 Results

Tables 2, 3 and 4 show the results obtained by each method in each collection. For each collection, the best results are highlighted.

In addition, Table 5 shows the number of clusters k estimated by each method in each collection. We include two numbers for each method, the total number of clusters (**All**), and the number of *relevant* clusters (**Rel**). The reason for this is that we have found that the AGGLOMERATIVE algorithm tends to find a high number of clusters, but many of them are small, possibly corresponding to outliers among the data.

Given that these small clusters are not relevant to the evaluation (and their detection as outliers is, in fact, an advantageous byproduct of the method), to obtain a more useful measure we have filtered those clusters smaller than a fourth of the average category size in the collection. The remaining ones are considered *relevant*, and their number is the figure appearing in the table. The number of categories (**Cats**) in each collection is also included in the table.

Following sections discuss the obtained results.

Overall Comparison. It can be seen how the **Major** approach outperforms the rest of the approaches in almost all collections in terms of F_1, and is also the

Table 2. F_1 values for all methods and collections

	Geo	Hi	IT	Major	Minor
APW	0.75	0.74	0.63	0.75	0.72
EFE	0.61	0.61	0.58	0.62	0.60
LAT	0.67	0.67	0.67	0.75	0.67
REU	0.88	0.79	0.76	0.88	0.88
SMT	0.85	0.82	0.71	0.93	0.91
SWB	0.79	0.26	0.53	0.44	0.66

Table 3. Purity values for all methods and collections

	Geo	Hi	IT	Major	Minor
APW	0.78	0.63	0.72	0.80	0.74
EFE	0.73	0.60	0.64	0.75	0.70
LAT	0.78	0.66	0.75	0.73	0.79
REU	0.84	0.73	0.77	0.86	0.85
SMT	0.92	0.71	0.89	0.93	0.93
SWB	0.69	0.15	0.38	0.29	0.53

Table 4. Inverse purity values for all methods and collections

	Geo	Hi	IT	Major	Minor
APW	0.73	0.88	0.56	0.70	0.70
EFE	0.52	0.63	0.53	0.53	0.53
LAT	0.59	0.68	0.61	0.79	0.59
REU	0.92	0.86	0.76	0.90	0.89
SMT	0.80	0.97	0.58	0.92	0.90
SWB	0.94	0.92	0.91	0.97	0.89

Table 5. Number of clusters k for all methods and collections

	Cats	Geo		Hi		IT		Major		Minor	
		All	Rel	All	Rel	All	Rel	All	Rel	All	Rel
APW	11	10	9	3	3	8	8	60.6	7.0	19	7
EFE	6	12	7	4	4	5	5	69.0	6.2	14	7
LAT	8	14	9	6	6	7	7	27.2	4.8	40	7
REU	10	6	6	4	4	6	6	18.2	5.2	13	6
SMT	4	6	5	3	3	9	7	20.6	4.0	18	4
SWB	22	15	15	3	3	8	8	10.4	5.8	22	12

best approach in terms of purity in four of the six collections. Its performance in terms of inverse purity is not always the best, but it is always comparable to that of the rest of the methods.

The performance of **Minor** and **Geo** is quite similar in terms of purity, but **Minor** suffers from lower inverse purity, so overall its F_1 is also lower. The **Hi** method usually gives solutions with a high inverse purity but a low purity, so in many cases the global F_1 scores are lower than other approaches. Lastly, the results of **IT** do not stand out in any aspect, and its utility outside the **Minor** combination seems limited, at least at the light of these results.

Nevertheless, we have applied a Friedman test, followed by pairwise Nemenyi tests, to account for statistical significance of these differences [19]. We only found that **Hi** is worse than **Major**, **Minor** and **Geo** in terms of purity; and that **IT** is worse than **Major** in terms of F_1. No other significant differences were found. This is relevant, because it means there is no empirical evidence supporting the rejection of any of the **Geo**, **Major** or **Minor** methods as less suitable to the task than the others, in terms of purity, inverse purity or F_1 score.

Estimation of the Number of Clusters. Concerning the estimated number of clusters, we can see how the ensemble-based approaches greatly overestimate the total number of clusters (**All**). As explained in Section 5.4, this is caused by the presence of a large number of small clusters, and the figures for the number of relevant clusters (**Rel**) are much closer to the actual number of categories (**Cats**).

However, it can be seen that the estimation of the total number of clusters by **Minor** is more accurate than that by **Major** in all but the LAT collection. **Major** shows a bias for purity, and shows a slightly displeasing tendency to disgregation.

Regarding the individual methods, whereas the estimation by **Geo** and **IT** is fairly accurate; **Hi** shows a tendency to underestimation, which explains its high inverse purity values and low purity values. The individual methods do not present such a large number of small clusters, which on the one hand means there is not such a risk of disgregation, but on the other one can mean a more limited capability to detect outliers.

Minor Method. As mentioned before, the performance of **Minor** method is only significantly better than that of **Hi** in terms of purity. Nevertheless, the results of the combination seem comparable to those of **Geo**, and better than those of **IT**.

Overall, **Minor** offers a greater stability across document collections than its components **Hi** and **IT**. Moreover, the fact that neither **Hi** nor **IT** do not perform competitively on document collections (particularly on SWB) suggests that using some other algorithm more suitable for this kind of data the performance of **Minor** could be boosted, and more competitive results could be obtained.

For this reason, together with the facts that its performance is not significantly worse than that of **Major**; that it gives a better estimation of the number of clusters; and that it has no parameters needing to be tuned, whereas **Major** requires the values of k_{max} and r have to be determined (see Section 3); we believe that the **Minor** method remains an attractive approach, and that more research should be carried on the topic of small ensembles of heterogeneous clusterings.

SWB Collection. The main exception to the general behaviour seems to be the SWB collection. Almost all methods experiment a considerable decrease in purity when applied to this data set. We believe this comes from the fact that, the size of all categories in SWB is quite similar, whereas for the rest of collections a few large categories cover most of the documents. This makes the SWB collection harder than the rest, and specially sensitive to underestimation of the number of clusters.

The fact that all the considered methods do underestimate the number of clusters (as can be seen in the **Rel** columns of Table 5), causes low values of purity (in some causes dramatically low, e.g. **Hi**), and hence of F_1. Only **Geo** and, to a lesser extent, **Minor** seem able to find a reasonable (even if still underestimated) number of relevant clusters (column **Rel**) in this collection.

6 Conclusions

We have studied the application of a non-parametric ensemble clustering approach to document collections, and considered two different strategies for the generation of the clustering ensembles. Lastly, we have carried a set of experiments with real-world data.

At the light of the results, we can conclude that non-parametric ensemble methods do perform competitively for clustering of document collections. Regarding the two considered strategies, whereas the **Major** approach gives better figures of purity and F_1 score, the differences with **Minor** are not statistically significant, its estimation of the number of clusters is worse, and it has a number of parameters to be tuned.

For these reasons, and because there is further room for improvement of the individual components of **Minor**, we believe that the results of this heterogeneous approach can be boosted, and that it remains an attractive approach for the task.

Acknowledgments

This work has been partially funded by the Spanish Text-Mess Project (TIN2006-15265-C06); the Commissionate for Universities and Research of the Department of Innovation, Universities and Enterprises of the Catalan Government; and the European Social Fund.

References

1. Zhao, Y., Karypis, G.: Evaluation of hierarchical clustering algorithms for document datasets. In: Proc. of CIKM (2002)
2. Zhao, Y., Karypis, G.: Empirical and theoretical comparisons of selected criterion functions for document clustering. Machine Learning 55(3) (2004)
3. Milligan, G.W., Cooper, M.C.: An examination of procedures for determining the number of clusters in a data set. Psychometrica 50 (1985)

4. Li, T., Ma, S., Ogihara, M.: Document clustering via adaptive subspace iteration. In: Proc. of SIGIR (2004)
5. Tibshirani, R., Walther, G., Hastie, T.: Estimating the number of clusters in a data set via the gap statistic. Journal of the Royal Statistical Society Series B 63(2) (2001)
6. Fraley, C., Raftery, A.: How many clusters? Which clustering method? Answers via model-based cluster analysis. The Computer Journal 41(8) (1998)
7. Surdeanu, M., Turmo, J., Ageno, A.: A hybrid unsupervised approach for document clustering. In: Proc. of KDD (2005)
8. Topchy, A., Jain, A.K., Punch, W.: Clustering ensembles: Models of consensus and weak partitions. IEEE Transactions on Pattern Analysis and Machine Intelligence 27(12) (2005)
9. Strehl, A., Ghosh, J.: Cluster ensembles - A knowledge reuse framework for combining multiple partitions. Journal of Machine Learning Research 3 (2002)
10. Siersdorfer, S., Sizov, S.: Restrictive clustering and metaclustering for self-organizing document collections. In: Proc. of SIGIR (2004)
11. Greene, D., Cunningham, P.: Efficient ensemble methods for document clustering. Technical report, Department of Computer Science, Trinity College Dublin (2006)
12. Gionis, A., Mannila, H., Tsaparas, P.: Clustering aggregation. In: Proc. of ICDE (2005)
13. Fred, A., Jain, A.: Robust data clustering. In: Proc. of CVPR (2003)
14. Li, T., Ogihara, M., Ma, S.: On combining multiple clusterings. In: Proc. of CIKM (2004)
15. Nigam, K., McCallum, A., Thrun, S., Mitchell, T.: Text classification from labeled and unlabeled documents using EM. Machine Learning 39(2/3) (2000)
16. Slonim, N.: The Information Bottleneck: Theory and Applications. PhD thesis, The Hebrew University (2003)
17. Calinski, T., Harabasz, J.: A dendrite method for cluster analysis. Communications in Statistics 3 (1974)
18. Dhillon, I., Guan, Y.: Information theoretic clustering of sparse co-occurrence data. In: Proc. of ICDM (2003)
19. Demsar, J.: Statistical comparisons of classifiers over multiple data sets. Journal of Machine Learning Research 7 (2006)

A Language Modelling Approach to Linking Criminal Styles with Offender Characteristics

Richard Bache[1], Fabio Crestani[1], David Canter[2], and Donna Youngs[2]

[1] Department of Computer and Information Science,
University of Strathclyde, Scotland
{r.bache,f.crestani}@cis.strath.ac.uk
[2] Centre for Investigative Psychology, England
{d.canter,d.e.youngs}@liverpool.co.uk

Abstract. The ability to infer the characteristics of offenders from their criminal behaviour ('offender profiling') has only been partially successful since it has relied on subjective judgments based on limited data. Words and structured data used in crime descriptions recorded by the police relate to behavioural features. Thus Language Modelling was applied to an existing police archive to link behavioural features with significant characteristics of offenders. Both multinomial and multiple Bernoulli models were used. Although categories selected are gender and age group, in principle this can be applied to any characteristic recorded. Results indicate that statistically significant relationships exist between both age and sex in certain types of crime. Both types of language model perform with similar effectiveness. It is also possible to identify automatically specific terms which when taken together give insight into the style of offending related to a particular group.

Keywords: Text Data Mining, Language Models, Crime Data, Investigative Psychology, Offender Profiling.

1 Introduction

Since the earliest criminological studies it has been clear that broadly speaking criminals have characteristics that distinguish them from the general population. There have also been attempts to demonstrate that certain classes of crime are typically committed by people who have similar characteristics. It has also been claimed that what may be called the 'style' of the crime or the pattern of behaviour, typical of any set of crimes, relates directly to subsets of characteristics of offenders. This process of making inferences about significant features of an offender on the basis of the kinds of people who commit crimes in that style has often been called 'offender profiling'. In general such 'profiles' are drawn from the subjective judgement and experience of putative experts with little empirical basis for their claims. In a few studies, most notably Canter and Fritzon's study of arson [1], it has been demonstrated that there are empirically sound relationships between, *inter alia*, the age, psychiatric background and personal relationships of offenders and dominant

E. Kapetanios, V. Sugumaran, M. Spiliopoulou (Eds.): NLDB 2008, LNCS 5039, pp. 257–268, 2008.

features of the crime, such as the nature of the target and whether there was more than one linked incident.

However, the few empirical studies that have been carried out to develop models relating offence style to offender characteristics have relied on intensive content analysis procedures that derive categories from open-ended police and related data sources. The labour intensity of the work as well as problems of access has meant that only limited archives have been examined. Furthermore the procedure of deriving content categories and assigning cases to those categories requires a high level of expertise and even with extensive training the content analysis can suffer from the unreliability inherent in subjective judgments. Although the application of data mining to criminal data is not new [2], the use of generative probabilistic models to unstructured data presented here departs significantly from typical data mining techniques.

Information Retrieval (IR) techniques and in particular Language Modelling have been used to link solved crimes to unsolved crimes with the purpose of prioritising suspects from a list of known offenders that have committed crimes of the same type [3]. Such techniques rely on the words and structured data used in the descriptions relating to features of the offenders' behaviour. In this study, we wished to determine whether Language Modelling could identify the group to which an unknown offender belonged from analysing common patterns of behaviour and thus relate offender characteristics to criminal style. Although the mathematical approach is similar, from a psychological point of view it is quite a different problem. We therefore looked at two crucial characteristics of an offender, age and gender, to determine if automated analysis of free text could identify differences between the terms used to describe crimes committed by offenders in different groups. Of course, in principle, this approach can be extended to any other characteristics of an offender that may be of interest, such as ethnic group, known usage of drugs, how far the offender has traveled to the crime site, a history of committing crimes of a different type (e.g. sex offenders with or without a history of property crime), marital status, sexual orientation or even personality characteristics. The limitation on such possible studies is a function of the availability of the criterion data about the characteristics of an offender, which must be derived from solved crimes where such information is recorded.

Age and gender are readily available in a reasonably (although it must be admitted not totally) reliable form in all police records and are of great theoretical and tactical interest and were thus a useful starting point for exploring the applicability of Language Models to the 'offender profiling' problem. There were four purposes of the investigation. Firstly, we wanted to see whether differences existed between groups defined by age and sex. Secondly we wished to show which type of language model would perform better for this data since there are two types that may be applied to this kind of data: multinomial and multiple Bernoulli (explained below). Thirdly, it would be useful to determine if the language models could give us insight as to what the differences in style between the groups were. Fourthly, a possible use of this approach is attempting to predict the age and sex of an offender from the description of an unsolved crime. Although eye-witness descriptions of offenders can be unreliable they are unlikely to mistake sex or very broad age group. However, there are areas where such a predictive model would be useful, for example where no witnesses were present such as certain burglaries, thefts and acts of criminal damage. We wished to establish if the models had any predictive power.

The rest of the paper is organised as follows: Section 2 describes the nature of the police digital archives and of the data to be analysed. In Section 3 we show how Language Modelling may be used to characterise criminal behaviour and provide a novel justification for its use. We also explain the two types of language model used here. Section 4 describes how the crime descriptions, recorded as a mixture of structured data and free text, are indexed for the purposes of analysis. In Section 5 we apply the models to 8 datasets representing different crimes types and show that there is a statistically significant difference between groups defined by age and sex. Section 6 looks more specifically at the features associated with each sex and age group; this takes us beyond a traditional classification problem. In Section 7, we offer some conclusions.

2 Criminal Data Used for Analysis

The data used in this study were extracted from a police distributed digital archive containing crimes committed over a four-year period in an inner city district. The three fields which gave information about the offenders' behaviour were:

- Free text describing the method employed by the offender in an informal note form, typically consisting of 20 words;
- Zero or more feature codes (typically 1 or 2) where each code represents the presence of a specific aspect of behaviour from a predefined list;
- A single allegation code describing the type or sub type of offence.

Henceforth we shall consider this information to be a single document.

Crimes were grouped according to allegation code with similar offences placed in the same set. For example, there are 8 possible allegation codes which describe different sub-types of burglary and these offences were grouped as burglaries. Only solved crimes were considered for the study and then only offences with a single offender were used. For each crime therefore there was an offender with an age (as of when offence was committed) and a gender. Note that although it is the eventual intention to apply the data to actual unsolved crimes, for the purpose of analysis some solved crimes are treated as 'unsolved' to test the models proposed here.

3 Language Models of Behaviour

The rationale for being able to distinguish sex or age of offenders from their actions rests on there being significant behavioural differences between these groups and these differences being revealed in the vocabulary used to record the crime within the criminal records. We argue in the rest of this section that Language Modelling provides a theoretically principled way of relating a document describing an offence to the group most likely to contain the culprit.

3.1 Features and Terms

From a crime description it is possible for researchers to identify salient features relating to aspects of the offenders behaviour. In previous studies [4, 5], features have been identified by means of content analysis of the text. Such a procedure necessarily requires human analysis of each crime description and is both time consuming and requires a degree of expertise by the indexer. An alternative approach, employed here, is to perform automatic indexing of the descriptions using standard Information Retrieval techniques. Thus each document is reduced to a number of tokens (words, phrases or codes) and these are counted to yield a vector of terms.

Many words used to describe the crime will reveal features of that offence. For example words such as *punch*, *argument* and *drunk* will reveal the behaviour of the offender when committing an assault. However, the relationship between automatically derived terms and behavioural features that would be identified by humans is complex and still not fully understood. There is certainly no one-to-one correspondence. Many terms in the document give no information about the offenders' behaviour and thus do not relate to any identifiable feature, e.g. *suspect, unlawfully, address*. There will also be sets of terms which may all relate to the same feature since identical behaviour can be recorded in different words. For example, where an assault has the feature of being racist in nature the terms *racist*, *racial*, *racially* or indeed quoted terms of racist abuse would all indicate the presence of this feature. Nevertheless, by using a Language Modelling approach the underlying relationship between terms and behavioural features can be exploited. This was demonstrated by linking individual offenders to unsolved crimes from past offending history [3] and provides evidence that vectors of terms contain feature information. Indeed, if there were not some identifiable relationship between the language used to describe actions and those actions then language would have little utility. The task, though, is to demonstrate that specific, automatable Language Modelling approaches do actually generate fruitful and reliable distinctions.

3.2 Document Generation

The (unintended) consequence of an offender committing a crime is that, if reported to the police, a document will be created to describe it. Such a document will be the result of some investigation and will describe features of the offender's behaviour. We therefore argue that the offender's behaviour generates a document even though the putative author is actually a police officer. In fact the police officer is only partially free in the description s/he gives. Training will have limited the language on which the police officer can draw as will the particular data management system that s/he will be required to use. The police officer is thus an 'author' in a very restricted sense, but the document is definitely initiated and its content greatly influenced by what the offender does.

A language model [6, 7] is a process which randomly emits terms (e.g. words) drawn from a predefined vocabulary. Each model will emit terms with differing frequencies and indeed a language model can be wholly characterised by the probability assigned to each term in the vocabulary. We argue that an offender generates behavioral features when committing an offence. Thus we can model an individual offender as having probabilities associated with each feature. Categories of offenders too will have probabilities associated with them although differences may be less pronounced. This will then be reflected in the probability of occurrence of terms in the documents generated by criminal activity.

Assigning crime reports to offender categories is a form of document classification. Researchers [8, 9, 10] have previously used language models for document classification and such an approach was essentially Bayesian. We too adopt a Bayesian approach but, in common with most IR applications, apply models that are *unigram* in that they consider each term independently and do not take account of the preceding tokens. The fact that we are using a mixture of terms derived from free text and structured data means that *n*-gram models which consider preceding terms would not be appropriate. Furthermore *n*-gram models require considerable training data both in terms of number and length of the documents which is not the case here.

The language model is thus a stochastic process which creates some vector of terms which corresponds to that created by automatic indexing. Strictly speaking, therefore, we should say that the language model generates the *index* of the document, but for brevity we shall say that we are generating the document. For each category of offender we will have one language model and for an unsolved crime we calculate which of these models was most likely to have generated it.

There are two types of unigram language model: multinomial and multiple Bernoulli. The former considers the number of occurrences each term has in a document and produces a non-negative integer-valued vector. We can consider multinomial model to be like drawing balls from an urn with replacement where many balls may have the same term written on them. The Bernoulli considers only whether a term is present of absent and so yields a Boolean vector where entries are 0 or 1. It can be modelled as someone systematically running through each term in the vocabulary and throwing dice to determine if it should be present or not where the dice score to trigger inclusion will vary for each term. The original application of Language Modelling to Information Retrieval applied the Bernoulli model [6] although multinomial models have become more frequently used in IR [7]. There are a number of reasons for believing *a priori* that Bernoulli models more closely match the properties of these data.

1. Terms derived from codes are either present or absent.
2. Terms from free text rarely appear more than once except stopwords (e.g. *the, by, and*) which were removed from the analysis anyhow.
3. Losada [11] shows that for question answering (QA), Bernoulli models work better. QA deals with very small sub-documents of a size not unlike the ones which are the subject of this analysis.

However, McCallum and Nigam [12] show that multinomial models start to outperform Bernoulli in document classification where the vocabulary is over 1000. The vocabulary sizes vary between datasets (see Table 1) some being greater and others less than 1000. Therefore we applied both types of model for the purposes of comparison.

3.3 Probability of Generation

Without loss of generality we consider categorisation by gender. Thus we will have two language models: male (m) and female (f). For multinomial models, the probability of a given document, d, being created given the male model is:

$$p(d\mid m)=\prod_{t\in d}p(t\mid m)^{c(t,d)} \qquad (1)$$

where $p(t\mid m)$ is the probability that the male model will emit term t and $c(t,d)$ is the count of occurrences of t in document d. For multiple Bernoulli models

$$p(d\mid m)=\prod_{t\in d}p(t\mid m)\times\prod_{t\notin d}(1-p(t\mid m)) \qquad (2)$$

We actually want to calculate the probability that an offender is male given we have seen the description of that crime, so we use Bayes' theorem.

$$p(m\mid d)=\frac{p(m)\cdot p(d\mid m)}{p(d)} \qquad (3)$$

where $p(m)$ is the prior probability that the offender is male. For most crime types, males are much more likely to offend than females. So, this can be estimated from the solved crimes as simply the proportion of offences committed by males. The denominator $p(d)$ is the probability that the document could have been generated at all. It can be eliminated by requiring that the sum of all probabilities is unity.

The solved crimes are used to calibrate the language models. We can estimate the probability that a language model (say the male one) will emit a given term t from the frequency with which it was emitted when generating all the solved male crimes. Thus for a multinomial model we have:

$$p(t\mid m)=\frac{\sum_{d\in M}c(t,d)}{\sum_{d\in M}|d|} \qquad (4)$$

where M is the set of offences known to have been committed by males and $|d|$ is the size of document d. For a Bernoulli model

$$p(t\mid m)=\frac{|S_t|}{|M|} \qquad (5)$$

where $S_t\subseteq M$ is the subset of documents containing the term t. These formulae define the *unsmoothed* language models, which if used to calibrate the language models, will result in poor performance because of the Zero Propagatability Problem (ZPP) as now explained.

3.4 Smoothing

The ZPP was identified as a problem in the application of Language Models [6, 13] to Information Retrieval. It refers to the problem of any term not appearing in the training data for a given category. The unsmoothed model will calculate the probability of any such term being emitted to be zero, and thus assign a probability of zero to any document containing it. For language models of behaviour, this has a clear interpretation.

A fundamental assumption of Investigative Psychology is that behaviour observed in the past is more likely to be observed in the future. But, if it has not been observed in the past, it *cannot* be assumed that it will *never* occur in the future – it is just less likely. The unfortunate consequence of an unsmoothed model is that it forbids novel behaviour. Furthermore, given the imprecise relationship between terms and features, the same behavioural feature might be described differently with new terms and this would mean a language model would also have a zero probability of generating it.

The standard solution to this problem is smoothing. This means adjusting the language model so that no term has a zero probability. There are numerous smoothing strategies. Jelinek-Mercer [14] smoothing was shown previously to work well with crime data [3] and this also has a natural interpretation. We define one extra language model - a universal model which models the behaviour of *every* offender. This is calibrated using the entire dataset, including both solved and unsolved crimes. Thus the probability of any term being emitted will, by definition, never be zero. So, by mixing the past behaviour of male offenders with those of the universal offender we create a model which implies any future behaviour is possible, although past male behaviour is more likely. This can be expressed as:

$$p_{smoothed}(t\,|\,m) = (1-\lambda)p(t\,|\,m) + \lambda p_{universal}(t) \tag{6}$$

where λ is the smoothing parameter set to 0.5.

4 Automatic Indexing

In order to calibrate the language models, we require that each document is reduced to a vector of terms. A (context free) lemmatiser was used rather than a stemmer so that words would be more readable and not have the ends mutilated. This was built using WordNet [15]. Stopwords from a standard list were also removed from the text.

The structured data (feature and allegation codes) appears in the police archive as two letter codes. These were mapped to meaningful hyphenated phrases and preceded by a non-alphabetic character ($) so that they could be identified from terms which originated from the free text e.g. *$victim-harassed*.

There was also a problem of gender and age specific vocabulary used in the description. In many cases the age and sex of the offender would be known when the document was created, e.g. if the offender was caught while committing the offence or a clear description was given by an eye witness. If there were words in the text which identified the age or sex of the offender, the language models would pick up on this so that would seriously bias the model. Therefore, sex and age specific vocabulary were either removed or converted to gender and age neutral proxies e.g., *girlfriend* and *boyfriend* became *partner*.

5 Experimental Procedure and Results

The approach described here will work for any finite number of categories. Clearly when using gender as a characteristic, there will only be 2. For age there could, in principle be any number although to create one for each year of age would spread the training data too thinly but 3 or 4 categories would be sensible. Nevertheless it was decided that for the purposes of the experiment there would be just 2, above and below the median age henceforth known as *older* and *younger*.

Within each crime set there were a number of serial crimes – two or more crimes by the same offender. Including these in the analysis would create a possible source of bias. If, for example, a prolific female offender committed a number of crimes with a distinct Modus Operandi, this might be seen as defining female features when in fact it was just the behaviour of one female. We term this the serial crime problem.

An adaptation of the Leave-One-Out (LOO) method was adopted to validate the models. Usually, this would involve removing each data point in turn and using all others to train the models, simulating a scenario where there was already an archive of solved crimes and one new unsolved crime. However, to avoid the serial crime problem, we remove all the crime committed by each offender in turn and use all crimes not committed by that offender to train the models. Table 1 shows the size of the 8 data sets with categorisation of the crimes. Because there were so few female offenders for some datasets it was not possible to validate the gender model in these cases.

Table 1. Numbers of Crimes, Offenders, Vocabulary Size etc

Crime Type	Number of Crimes	Male Crimes	Female Crimes	Number of Offenders	Median Age	Vocabulary Size
Theft from Vehicles	248	244	4	159	26	367
Other Theft	326	260	66	284	25	736
Shoplifting	2060	1399	661	1618	29	968
Assault	2073	1691	382	1881	31	1507
Criminal Damage	849	725	124	724	26	1114
Damage to Vehicles	220	193	27	207	27	438
Burglary	1126	1060	66	556	27	1136
Robbery	263	252	11	210	18	544

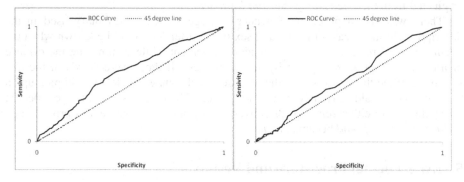

Fig. 1. ROC Curve for Age Model (left) and Sex (right) Applied to Criminal Damage

5.1 ROC Analysis

This form of analysis is applicable for any continuous variable being used to predict an item belonging to one of two groups. Here, we have the estimated probability that the offender for a given crime is either male or adult respectively. By setting some cut-off point, the offences can be partitioned into two sets: upper and lower. For each possible cut-off point between 0 and 1, we expect more males (respectively older) to be in upper set. A ROC graph shows the sensitivity of the model, the proportion of males (older) in the upper set, against the specificity, the proportion of females (younger) in the lower set. A straight line implies that the model has no predictive power and is no better than random. A curve bowed above the 45° line implies that the model does have predictive power. A measure of the bowedness of this curve is AUC (Area Under the Curve) and values above 0.5 imply some predictive power. Figure 1 shows the ROC curves for criminal damage for the age and sex models respectively using the Bernoulli formulation (The multinomial version would appear very similar). It demonstrates that whereas there are distinct behavioural differences based on age, these are weaker when considering male and female offenders. Table 2 and 3 shows AUC for both models applied to the 8 datasets. For 2 data sets there were not enough crimes committed by females to make analysis meaningful. Note that all models except damage to vehicles by sex show AUC at greater than 0.5 showing measurable differences in behaviour between age and sex categories. Note also that if the results were purely random, we would expect AUC to be below 0.5 as often as above it. Applying the Wilcoxon Ranked Sign Test to the age and sex models allows us to conclude a relationship with age to 1% significance and with sex to 5% significance. There are no striking differences between the performance of both types of models. However, if we use AUC as a measure of the performance of each model then we can use the two-sided Ranked Sign Test to determine if there were significant differences in performance of the models. We can conclude with 5% significance that Bernoulli models work better than multinomial ones.

5.2 Statistical Significance of Predictions

We attempted to use the model to perform actual classification by choosing the optimal category – this being the one which maximises the posterior probabilities. Given that the models estimate the probability that the offender will be in one of two categories, we can set the cut off point at 0.5. So if a crime shows a probability of

Table 2. Results of ROC Analysis and Chi-squared Test for Sex Model

Crime	ROC Analysis - AUC		Chi-Squared Test - significance	
	Multinomial	Bernoulli	Multinomial	Bernoulli
Other Theft	0.558	0.564	0.483	0.269
Shoplifting	0.653	0.655	**<0.001**	**<0.001**
Assault	0.697	0.705	**<0.001**	**<0.001**
Criminal Damage	0.557	0.558	0.219	0.455
Damage to Vehicles	0.453	0.474	0.141	0.199
Burglary	0.584	0.588	**<0.001**	**<0.001**

Table 3. Results of ROC Analysis and Chi-squared Test for Age Model

Crime	ROC Analysis - AUC		Chi-Squared Test - significance	
	Multinomial	Bernoulli	Multinomial	Bernoulli
Theft from Vehicles	0.638	0.623	**0.015**	**0.002**
Other Theft	0.569	0.574	**0.046**	**0.022**
Shoplifting	0.586	0.587	**<0.001**	**<0.001**
Assault	0.564	0.561	**<0.001**	**<0.001**
Criminal Damage	0.586	0.609	**<0.001**	**<0.001**
Damage to Vehicles	0.586	0.610	**0.006**	**0.002**
Burglary	0.527	0.525	0.155	0.081
Robbery	0.677	0.690	**<0.001**	**<0.001**

more than 0.5 of being a male we assume the model predicts it to be a male. Otherwise we assume that we are predicting a female. A similar technique may be used for the age models. We can compare the predictions with the known category by using the chi-squared test. Tables 2 and 3 show the level of significance with bold indicating significance at 5% for a one sided Chi-squared test.

We can see that shoplifting and assault show significance differences for both sex and age. Criminal damage and other theft show significance for age but not sex. Robbery and theft from vehicles, which are overwhelmingly male crimes, also show significance for age. Burglary shows significance for sex but not age. The fact that some datasets do not show significance has two possible explanations. It may be that there was insufficient data and a larger dataset would pick up a (probably weak) correlation. However it may also be that there are no observable behavioural differences between the groups in the first place. Note that again that multinomial and Bernoulli models perform similarly.

6   Analysis of Vocabulary

For each term in the vocabulary, we can determine to what extent it is associated with each category. Terms with a behavioural significance can be used to signify features more commonly associated with particular characteristics of offenders. So, we can define a measure of the *maleness* of a term as

$$p(t \mid m) - p(t \mid f) \tag{7}$$

where these are the probabilities of a given term being emitted from the unsmoothed language models of either type. A measure of *maturity* can be defined similarly using the younger and older models. We can then rank the terms by this measure and identify the terms most associated with male or female offenders (younger and older offenders respectively). Table 4 shows the top ten terms associated with offender characteristics for assault and criminal damage using here the multinomial models. Each model type will rank vocabulary slightly differently but the results are broadly similar. Note that some terms are proxies to avoid gender-specific terms. Examples are *parent*, *partner* and *spouse*. Note also that terms beginning with a $ sign are feature or allegation codes. All others come from the free text.

Table 4. Most Common Terms Associated with Each Language Model

Rank	Assault		Criminal Damage	
	Most Male-related Terms	Most Female-related Terms	Most Older-related Terms	Most Younger-related Terms
1	punch	$common-assault	door	stone
2	$victim-punched	parent	victim	$criminal-damage-not-high-value
3	$actual-bodily-harm	slap	smash	cause
4	suspect	victim	front	$graffiti
5	assault	$attack	arrest	graffito
6	head	accuse	entry	property
7	partner	scratch	make	spray
8	$victim-pushed	hit	spouse	break
9	$victim-threatened	face	drive	wall
10	$assault-section-eighteen	injury	suspect	accuse

Inspection of the terms alone shows some distinct patterns. A typical style of assault by a male will involve *punching* and injury to the *head* as well as *pushing* and *threatening,* whereas females *scratch* and *slap*. Males are more likely to assault a girlfriend (*partner*) whereas for females the violence is inter-generational. It is not clear from the terms whether the assaults are of *parents* or by *parents* although further inspection of the actual text reveals the latter is actually more likely. Male assaults appear to be more serious – *section 18 assault* and *actual bodily harm* are more serious allegations than *common assault*. For criminal damage, typical younger behaviour appears to be throwing *stones* and *spraying walls* with *graffiti*. Adults are more likely to *smash windows* and *kick* in *doors*. The damage appears to be directed at a specific *victim*. Smashing up a bus shelter is not a crime directed at an individual. However, criminal damage can be directed at a particular person and this appears to be a feature of older offenders. The presence of *spouse* implies that these are often domestic incidents. Clearly a more detailed analysis of all the terms, of which there are several hundred, over all the datasets would reveal further patterns. For example, older male shoplifters are more likely to steal spirits evidenced by words such as *bottle*, *whisky* and *vodka*. Also this is, of course, only one dataset from one locality and does not mean that these patterns are universal.

7 Conclusions

The results show that Language Models are capable of identifying differences in criminal styles between males and females and between older and younger offenders. It thus shows the power of an approach that could be applied more widely to other categories. Bernoulli models outperform multinomial models although the difference in performance is not great. For offences where there may be no eye-witnesses such as criminal damage, burglary and damage to vehicles the approach can indicate the likely age or sex of the offender. Furthermore the models can be used to characterise styles of behaviour. The analysis of the most dominant terms in each language model does reveal interesting characterisations of behavioural styles. This takes the use of language models beyond mere classification in that it can shed light on the material differences in behaviours between categories of offender.

References

1. Canter, D., Fritzon, K.: Differentiating Arsonists: A Model of Firesetting Actions and Characteristics. Legal and Criminal Psychology 3, 73–96 (1998)
2. Chen, H., Chung, W., Xu, J.J., Qin, G.W.Y., Chau, M.: Crime Data Mining: A General Framework and Some Examples. Computer 37(4), 50–56 (2004)
3. Bache, R., Crestani, F., Canter, D., Youngs, D.: Application of Language Models to Suspect Prioritisation and Suspect Likelihood in Serial Crimes. In: International Workshop on Computer Forensics (2007)
4. Canter, D.: Offender Profiling and Criminal Differentiation. Legal and Criminological Psychology 5, 23–46 (2000)
5. Canter, D., Bennell, C., Laurance, A.: Differentiating Sex Offences: A Behaviorally Based Thematic Classification of Stranger Rapes. Behavioral Sciences and the Law 21, 157–174 (2003)
6. Ponte, J.M., Croft, W.B.: A Language Modeling Approach to Information Retrieval. In: Proceedings of the Twenty First ACM-SIGIR, Melbourne, Australia, pp. 275–281 (1998)
7. Lafferety, J., Cheng-Xiang, Z.: Probabilistic Relevance Models based on Document and Language Generation. In: Croft, W.B., Lafferty, J. (eds.) Language Modeling for Information Retrieval. Kluwer Academic Publishers, Dordrecht (2003)
8. Bai, J., Nie, J., Paradis, F.: Text Classification Using Language Models. In: Asia Information Retrieval Symposium, Poster Session, Beijing (2004)
9. Peng, F., Schuurmans, D.: Combining Naive Bayes and n-Gram Language Models for Text Classification. In: Twenty-Fifth European Conference on Information Retrieval Research (2003)
10. Peng, F., Schuurmans, D., Wang, S.: Augmenting Naive Bayes classifiers with statistical language models. Information Retrieval 7(3), 317–345 (2003)
11. Zhai, C., Lafferty, J.: A Study of Smoothing Methods for Language Models Applied to Ad Hoc Information Retrieval. In: Proceedings of SIGIR, pp. 334–342 (2001)
12. Losada, D.: Language Modeling for Sentence Retrieval: A Comparison between Multiple-Bernoulli Models and Multinomial Models. In: Information Retrieval Workshop. Glasgow (July 2005)
13. McCallum, A., Nigam, K.: A Comparison of Event Models for Naïve Bayes Text Classification. In: Proc. AAAI/ICML-98 Workshop on Learning for Text Categorisation, pp. 41–48. AAAI Press, Menlo Park (1998)
14. Jelinek, F., Mercer, R.: Interpolation estimation of Markov source parameters from sparse data. In: Workshop on Pattern Recognition in Practice, Amsterdam, The Netherlands (1980)
15. Fellbaum, C. (ed.): WordNet – An Electronic Lexical Database. MIT Press, Cambridge (1998)

Software (Requirements) Engineering and Specification

Towards Designing Operationalizable Models of Man-Machine Interaction Based on Concepts from Human Dialog Systems*

Helmut Horacek[1,2]

[1] Universität des Saarlandes
FB 14 Informatik, Postfach 1150
D-66041 Saarbrücken, Germany
horacek@cs.uni-sb.de
[2] Technische Universität Wien
Institut für Computertechnologie
A-1040 Wien, Austria
horacek@ict.tuwien.ac.at

Abstract. Designing interfaces for human-computer interaction components is time-consuming, error-prone, and requires substantial programming skills. Therefore, a variety of models for designing interfaces on a more abstract and intuitive level have been proposed, a few of them being based on ingredients of human discourse processing techniques. In this paper, we compare and contrast two of these approaches, and we develop essentials of a joint model which profits from their complementary advantages. Essential factors in this model are an adequate distribution of labor between the design made by the user and the operationalization capabilities of the system, suitable elements for the design language available to the user, and techniques that allow the specification of effective interaction sequences. The model is intended as the basis for an interaction design tool that allows the specification of graphical user interfaces as well as multi-modal interaction with a robot in a restricted real-world environment.

1 Introduction

Designing interfaces for human-computer interaction (HCI) components tends to require increasing resources in developing software systems. Moreover, existing tools do not well support the designer in avoiding errors, and developing interfaces requires substantial programming skills. The last two factors are also characteristic for the development of the proper interaction part in natural language dialog systems. Because of this bottleneck in software engineering, a variety of models for supporting the design of human-computer interaction components on a symbolic level have been proposed, and the results are ultimately compiled into the proper interface software. However, only a few of these appraoches operate on a more abstract and intuitive level in terms of the semantics of the underlying communication rather than the

---

* This research has been carried out in the CommRob project and is partially funded by the EU (contract number IST-045411 under 6th framework programme).

E. Kapetanios, V. Sugumaran, M. Spiliopoulou (Eds.): NLDB 2008, LNCS 5039, pp. 271–286, 2008.

syntax of widgets. Among these, only two [3,7] are based on ingredients of human discourse processing techniques, so that also communication in natural language is supported, which makes them highly interesting for our purposes.

In this paper, we compare and contrast these two approaches. The first one is based on concepts of rhetorical relations and communcative acts, and it comes with a graphical model development tool and operationalizations in terms of finite-state machines. The second one is based on concepts of information state-based techniques for modeling the effects of natural language discourse, applied to human-computer communication situations; it is only a theoretical model. On that basis, we develop essentials of a joint model which profits from their complementary advantages. Essential factors in this model are an adequate distribution of labor between the design made by the user and the operationalization capabilities of the system, suitable elements for the design language available to the user, and techniques that allow the specification of effective interaction sequences.

This paper is organized as follows. We first motivate our approach. Then we describe the two complementary models, and we compare and contrast them. We follow with stating and justifying desiderata for a joint model. Next, we describe and explain some central concepts in a joint model that meets these desiderata, and we discuss some examples. Finally, we describe the intended application of the joint model and discuss some future extensions that strongly increase its usefulness.

2 Motivation

In the introduction, we have stated that designing interfaces for human-computer interaction components is time-consuming, error-prone, and requires substantial programming skills. In our view, a fundamental reason for this unsatisfactory situation lies in the way the development process is organized, which widely works on a syntactic level in terms of sets of widgets, rather than on a semantic level capturing the task-relevant flow of information. Thus, we believe that specifying interfaces on a more abstract level is easier and more intuitive for developers and it can be done faster, since it discharges them from the burden of manipulating low-level technicalities. Consequently, it allows one to concentrate on conceptual issues of the interface development.

An essential hope to make the development of human-computer interaction components easier and less error-prone lies in using descriptions of components, functionality, and behavior of these interfaces that capture the semantics of the communication involved. An important impact of such a specification technique lies in avoiding the need for technical expertise. This would make the interface development process accessible to the domain experts as well, which is beneficial since these people have a better understanding of the required system functionality. Opening up the task of interface development to a larger segment of potential users is the essential social and organizational motivation for the approaches addressed in this paper.

Besides increasing the ease of development and its comprehensibility, there are also a number of technical motivations that have impact on how interfaces can be designed, tested, maintained, and even applied in varying contexts. Some of these issues are related to the correctness aspect:

- *Consistency*
  In principle, abstract models have the potential of supporting a developer significantly in this respect, when generalities can be expressed at an appropriate place. For example, navigation options can be specified once for an application fragment so that they are percolated systematically to all relevant places. Another factor are requirements of being conform to norms or conventions, some of which can only be checked at an abstract description level. Handling such requirements effectively is also considered an advantage of natural language generation systems.
- *Completeness*
  The development of a human-computer interaction component may turn out to be problematic when applications get larger and taking into account all possible constellations gets increasingly harder for the designer. Consequently, it may easily happen that some situation is forgotten, which is very expensive to uncover by extensive testing.

Further advantages concern the development over time and its usage in divergent environments:

- *Extendability*
  This issue essentially addresses the first two items above, from the perspective of incremental development. Hence, every support mentioned there is likely to pertain to supporting extendability as well. Specifically, the use of an abstract level of design pays off in cases where the additional properties are a simple adds-on in conceptual terms, but require considerable recasting in the appearance of the interaction elements. Moreover, incremental development with inspection facilities of the state of completion are even beneficial for a first-shot development.
- *Adaptivity*
  The demand of using an application program under varying devices is best met by a separation into *abstract* and *physical* description levels of an interface, which is typically underlying linguistic approaches [1, 9]. Such a separation essentially corresponds to factoring out device-independent issues from those which at least potentially depend on given presentation facilities.

In the following descriptions, we will refer to some of these motivations.

3 Two Methods for Interface Design

In this section, we describe the two methods that model human-computer interaction on a more abstract semantic level, and we compare and contrast them in terms of their contribution and potential cross-fertilization. These models are (1) an operationalized system mostly based on communicative acts and rhetorical relations [2,3] that comes with a modeling tool (subsection 3.1), and (2) a theoretical development [7] that is based on the technique of representing and processing "information states" (subsection 3.2). The comparative assessment is made in the last subsection (3.3.).

3.1   The Model Based on Communicative Acts and Rhetorical Relations

The model described in this subsection is based on results from *Speech act theory* [14], *Rhetorical Structure Theory* (RST) [12], and *Conversation analysis* [11], and it is used for modeling interaction design. From *Speech act theory*, this model borrows communicative acts that express intentions in the sense of desired effects of the environment. Communicative acts are generalizations of speech acts in the sense that they abstract from the form in which the information associated is transmitted, where they refer to elements of the domain of discourse. This may not only be done in natural language, but also by some other medium, such as some kind of widgets in a Graphical User Interface (GUI), or even some conventionally interpretable gesture. Some of the communicative acts used are *Question, Answer, Offer, Accept, Reject, Request,* and *Confirm. Rhetorical Structure Theory* which is a linguistic theory focusing on the function of text and widely applied to natural language generation, provides internal relationships among text portions and associated constraints and effects. These relations are used to put together related portions of information in the interaction model, and the associated constraints and effects are exploited in the operationalization, for example, to motivate decisions about how to organize the presentation of semantically related information in the later applied rendering process. Rhetorical relations used comprise a subset of the *ideational* subcategory (essentially relations that express a semantic view), including JOINT, CONDITION, ELABORATION, and BACKGROUND. *Conversation analysis*, finally, provides evidence about sequences of naturally occurring talk "turns", which constitute patterns regarded familiar to humans. Some of the simpler patterns are used for modeling interaction design, such as "adjacency pairs" (e.g., an *Offer* and an *Accept*), and "inserted sequence".

In structural terms, the model combines *communicative acts* into *adjacency pairs* so that the communicative acts form meaningful patterns. To account to some extent for the diversity of possible reactions in a discourse, alternatives are allowed for the second component of such a pair, such as *Accept* and *Reject* going together with an *Offer*. These *adjacency pairs* (sometimes also two related *adjacency pairs* sharing the first component) form the leaf nodes in a tree which is composed by rhetorical relations. Thus, *rhetorical relations* recursively build a discourse tree, as in *Rhetorical Structure Theory*, but the leaf nodes are populated by small interaction patterns in the form of adjacency pairs instead of text portions. In that sense, the model presents a combination of dialogic elements with rhetorical relations that are typically used to describe internal relations in monologic discourse. A graphical tool has been implemented that allows the user to incrementally design a task-oriented interaction model. Thereby, the user is suitably guided in his actions insofar as he is only allowed to build models that conform to the underlying definitions – for example, he is forced to obey to the arity of rhetorical relations, and he can only put together communicative acts that form an *adjacency pair*. Figure 1 shows a fragment of a discourse tree for this approach. Rhetorical relations are enclosed in rectangles, joining their nucleus (N) and satellite (S). Communicative acts appear in ovals, with the category of the communicative act on top, and content specifications below. The small triangles link together adjacency pairs. The diagram models a simplified flight booking dialog: the system asks the user about date, departure and arrival localions, connected by JOINT relations. Provided this worked out successfully (according to the

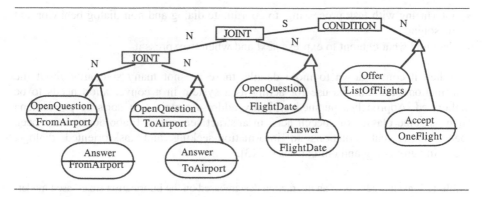

Fig. 1. A fragment of an interaction design based on the model based on rhetorical relations

rhetorical relation CONDITION), the system offers the user a set of flights that meet these specifications, so that the user can finally accept this offer by picking one of these flights.

The interaction design built this way can be conceived as design rationale for the interface. This tree represents a kind of typical interaction, strictly speaking one straightforward example of a conversation where no extra or repetitive turns of interaction occur. Since this is both unrealistic and inflexible, some reasonable form of "extended" interpretation is implicitly assumed, such as repeating a question until answered satisfactorily. Consequently, a rhetorical tree in the discourse model somehow represents a set of discourses where each member is conform to one of the options implicitly contained in the rhetorical tree specifications. So far, a member of this set differs from other members only in terms of user answers (acceptable or inacceptable ones), which may or may not lead to the repetitions of a subdialog in dependency of the answer category.

Based on a number of tutorials and a few experiments, there is evidence that moderately trained people are able to specify reasonable interaction models by using the tool. Fully specified models can be operationalized in terms of an abstract user interface model [4], which is further transformed into a concrete interface with suitably rendered widgets. For this transformation step, techniques have been developed to adapt the interface to the situational context, such as device constraints [5].

3.2 The Model Based on the Information State Approach

Before describing the model based on the information state approach, we give some background information about this technique. According to Traum and Larsson [10, 15], the purpose of dialog modeling with information states includes the following functionalities:

- updating the dialog context on the basis of interpreted utterances
- providing context-dependent expectations for interpreting observed signals

- interfacing with task processing, to coordinate dialog and non-dialog behavior and reasoning
- deciding what content to express next and when to express it

When it comes down to more details, there are not many standards about the information state, and its use for acting as a system in a conversation needs to be elaborated. Approaches pursued typically address certain text sorts or phenomena such as some classes of speech acts, in abstract semantics. Elaborations have been made for typical situations in information-seeking and task-oriented dialogs, including grounding and obligations [8, 13].

Table 1. Some ingredients of an interaction design based on the information state-based model

Rule trigger	Force of the rule and some preconditions
ask(S,U,x)	introduces a user obligation to answer the question eventually
answer(U,S,x)	requires an obligation for conveying x, adds new information to the common ground, and propagates constraints
inform(S,U,x)	effects are not modeled; used for error messages
accept(U,x)	requires an active user obligation regarding an offer with regard to x, drops the obligation, and activates x
accept(S,x)	retracts all user obligations within the scope of x, pops previous user communicative acts from the stack, and also preceeding system communicative act, unless corrections of specifications are permitted
reject(S,x)	retracts preceeding common ground update, pops previous user communicative act from the stack

The model based on the information state approach is built around four central concepts: 1) communicative acts, 2) the discourse history, 3) discourse objects, and 4) system functionalities. These concepts are quite standard except to the system functionalities, which are the concept particular to this model. Communicative acts are widely in accordance with the other model described in this section, with categories chosen/elaborated according to the needs of HCI. The discourse history constitutes a representation of the active part of the course of interaction, as far as it is relevant for references to discourse objects and reasoning about communicative acts, organized in terms of focus spaces in the sense of Grosz and Sidner [6]. Discourse objects represent abstractions of domain objects, in terms of partial descriptions of the properties of domain objects. They play the role of binding several pieces of information that are possibly communicated at different times. Hence, these objects are incrementally made more precise over the course of communication. System functionalities represent purely functional views of subcomponents of the application system such that these functionalities are accessible by appropriate procedures. The system functionalities are conceived as a set of actions performed by the application system, reduced to their input/output behavior. The input is defined in terms of a set of discourse objects. The output expresses changes imposed on discourse objects or the creation of new ones, accompanied by the state of the action performed. In addition, dependencies on system functionalities are expressed, which is used for determining the state of a task on the basis of the state of those tasks it is depending on. This way, task completion is checked, which enables adequate bookkeeping in the discourse history.

On the basis of these four concepts, information state-based update rules are defined. These rules encapsulate the effect of communicative acts, and they are invoked to determine a reaction when it is the system's turn. Obligations and commitments carried by communicative acts are put on the stack that represents the discourse history. Simultaneously, the state of discourse objects is maintained; once specified to a sufficient degree, access conditions of system functionalities may be accomplished, so that they are invoked by update rules. The results of system functionalities, in turn, are typically such that some of the obligations and commitments are retracted again as a consequence of the corresponding rules. Table 1 shows descriptions of the effects of some communicative acts on the information state. In Figure 2, a fragment of interaction sequences for this approach is shown, elaborated for the task of flight booking. It widely covers the same task as Figure 1 does for the other model. In Figure 2, system functionalities are enclosed in triangles, and a rectangles is used to put together the set of communicative acts that the user can address during the flight specification, which can be considered the abstract counterpart to some portion of a form that allows users to specify their demands in a GUI. The system functionality *Check Inform* takes care of verifying the correctness of individual user specifications, which may be influenced by the information state – for example, the specification of the second airport location must be such that a connection exists from the first location. The system functionality *Perform Action* searches the database for the given specifications and delivers the results in terms of a list of flights.

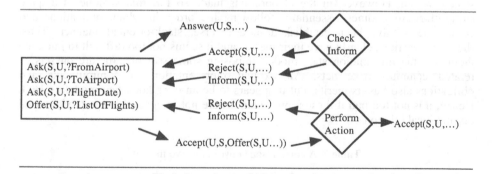

Fig. 2. A fragment of an interaction design based on the information state-based model

3.3 Comparing and Constrasting the Two Models

The preceeding two subsections have illustrated that the two models have a widely comparable functionality, as far as the abstract perspective of the models is concerned. Obvious commonalities are in some sense the "observables", that is, the communicative acts and the role of the domain model. Compared to the system functionalities, the incorporation of the domain model elements is specified in the model based on rhetorical relations in a less formal manner. Another commonality is the primary effect that both models achieve: they regulate the sensible composition of communicative acts, both in terms of sequences that are possible in principle, as well as with regard to the development of the discourse in dependency of the content of communicative acts, although the model based on rhetorical relations does this in an implicit form – repeating a communicative act until some condition is met. However, the techniques to regulate compositions are quite different.

There is one component in each model which has no counterpart in the other one: rhetorical relations in the model presented first, and obligations and commitments in the other model. Each of these components has a specific benefit. Rhetorical relations do not only indicate constraints for the course of the conversation, such as a CONDITION, which requires the subdialog rooted under its satellite to provide suitably instantiated domain model elements (such as flight data in Figure 1), in order to process its nucleus (such as the list of flights matching these specifications in Figure 1). Moreover, rhetorical relations carry information about how components of closely related subtrees, such as some connected information-seeking adjacency pairs, are semantically related. In Figure 1, for instance, the two subdialogs about the from- and to-airport built one substructure, which is then connected to a subdialog about the flight date; in Figure 2, the questions associated with these items appear in a simple list, together with the offer to list suitable flights. In the model based on rhetorical relations, this information is exploited by the subsequent rendering process to support a cognitively adequate grouping of the associated communicative acts. The role of obligations and commitments lies more on a meta-level. They constitute explicit representations of open issues in a discourse, which in the model based on rhetorical relations resides implicitly in the composition of subdialog structures. Consequently, the model based on information state allows for reasoning about open obligations and commitments, which can be incorporated in update rules that determine the next system action. However, in the elaborations made so far, this is done in a quite straightforward manner, essentially following the order in which obligations and commitments have been established in a reverse, stack-oriented manner. Thus, rhetorical relations clearly have their merits, and it seems not too difficult to integrate them into the information state model, imposing some sort of structure on sets of related communicative acts. An explicit representation of commitments and obligations also has its merits, but it appears to be an overshot when modeling HCI. Hence, it is not too bad if these representations are not integrated in the model based on rhetorical relations.

Table 2. A comparison between the two models

Model based on	Rhetorical relations	Information state
control over correctness	high, especially in the form	in general low
flexibility in specifications	relatively low	relatively high
coverage so far	basic	medium
ease to build specifications	reasonably good	requires rule building skills

A crucial difference between the models are the methods by which results are achieved. In the model based on rhetorical relations, a combination of structural and functional elements captured in the composition of rhetorical relations, communicative acts, and the interpretation of alternatives and repetitions due to still unmet constraints are the driving forces for the operationalization of the model. In the model based on information state, the operationalization lies entirely in the update rules and their coordination. The structural backbone of the model based on rhetorical relations gives the designer much control, but the repertoire of the defined structural compositions somehow limits the flexibility in modeling. In contrast, the set of update rules in the information state-based model gives the designer much flexibility, but it also leaves him with the burdon of controling the effects of the defined set of rules.

Further differences concern the coverage achieved so far, and the ease to build specifications. At present, the coverage of the information state-based model seems to be higher insofar as typical patterns of subdialogs are handled, such as explicit rejection and associated information, and repeated options of making additional specifications or changing already given ones. The specification of these patterns can also be done quicker than in the model based on rhetorical relations, but it requires familiarity with formulating updates rules and coordinating them with the existing set of rules. As mentioned earlier, there is evidence for the ease and intuitiveness in modeling with rhetorical relations and communicative acts, which enables users to produce something meaningful after a short introduction. For the information state-based model, developing a rule-based system is the task to be accomplished, which requires some sort of programming skill. Hence, the two models can in some sense be contrasted on a spectrum ranging from good control which is increasingly compensated by flexibility along this scale. A similar contrast appears between coverage, which is in some sense associated with quick development facilities, and ease of handling and understanding the modeling facilities. The compensative effects are listed in Table 2.

Consequently, when aiming for a joint model there is a crucial choice whether to put the emphasis on control or on development flexibility. Depending on what the preference among these choices is, the model favored should form the basis in the joint model, and extra components of the other model should be re-interpreted in the scope of the model that forms the basis. When referring back to the motivations for building HCI specifications on the lines of a model for human interaction, one of the major arguments was the goal of having domain experts design the interface in a way that is intuitive and does not require any kind of programming skills. Therefore, the option of using the information state-based model as the major starting components appears less suitable, since it requires familiarity with developing rule-based systems. Moreover, it is far from clear how the support for correctness of compositions, which resides in the GUI of the tool associated with the model based on rhetorical relations, can be rebuilt in a rule-based system which puts little restrictions other than accordance with domain ontology on how precisely the rules are formed. Hence, we suggest that the model based on rhetorical relations forms the basis for a joint model. The model based on information states constitutes then some sort of platform for exploring communication situations suitable for HCI, thereby formulating the underlying constraints and effects.

4 Desiderata for a Joint Model

Following the considerations raised in the previous section, we formulate desiderata for a joint model, which amounts to suitably extending the system based on rhetorical relations to incorporate functionalities of the information state-based model, and probably more than that. These extensions concern two perspectives: the coverage achieved by the functionalities of the model (1-3), and the proper interaction design in terms the kind of specification tasks that the user has to perform (4):

1. *Clarification, confirmation and correction subdialogs*
   Dealing with situations where the communication medium is associated with some kind of uncertainties, such as speech input, raises the need for checking the content of messages to reduce the risk of misunderstandings. Another more advanced and content-related motivation for checking operations is present if there is some doubt on the plausibility of specifications made by the user. Moreover, allowing the user to correct his specifications in the context of such subdialogs is likely to contribute to increasing the effectiveness of the interaction.

2. *Communication to jointly perform composed tasks*
   In moderately complex application domains, the accomplishment of some tasks may be achievable by the accomplishment of a set of related tasks which the original one is composed of. In communicating about these tasks, addressing all related concepts, specifically those coming from embedding tasks in addition to those joined with the subtask at hand significantly contributes to the flexibility of the overall conversation.

3. *Flexibly navigating between tasks*
   If the system's domain comprises a number of tasks, some of which can be carried out alternatively or where the order of the tasks can be freely interchanged, the user should be given considerable degree of freedom in the order in which these tasks are accomplished; hence, some task might be interrupted, interleaved with another one, and resumed again later.

4. *Accordance between what the user must specify and what he is able to do*
   There are a couple of properties of the application which the user must specify in some way in order to contribute to the extended functionalities as stated above. The means to achieve this desideratum essentially lies in letting the user make his specifications in a declarative manner, referring to rhetorical relations, communicative acts, and elements of the domain ontology.

5 Essential Concepts in a Joint Model

In order to meet these desiderata, we introduce some new concepts in the interaction specification:

- *Discourse variables*
  In the model based on rhetorical relations, the user can only refer to elements of the domain model when specifying the content of communicative acts. Beyond that, there are no expressive means that concern their effects, which is quite restricted compared to the general information state-based approach. In order to express the effects of some interaction properly, discourse variables are introduced, which change their values according to specifications attached to communicative acts. Among others, this concept may be used to regulate the sequence of interactions within clarification/confirmation subdialogs. For example, the information

conveyed through some communicative act, typically an answer or a request, may be associated with some degree of uncertainty, especially for speech. To express this assessment, the information conveyed is associated with a discourse variable, which gets assigned a value that expresses the state of this assessment, for example "uncertain". Another example of a discourse variable is shown in Figure 3, with a property of the "current action" being referred to as "ambiguous". Through additionally associating constraints that access such discourse variables with optional parts or alternatives in the interaction specifications, the circumstances under which specific interactions are pursued are made explicit, for example, whether or not a clarification subdialog is invoked

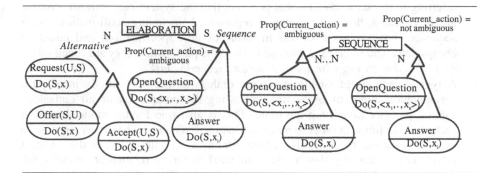

Fig. 3. An alternation and a compact expression (left side) and the expansion of the latter (right side)

(see also Figure 3), or a confirmation is asked for. In addition, effects manifest themselves in changes of the discourse variables involved, such as changing the state from "uncertain" to "certain", and later on to "confirmed". Another category of situations where discourse variables are beneficial, concerns iterative repetitions of interactions. Examples again include clarification subdialogs, which may be repeated until the assessment of the state of the associated information changes, as well as sequences of tasks, through making a new offer to the user once the current task is finished. Through making the primary effect of the response accessible (e.g., the user wants to stop) the condition for exiting this iteration is fulfilled. In some sense, the discourse variables used in this way represent a limited and formally constrained version of the information state.

- *Compactified representations of discourse subtrees*
There are a number of discourse patterns which obey a quite regular structure but cannot easily be visualized in a compact and intuitively comprehensible form. One of these patterns concerns the kind of iterations discussed in the previous paragraph. In terms of rhetorical relations, such a pattern amounts to an arbitrary number of chained SEQUENCE relations, where each of the subtrees rooted in individual SEQUENCE relations has the same generic specifications – see Figure 3 for the repeated performance of a clarification subdialog until the need for clarification is eliminated. Thus, it seems desirable to abstract from the whole sequence in the context of the user specification, and restricting the notation to the

subtree common to all chained elements in the sequence. In order to do this properly, this tree should be encapsulated to make it referable as a unit, a "sequence" marker should be attached to it, and a constraint over relevant discourse variables should be formulated which regulates continuation and termination of the iterating process. Figure 3 shows an example (the SEQUENCE part on the left), with an expanded version of this fragment on the right: an arbitrary number of elements of the sequence associated with the entry condition, and a final element associated with the negated condition. Another category of situations where encapsulating compositions of rhetorical relations is beneficial concerns alternations for achieving identical effects. An example are situations with mixed initiative, e.g., an offer made by the system and a request by the user referring to the same content, that is a task that the system can perform – on the left of Figure 3, this constellation is represented. Including confirmations, such sequences have identical effects in terms of task specifications and could be encapsulated and referred to compactly from the place where they are specified.

- *An interaction strategy for interrupting and resuming tasks*
  A typical property of simply structured dialogs is that they strictly insist on a chronological order of processing, including the completion of all embedded subdialogs prior to shifting the topic. This is also true for the model based on rhetorical relations. In order to make the interaction more effective, we propose to give the user means to specify where continuations alternative to the adjacent communcative act or following the connected rhetorical relation are meaningful. The set of candidate continuation points amounts to the path of embeddings from the root of the discourse tree to the currently modeled subdialog – new subdialogs may be started at any of these levels: for instance, a flight booking task may be interrupted to start a subdialog about train booking, or a new dialog may be started at the entry point of the whole application. In addition, embedding of a subdialog in the current position can be made if the newly started one does not interfer with the present one (e.g., activating the help function), so that the previous task can be resumed when the newly started one is finished (see the categories in Table 3). The user's task essentially lies in putting restrictions on the set of possible continuation points, such as choosing from task lists offered or specifying constraints on feasible continuation points. For example, the user may disallow going back over some compound task in the discourse tree, so that canceling this task within the interaction requires an explicit communicative act that signals this cancelation.

Table 3. Handling user options for specifying dialog interruption and continuation points

Category of interruption and continuation point	Specifcation made by user	Examples
Goes back to an embedding task	restrictions on levels or tasks	restarting at the introduction
Task parrallel to the present one	selection from the set of tasks	switching from flights to trains
Dialog about an embedded task	selecting tasks from offered list	activating the help function

These concepts are tailored in such a way that the user can specify the associated information in a declarative manner. Consequently, the system has to perform a number of additional functionalities to provide operationalizations for the extended specifications. They comprise percolating the state of discourse variables across subdialogs as context settings, expanding compactified subtree notations into its components joined by the associated constraints (as shown in Figure 3), and handling interruption and continuation policies for subdialogs. The last point comprises:

- *Making use of domain task definitions*
  In order to offer choices for continuation points, the system must be acquainted with relations among domain tasks, such as decomposition and dependency. On the basis of that, lists for being offered to the user are computed (e.g., embedding tasks according to task compositions).
- *Handling the structure of subdialogs*
  This task essentially is the administration of a stack, with some of the discourse variables duplicated for each new pile on the stack – for example, the state of the local goal specification. In dependency of chosen continuation points, potentially unfinished subdialogs must be closed and new subdialogs must be opened at the right level of embedding the discourse structure.
- *Taking the interaction mode into account*
  The interruption and continuation handling is slightly different when the interation is made via a GUI, as opposed to interaction in natural language. For a GUI, interruption must be made possible by providing facilities to access predefined continuation points. They are ultimately converted into widgets which announce, for example, "start" or "train booking". In the context of NL interaction, communicative acts unsuitable for a specified subdialog must be tested as to whether they could be interpreted as possible interruption and continuation messages. Hence, such a communicative act opens a subdialog consistent with a possible continuation point, it is interpreted in this manner; otherwise the contextually improper contribution is refused.

6 Future Developments

The model discussed in the previous section will be applied within a real-world environment where the communication platform to be developed is supposed to regulate the interaction between a robot and a user (EU project CommRob). The application situation is a supermarket environment where the robot has the form of a trolley. The trolley supports the user in navigating through the shopping area, and it helps the user in processing a list of items which he intends to inspect and potentially buy. For this activity, multi-modal interaction is foreseen, including a traditional GUI, a restricted speech interaction, and even some form of gesture interpretation. The interface component presented here abstracts from these forms of interaction, but it takes into account the varying degrees of uncertainty associated with the information conveyed in dependency of the medium used (for example, GUI versus speech), which influences the need for clarification and confirmation subdialogs. In addition, the application environment is complex enough so that extended functionalites as described in the preceeding sections are needed: a compound task, implicitly specified in terms of a shopping list is interpreted in terms of a set of embedded navigation tasks, and carrying out these subtasks may be interrupted or changed by further user specifications, such as stopping for some time or performing another navigation task in between, or corrections in the shopping list. In order to achieve such a behavior with the robot, the communication must enable flexible task interaction, which is in accordance between speech and the GUI, that is the user may utter a command to start a new task interleaving with a still running one, and the GUI must also foresee the possibility of such an interruption and embedding.

In the future, we envision further extensions of the model by a couple of advanced capabilities. They should enable the user to work with the system more effectively, help him to reduce sources of errors, and support him to increase the quality and effectiveness of the ultimately resulting interface. The following quite challenging capabilities contribute to this goal:

- *Making use of a catalog of reusable subdialogs*
  A number of interaction patterns in HCI recur across domains and applications, with adaptations to specific situations. Examples are clarification and correction subdialogs, which follow stereotype sequences of communicative acts, in dependency of the number of items and the type of information to be conveyed and the associated degree of uncertainty. Abstracting such a pattern into a unit where the user only has to specify the ingredients, and their structural interplay is then determined by the tool would greatly improve design effectiveness.
- *Checking the discourse model in terms of completeness*
  When building interface specifications, especially with the extended expressiveness obtained by discourse variables, it can always be the case that some constellation in terms of a set of values assigned to discourse variables is not handled or mistreated in some other way. Checking the completeness in terms of the presence of situational specifications for each value that a discourse variable is assigned through one or another component of the interface specifications would enable one to discover such unaddressed situations. However, developing a method that is able to perform such an analysis is quite challenging.
- *Critiqueing a discourse model in terms of potential communicative weaknesses*
  Another candidate for an automated analysis of interface specifications lies in reasoning about the resulting flow of control, by integrating changes in the values of discourse variables and updates in the information that constitutes the common ground. By comparing the effect of the interactions specified with existing building blocks and their functionality, proposals about a possibly more direct or otherwise preferable interaction composition can be made.

The last item, as opposed to the other two, has the potential of going beyond a mere abstract model of communication. In addition, issues of rendering may be considered in the selection among competing alternatives of communicative patterns. Some of these issues may be device-dependent to a certain extent, but not all – for instance, guidelines about suitable numbers of menu items.

7 Conclusion

In this paper, we have compared and contrasted two approaches to building interface specifications, motivated by models of human communication that can be operationalized for building HCI components. We have argued in favor of essential parts and further extensions in a joint model which profits from the complementary advantages of these two models. Essential factors in this model are an adequate distribution of labor between the design made by the user and the operationalization capabilities of the system, suitable elements for the design language available to the user, and techniques that allow the specification of effective interaction sequences. The model to be developed will form the basis for an interaction design tool that allows the specification of graphical user interfaces as well as multi-modal interaction, and it will be applied in the restricted real-world environment of the project CommRob to allow a user to communicate with a robot.

References

1. Bateman, J., Kleinz, J., Kamps, T., Reichenberger, K.: Towards Constructive Text, Diagram, and Layout Generation for Information Presentation. Computational Linguistics 27(3), 409–449 (2001)
2. Falb, J., Popp, R., Röck, T., Jelinek, H., Arnautovic, E., Kaindl, H.: Using Communicative Acts in High-Level Specifications of User Interfaces for their Automated Synthesis. In: Proceedings of the 20th IEEE/ACM International Conference on Automated Software Engineering (ASE 2005) (tool demo paper), pp. 429–430. ACM Press, New York (2005)
3. Falb, J., Kaindl, H., Horacek, H., Bogdan, C., Popp, R., Arnautovic, E.: A Discourse Model for Interaction Design Based on Theories of Human Communication. In: CHI 2006, Extended Abstracts on Human Factors in Computing Systems, pp. 754–759 (2006)
4. Falb, J., Popp, R., Röck, T., Jelinek, H., Arnautovic, E., Kaindl, H.: Using Communicative Acts in Interface Design Specfications for Automated Synthesis of User Interfaces. In: Proceedings of the 21th IEEE/ACM International Conference on Automated Software Engineering (ASE 2005) (tool demo paper), pp. 261–264. ACM Press, New York (2006)
5. Falb, J., Popp, R., Röck, T., Jelinek, H., Arnautovic, E., Kaindl, H.: Fully Automatic Generation of User Interfaces for Multiple Devices from a High-Level Model Based in Communicative Acts. In: Proceedings of the 40th Annual Hawaii International Conference on System Sciences (HICSS-40), Hawaii, p. 10 (2007)
6. Grosz, B.J., Sidner, C.L.: Attention, Intention, and the Structure of Discourse. Computational Linguistics 12(3), 175–204 (1986)
7. Horacek, H.: An Abstract Model of Man-Machine Interaction Based on Concepts from NL Dialog Processing. In: Kop, C., Fliedl, G., Mayr, H., Métais, E. (eds.) NLDB 2006. LNCS, vol. 3999, pp. 129–140. Springer, Heidelberg (2006)
8. Kreutel, J., Matheson, C.: Incremental Information State Updates in an Obligation-Driven Dialogue Model. Logic Journal of the IGPL 11(4), 485–511 (2003)
9. Lampert, A., Paris, C.: Information Assembly for Adaptive Display. In: Proceedings of the 2004 Australasian Language Technology Association Workshop (ALTW 2004), Macquarie University, pp. 63–70 (2004)
10. Larsson, S.: Issue-Based Dialogue Management. PhD thesis, Gothenburg University (2002)

11. Luff, P., Gilbert, N., Frohlich, D.: Computers and Conversation. Academic Press, London (1990)
12. Mann, B., Thompson, S.: Rhetorical Structure Theory: Toward a Functional Theory of Text Organization. Text 8(3), 243–281 (1988)
13. Matheson, C., Poesio, M., Traum, D.: Modelling Grounding and Discourse Obligations Using Update Rules. In: Proceedings of the 1st Annual Meeting of the North American Association for Computational Linguistics (NAACL 2000), pp. 1–8 (2000)
14. Searle, J.: Speech Acts: An Essay in the Philosophy of Language. Cambridge University Press, Cambridge (1969)
15. Traum, D., Larsson, S.: The Information State Approach to Dialogue Management. In: Current and New Directions in Discourse and Dialogue. Kluwer, Dordrecht (2003)

Using Linguistic Knowledge to Classify Non-functional Requirements in SRS documents

Ishrar Hussain, Leila Kosseim, and Olga Ormandjieva

Department of Computer Science and Software Engineering,
Concordia University, Montreal, Quebec, Canada
{h_hussa,kosseim,ormandj}@cse.concordia.ca

Abstract. Non-functional Requirements (NFRs) such as software quality attributes, software design constraints and software interface requirements hold crucial information about the constraints on the software system under development and its behavior. NFRs are subjective in nature and have a broad impact on the system as a whole. Being distinct from Functional Requirements (FR), NFRs are dealt with special attention, as they play an integral role during software modeling and development. However, since Software Requirements Specification (SRS) documents, in practice, are written in natural language, solely holding the perspectives of the clients, the documents often end up with FR and NFR statements mixed together in the same paragraphs. It is, therefore, left upon the software analysts to classify and separate them manually. The research, presented in this paper, aims to automate the process of detecting NFR sentences by using a text classifier equipped with a part-of-speech (POS) tagger. The results reported in this paper outperform the recent work in the field, and achieved a higher accuracy of 98.56% using 10-folds-cross-validation over the same data used in the literature. The research reported in this paper is part of a larger project aimed at applying Natural Language Processing techniques in Software Requirements Engineering.

1 Introduction

Software systems are characterized both by their functionality (what the system does) and by their non-functionality (how the system behaves with respect to some observable attributes like reliability, reusability, maintainability, etc.), captured in a Software Requirements Specification (SRS) document [10]. Software nonfunctional requirements (NFRs) are defined as requirements that constrain the design of software, but do not describe a service the software is to provide [11]. The functional requirements (FRs), on the other hand, are defined as a subset of the user requirements which represent the user practices and procedures that the software must perform to fulfill the users' needs; they exclude NFRs such as quality requirements and any technical requirements [11]. Consider the following example: "The System shall allow generation of Inventory Quantity Adjustment documents on demand. The System shall not require additional third party licenses resulting in royalty fees." [6] Here, the first

E. Kapetanios, V. Sugumaran, M. Spiliopoulou (Eds.): NLDB 2008, LNCS 5039, pp. 287–298, 2008.

sentence describes the behavior of a system, and therefore, is a functional requirement. In contrast, the second sentence explains a required quality of the system, and therefore, is a non-functional requirement.

Both FRs and NFRs are relevant to software development. However, non-functional issues have received little attention relative to functional ones for the reason that such requirements are difficult to address in many projects, since the NFRs for each system typically interact with one another, have a broad impact on the system, and may be subjective. Moreover, one of the software engineering problems faced in building SRS documentation is the inability to detect NFRs in user requirements written in natural language. Empirical reports consistently indicate that neglecting NFRs in the requirements elicitation and analysis phase leads to project failures, or at least to considerable delays, and, consequently, an escalation in the cost of software development [2,4]. The ability to detect NFRs in NL text at a very early stage of requirements elicitation has the potential to bring about significant industrial savings in cost and aggravation, as well as to prevent very costly misinterpretation.

The research reported in this paper addresses the problem of providing automated assistance to the elicitation of the NFRs from the user requirements text and is aimed at applying Natural Language Processing (NLP) techniques to Requirements Engineering (RE). We present our approach towards an effective method for automatic classification of textual requirements into two categories, namely FRs and NFRs, by means of a text classifier equipped with a part-of-speech (POS) tagger. Since the characteristics of FR and NFR remain within the scope of sentences, the classifier works only at the sentence-level.

The remainder of this paper is organized as follows: section 2 presents the related work. The methodology for automatic assessment of NFRs is introduced in section 3. Section 4 presents a discussion of the results and observations. Finally, our conclusions and directions for future work are outlined in section 5.

2   Related Work

The current processes to extract NFRs from SRS documents mostly rely on manual inspection, where an analyst reads the texts to identify a sentence manually as FR or NFR following different approaches (e.g. [5,7,8]). Research in this field to automate the process of extracting NFRs from SRS documents has been scarce.

A recent study by Cleland-Huang et al. [6] explored the use of text classification as an attempt to classify requirements statements into FR and NFR. As reported in their paper, their work attained a recall measure of 0.767 and a precision measure of 0.248 with their corpus. The authors used a stemmer to stem the words of the documents, and then selected keywords based on their high probability of occurrences in NFR statements. Their system then classified a statement as NFR, if the density of those selected keywords in that statement exceeds a particular threshold, else, otherwise.

The research work presented in this paper used the same corpus that was used by [6] and compares the performance of the resultant classifier with that of theirs. The results reported in this work outperform the recent work in this field [6], and attain a higher accuracy of 98.56% using 10-folds-cross-validation over the data used by [6].

3 Methodology

Ideally, classifying the requirements as FR or NFR should be performed, or at least assisted, automatically. This section introduces our methodology aimed at improving the NFRs detection in requirements documents.

3.1 The Corpus

The corpus used by [6] was made freely available for download via [1]. It contains 15 SRS problem statements, all from different domains, with a total of 765 sentences: 495 (65%) of them were annotated as "NFR", while 270 (35%) of them as "FR". These will be referred to as *CorpusN* and *CorpusF*, respectively.

According to [6], all these statements were manually annotated by fifteen graduate students of DePaul University (who also work in the software industry as professionals). The same corpus is used in this project to train our classifier and also for testing its performance.

3.2 Syntactic Features

By definition of NFR, it can be realized that some categories of words are better indicators of NFR by their occurrences in the sentences. For example, NFR sentences often explain quality attributes of a component or the system as a whole, and such sentences are likely to contain adjectives and adverbs. Again, NFR sentences that explain constraints of the system are likely to contain cardinals or numeric figures. Following these characteristics of NFR, as described in [4], we were motivated to choose a list of syntactic features as candidates and test their probabilities of occurrence in our collection NFR sentences (*CorpusN*), and thus, validate them to the most representative list of syntactic features.

We used the Stanford Parser [12] (equipped with Brill's POS tagger [3] and a morphological stemmer) to morphologically stem[1] the words and extract five syntactic features from each of the training instances (sentences) of the corpus. These features are:

- Number of Adjectives
    (e.g. "good", "bad", "efficient" etc.)
- Number of Adverbs
    (e.g. "very", "well", "properly" etc.)
- Number of Adverbs that modify Verbs
    (e.g. "well", "efficiently", "perfectly" etc.)

---

[1] The morphological stemmer comes built-in with the Stanford parser [12]. It stems with the prior knowledge of the morphology of a word, and stems only to the point at which it retains its original POS class.

- Number of Cardinals

   (*e.g. "1.56", "800x600", "twelve", "11" etc.*)

- Number of Degree Adjectives/Adverbs

   (*e.g. "better", "worse", "best", "more", "most" etc.*)

We only identified these features as candidates that can have some influence in the process of classifying a statement as NFR. Among these features, the ones that are valid for detecting NFR were to be selected automatically based on their ranks of the higher probabilities of their occurrences in NFR sentences of the training dataset. This probability measure for a candidate feature f_i is computed as follows:

$$Pr(f_i) = \frac{\text{Frequency of } f_i \text{ in CorpusN}}{(\text{Frequency of } f_i \text{ in CorpusN} + \text{Frequency of } f_f \text{ in CorpusF})} \quad (1)$$

Note that this probability is not normalized, while 61% of the total number words in both the corpora belonged to CorpusN. The probability values of all the syntactic features are shown in Table 1.

Table 1. Probability Ranking of Syntactic Features (the numbers, computed by formula (1), indicates the probability of a feature to appear in Non-functional Requirements (i.e. *CorpusN*)

Feature	Probability
Cardinals	0.8762
Degree Adjectives/Adverbs	0.8571
Adverbs that modify Verbs	0.8571
Cutoff Threshold = 0.8	
Adverbs	0.7387
Adjectives	0.6193

Here, we manually selected a cutoff threshold (>0.8, in this case), and all the features exceeding the cutoff threshold were selected as valid (or most discriminating) features.

3.3 Keyword Features

Previous work of [6,14] has shown that NFR statements are mostly identifiable by the use of specific keywords that belong to different part-of-speech categories. The work of [6] also identified specific keywords, but with no regards to their parts-of-speech group. This allowed many words of unwanted parts-of-speech group to be included in their final list.

Analyzing the types of most probable words used in NFR, as described in [4], we have considered keywords of 9 different parts-of-speech groups separately— the frequencies of each becoming a feature in our final feature list. These 9 part-of-speech based keyword groups are:

– Adjective-keywords (coded as: *JJ_kw*)
– Adverb-keywords (coded as: *RB_kw*)

- Modal-keywords (coded as: *MD_kw*)
- Determiner-keywords (coded as: *DT_kw*)
- Verb-keywords (coded as: *VB_kw*)
- Preposition-keywords (coded as: *IN_kw*)
- Common Noun-keywords that appeared in singular form (coded as: *NN_kw*)
- Common Noun-keywords that appeared in plural form (coded as: *NNS_kw*)

Similarly to ranking the features, two different probability measures have been considered for ranking the keywords in each of the groups: (a) Unsmoothed Probability Measure, and (b) Smoothed Probability Measure.

3.3.1 Unsmoothed Probability Measure (UPM)
Unlike [6], we considered taking the probability measure of a keyword occurring in a particular parts-of-speech (POS) category.

$$\Pr(Word, POS) = \frac{\text{Frequency of} < Word, POS > \text{in CorpusN}}{(\text{Frequency of} < Word, POS > \text{in CorpusN} + \text{Frequency of} < Word, POS > \text{in CorpusF})} \qquad (2)$$

Keywords of a POS category were then ranked according to the higher values of unsmoothed probability measure. A high cutoff threshold is individually set for each group to select the most discriminating keywords that attain a higher value than the threshold,

3.3.2 Smoothed Probability Measure (SPM)
While examining the values of the Unsmoothed Probability Measure, we found that some keywords, simply by chance, appeared only a few times in CorpusN, but never in CorpusF. This led them to have the highest probability value of 1 ranked on the top. On the other hand, some very discriminating keywords appeared in CorpusN many times, and also, by chance, appeared only a few times in CorpusF. UPM method ranks these keywords below the earlier ones, since they attain a probability value less than 1.

Also, with UPM, the keyword that appear, for example, once in CorpusN and never in CorpusF, and the keyword that appear, for example, 10 times in CorpusN, and never in CorpusF — both attain the same probability value of 1, and therefore, ranked together in the same place. But, clearly the second keyword is more discriminating than the first one.

To address these issues, we used a second measure that adds a small smoothing factor to all values of UPM. This factor takes into consideration how many more times the keyword in a particular POS group actually appears in CorpusN, than that of CorpusF. That is,

$$\text{Smoothed} \Pr(Word, POS) = \Pr(Word, POS) + \frac{\text{Frequency of} < Word, POS > \text{in CorpusN}}{(\text{Frequency of} < Word, POS > \text{in CorpusF} + 1) \times \alpha} \qquad (3)$$

Here, the constant α determines how much scaling one would like to add to the UPM value (the smaller the value of α, the higher the scaling). In our experiments, we used $\alpha = 10$.

Like UPM, keywords are also ranked according to their values of SPM, and a cutoff threshold is chosen to select the most discriminating keywords.

We used both UPM and SPM values to rank the keywords in their respective POS groups. Thus, we had the options, not only to manipulate the cutoff thresholds, but also to choose the best probability measures for each of the keyword groups. Table 2 illustrates the benefit of using SPM by showing an example from our corpora set.

Table 2. Choosing between UPM and SPM (here, ranking of keywords are based on SPM)

Keyword	POS	SPM	UPM	Frequency in CorpusN	Frequency in CorpusF
"every"	DT	1.741	0.941	16	1
"no"	DT	1.482	0.882	30	4
*Cutoff threshold set on **SPM** = 1.3*					
"this"	DT	1.225	0.875	7	1
"these"	DT	1.100	**1.000**	1	0

Table 2 shows the frequency values of the words "every" and "no", indicating that they are the most discriminating words in this category. It can also be guessed by understanding the nature of NFRs, which are most likely to contain descriptions of constraints that usually have determiner words, e.g. "every" and "no", as quantifiers. On the other hand, determiners like "this" and "these" simply appeared more in NFRs by chance. We find that, in this case, SPM successfully isolates the two most discriminating keywords of POS group DT[2], while UPM failed to create a proper ranking by setting its highest value (1.0) for the keyword "these". Therefore, here, we chose to rank the keywords by SPM values and set a cutoff threshold on them to select the most discriminating keywords. Thus, while training the system, for each of the keyword groups, either SPM or UPM can be chosen, along with the cut-off threshold to fine-tune the results. A slice of the list of the keywords selected in this way is shown in Table 3.

Table 3. Some of the keywords of different POS group, selected automatically by the keyword extractor program using both SPM and UPM set with different cut-off thesholds

JJ_kw	RB_kw	MD_kw	DT_kw	VB_kw	IN_kw	NN_kw	NNS_kw
acceptable	only	should	every	accommodate	within	abuse	answer
active	prior	may	no	accomplish	than	accordance	aspect
ad-hoc	adequately	might		achieve	between	animation	attack
additional	appropriately			activate	during	appearance	audience
adjacent	approximately			adhere	about	attention	button
african	currently			agree	alongside	auditor	calculation
appealing	especially			appeal	per	avoidance	condition
...	...			...	...	...	...

---

[2] The Penn-tree bank tags are used for the keyword groups. DT signifies Determiners.

3.4 Feature Extraction and Classification

Our final list of features, therefore, is as follows:

- Number of Cardinals
- Number of Degree Adjectives/Adverbs
- Number of Adverbs that modify Verbs
- Adjective-keywords (coded as: *JJ_kw*)
- Adverb-keywords (coded as: *RB_kw*)
- Modal-keywords (coded as: *MD_kw*)
- Determiner-keywords (coded as: *DT_kw*)
- Verb-keywords (coded as: *VB_kw*)
- Preposition-keywords (coded as: *IN_kw*)
- Common Noun-keywords that appeared in singular form (coded as: *NN_kw*)
- Common Noun-keywords that appeared in plural form (coded as: *NNS_kw*)

To classify the sentences, we developed a Java-based feature extraction program that parses the sentences from the corpora, and extracts the values of all the features mentioned above, and uses Weka [15] to train C4.5 decision tree learning algorithm [13]. We used the implementation of C4.5 (revision 8) that comes with Weka (as J48), setting its parameter for the minimum number of instances allowed in a leaf to 6 to counter possible chances of over-fitting. The results are discussed in the next section.

4 Results and Analysis

The results came out to be exceptionally well when using the whole dataset for training and testing. Since the dataset was not very large, we also used 10-fold-crossvalidation, and the results were very good as well. Table 4 shows the summary of the results.

Table 4. Summary of the results

	Scheme	Correctly Classified Sentences	Incorrectly Classified Sentences	Kappa	Comment
Concatenation of CorpusN & CorpusF (Size = 475 + 270 = 765)	Training + Testing on same set	760 (99.35%)	5 (7.63%)	0.9856	Tree is of desirable characteristics, not sparse, and also not flat. None of the branches are wrongly directed.
	Cross-validation (10 Folds)	754 (98.56%)	11 (1.44%)	0.9682	

The resultant decision tree after training on the complete dataset also came out well-formed. The tree is shown in Figure 3.

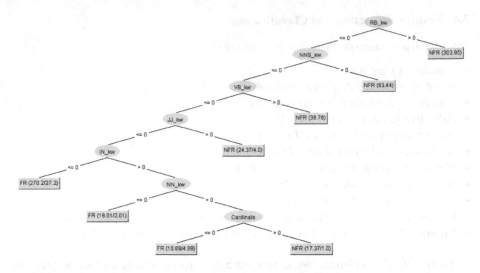

Fig. 1. The resultant C4.5 decision tree after training with the complete dataset

Detailed results of using 10-fold-crossvalidation with the final confusion matrix and standard deviation on all measures are shown in Table 5 and 6 respectively.

Table 5. Confusion matrix when using 10-fold-crossvalidation

	Classified as	
	FR	NFR
FR	259	11
NFR	0	495

Table 5 shows that the classifier retrieved all NFRs successfully (100% recall), without showing any sign of false negatives. Table 6, on the other hand, shows very low standard deviation over all the measurements taken during the iterations of crossvalidation (0.02 for precision and 0 for recall). This indicates that the results are likely to be robust.

Figures 2 and 3 show similar phenomena, where the curves hardly experienced any drastic change.

Table 7 compares our results of precision and recall using 10-fold-crossvalidation to the results obtained by the previous work [6]. Here, we have further broken down our results into steps of improvement, i.e. firstly using syntactic features only, and then using keyword features only, and then using both types of features, as documented previously.

Table 6. Detailed results of using 10-fold-crossvalidation

Test no.	Number of Training Instances	Number of Testing Instances	Number of Correctly Classified Instances	Number of Incorrectly Classified Instances	Total % of Correct	Total % of Incorrect	Kappa	Mean Absolute Error	RMS	For Classifying NFR			Characteristics of C4.5 Decision Tree		
										Precision	Recall	F-Measure	Total Nodes	Total Leaves	Total Rules
1	688	77	75	2	97.40	2.60	0.94	0.12	0.20	0.96	1.00	0.98	17	9	9
2	688	77	77	0	100.00	0.00	1.00	0.06	0.09	1.00	1.00	1.00	15	8	8
3	688	77	76	1	98.70	1.30	0.97	0.09	0.16	0.98	1.00	0.99	13	7	7
4	688	77	76	1	98.70	1.30	0.97	0.09	0.16	0.98	1.00	0.99	13	7	7
5	688	77	76	1	98.70	1.30	0.97	0.09	0.15	0.98	1.00	0.99	13	7	7
6	689	76	76	0	100.00	0.00	1.00	0.10	0.14	1.00	1.00	1.00	15	8	8
7	689	76	75	1	98.68	1.32	0.97	0.08	0.16	0.98	1.00	0.99	15	8	8
8	689	76	72	4	94.74	5.26	0.88	0.13	0.24	0.92	1.00	0.96	13	7	7
9	689	76	75	1	98.68	1.32	0.97	0.08	0.15	0.98	1.00	0.99	15	8	8
10	689	76	76	0	100.00	0.00	1.00	0.08	0.13	1.00	1.00	1.00	13	7	7
Mean			75.40	1.10	98.56	1.44	0.97	0.09	0.16	0.98	1.00	0.99	14.20	7.60	7.60
Standard Deviation(+/-)			1.35	1.20	1.57	1.57	0.04	0.02	0.04	0.02	0.00	0.01	1.40	0.70	0.70

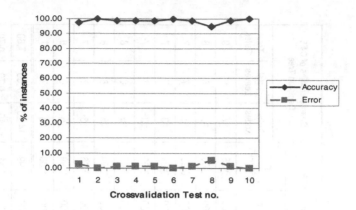

Fig. 2. Classifier's Accuracy/Error curve during 10-fold-crossvalidation

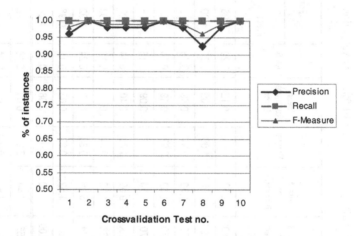

Fig. 3. Classifier's Precision/Recall/F-Measure curve for detecting NFR, during 10-fold-crossvalidation

Table 7. Comparison of our results to that of the previous work [6]

		Results	
		Precision	**Recall**
Work of Cleland-Huang *et al.* [6] - classification by density of keywords in the text		0.248	0.767
Our work: - classification by C4.5 decision tree learnner (10-fold-crossvalidation)	using Syntactic features Only	0.950	1.0
	using Keyword features Only	0.974	1.0
	using both Syntactic and Keyword features	0.978	1.0

The results of Table 7 show significant improvement over the most recent work [6] in the field. The classifier yielded high accuracy in performance, demonstrating results with 98.56% accuracy in the critical conditions of using 10-fold-crossvalidation. The precision and recall achieved by the classifier with 10-fold-cross-validation are 0.978 and 1.0 respectively, outperforming the work of [6] which attained the precision and recall as low as 0.248 and 0.767 respectively. Also, by using the syntactic features exclusively, and also by using our keyword features (that we selected based on their part-of-speech category) exclusively, yelled results which also surpass the performance of the work of [6] by a large margin.

5   Conclusions and Future Work

In this paper, a methodology for automatic classification of requirements by means of using a text classifier was presented. Our work extends the idea of [6] of using of Information Retrieval for classifying NFRs, and proved that using linguistic knowledge can help perform very well in this classification task. Ours research aimed at assisting the software analysts in highlighting the NFRs in the users' textual SRS documents to avoid their further oversight in the development process, which can lead to poor quality of the final product and eventually to project failure.

The research reported in this paper is a part of a larger NLP-driven Requirements Engineering project which intends at using Natural Language Processing techniques in Requirements Engineering [9]. The goal of the work presented here was to increase the quality of the requirements text by deriving a module for the aforesaid project that would explicitly flag the requirements statements into FR and NFR for further processing. The module presented here can also be run exclusively as a standalone program to perform the classification task on requirements text.

Our future work includes introducing more training and testing data, implementing a full-fledged prototype to demonstrate its use and a complete integration in our NLP-driven Requirements Engineering project.

Acknowledgements. The authors would like to thank Dr. Jane Cleland-Huang, co-author of [8], for making their corpus available online, and Dr. Doina Precup (McGill University, Montreal) for her valuable suggestions and feedback on the project. We are also grateful to the anonymous referees for their valuable comments on a previous version of the paper.

References

1. Boetticher, G., Menzies, T., Ostrand, T.: PROMISE Repository of empirical software engineering data, West Virginia University, Department of Computer Science (2007) (Last retrieved: December 10, 2007), http://promisedata.org/repository
2. Breitman, K., Leite, J., Finkelstein, A.: The world's a stage: a survey of requirements engineering using a real-life case study. Journal of the Brazilian Computer Society 6(1), 13–37 (1999)

3. Brill, E.: A simple rule-based part-of-speech tagger. In: Proceedings of the 3rd Conference on Applied Natural Language Processing (ANLP 1992), Trenton, Italy, April 1– 3, 1992, pp. 152–155 (1992)
4. Chung, L., Nixon, B.A., Yu, E., Mylopoulos, J.: Non-functional Requirements in Software Engiineering. Kluwer Academic Publishers, Dordrecht (2000)
5. Chung, L., Sapakkul, S.: Capturing and Reusing Functional and Non-functional Requirements Knowledge: A Goal-Object Pattern Approach. In: Proceedings of 2006 IEEE International Conference on Information Reuse and Integration, September 2006, pp. 539–544 (2006)
6. Cleland-Huang, J., Settimi, R., Zou, X., Solc, P.: The Detection and Classification of Non-Functional Requirements with Application to Early Aspects. In: Proceedings of 14th IEEE International Requirements Engineering Conference 2006 (RE 2006), pp. 36–45 (2006)
7. Cysneiros, L.M., Leite, J.C.S.: Non-functional requirements: from elicitation to modelling languages. In: Proceedings of the 24rd International Conference on Software Engineering (ICSE 2002), pp. 699–700 (2002)
8. Hill, R.: Quantifying Non-functional Requirements: A Process Oriented Approach, Requirements Engineering Conference. In: Proceedings of 12th IEEE International, September 6-11, 2004, pp. 352–353 (2004)
9. Hussain, I., Ormandjieva, O., Kosseim, L.: Automatic Quality Assessment of SRS Text by Means of a Decision-Tree-Based Text Classifier. In: Proceedings of the Seventh International Conference on Quality Software (QSIC 2007), pp. 209–218 (2007)
10. IEEE (1998). IEEE recommended practice for software requirements specifications (IEEE Std 830-1998). The Institute of Electrical and Electronics Engineers, Inc., New York (October 20, 1998) ISBN 0-7381-0332-2
11. International Standard ISO/IEC 14143-1:2007. Information technology - Software measurement - Functional size measurement - Part 1: Definition of concepts. International Organization for Standardization (2007)
12. Klein, D., Manning, C.D.: Accurate Unlexicalized Parsing. In: Proceedings of the 41st Meeting of the Association for Computational Linguistics (ACL 2003), pp. 423–430 (2003)
13. Quinlan, J.R.: C4.5: Programs for machine learning. Morgan Kaufmann, San Mateo (1993)
14. Rosenhainer, L.: Identifying Crosscutting Concerns in Requirements Specifications. In: Workshop on Early Aspects: Aspect-Oriented Requirements Engineering and Architecture Design, Vancouver, Canada (October 2004)
15. Witten, I.H., Frank, E.: Data mining: Practical machine learning tools and techniques. 2nd edn. Morgan Kaufman, San Francisco (2005)

A Preliminary Approach to the Automatic Extraction of Business Rules from Unrestricted Text in the Banking Industry

José L. Martínez-Fernández[1,2], José C. González[1,3],
Julio Villena[1], and Paloma Martínez[2]

[1] DAEDALUS – Data, Decisions and Language S.A.
Avda. de la Albufera, 321
28031 Madrid, Spain
{jmartinez,jgonzalez,jvillena}@daedalus.es
[2] Computer Science Department, Universidad Carlos III de Madrid
Avda. de la Universidad, s.n.
29811 Leganés, Madrid, Spain
{joseluis.martinez,paloma.martinez}@uc3m.es
[3] Dept. Telemática, Universidad Politécnica de Madrid
ETSI Telecomuncación
28040 Madrid, Spain

Abstract. This paper addresses the problem of extracting formal statements, in the form of business rules, from free text descriptions of financial products or services. This automatic process is integrated in the banking software factory, permitting business analysts the formal specification, direct implementation and fast deployment of new products. This system is fully integrated with the typical software methodologies and architectures used in the banking industry for conventional development of backoffice or online applications.

Keywords: Business rules, banking industry, natural language processing, financial ontologies.

1 Introduction

Current trends in software development are paying special attention to Business Rules Systems (BRS), especially useful in environments where specifications are changing everyday as a reaction to market evolutions. A Business Rule can be defined as a statement constraining some aspect of the business, and a BRS is the system in charge of verifying the correct application of these Business Rules.

One of the market sectors where needs and requirements are suffering continuous changes is the financial and insurance industry. The ITECBAN (Architecture for Core Banking Information Systems) project is an initiative funded by the Spanish Ministry of Industry (INGENIO 2010 initiative, CENIT programme). The aim of this project is to develop a new core banking distributed platform.

E. Kapetanios, V. Sugumaran, M. Spiliopoulou (Eds.): NLDB 2008, LNCS 5039, pp. 299–310, 2008.

In order to make it easy the integration of Business Rules in nowadays Information Technology infrastructures, BRS not only include inference engines to verify the compliance with a Business Rule but also different tools to manage the creation, maintenance and sharing of these rules. One of the major weaknesses of these BRSs is the need of specific technical knowledge to be able to define and develop Business Rules. Usually, software developers are in charge of creating and maintaining Business Rules, attending requirements of business analysts. There has been a lot of interest in developing tools to remove this technical dependence, including the ability for business analysts to use natural language expressions to define Business Rules. As already mentioned, the development of these rules must be integrated in the overall software development cycle implemented in the target organization.

The process of extracting Business Rules from free text is part of a software tool called K-Site® Rules, whose aim is to support the development of Business Rules. The tool pursues three main goals: make any information system independent from the specific rule engine used to implement Business Rules; facilitate integration of Business Rules development process into the whole software development process of the organization or company; and allow business analysts to develop and deploy Business Rules without the support of software developers.

K-Site Rules is developed under a Model Driven Architecture (MDA) [7] to make it easier the integration with the other services and software components constituting the ITECBAN platform. MDA considers a computational independent model (CIM) where domain specifics must be described and business logic is covered. Under this level, there is a platform independent model (PIM) where a software system to support the CIM is described using one or more platform independent models (in our case described through the Unified Modeling Language, UML, the Semantic Web Rules Language, SWRL, to describe rules, Web Ontology Language, OWL, to define the domain and the JSR-94 standard [10] for the interaction with the rule engine). Finally, several platform specific models (PSM) must be specified to support the PIM level; at this level, concrete products, rule engine vendors in our case, are considered. Fig. 1 depicts the languages and tools selected to define the corresponding models according to the MDA architecture.

This paper focuses on the first level, the computationally independent model, more specifically, on the process to obtain expressions in the Semantics of Business Vocabulary and Rules (SBVR) language from unrestricted text. After this, SBVR sentences must be transformed into SWRL expressions. This later step will constitute the transformation between the CIM model and the PIM model used to describe business rules to be implemented by the system. The next section describes the method designed to identify Business Rules from text, including details about standards and technologies applied. The third section includes a description of the architecture designed for the system, identifying main elements and software components used. The following section is centered on the linguistic processing integrated as part of the system, included in the STILUS linguistic software platform. The fifth section covers the semantic modules introduced to reduce the aforementioned ambiguity and to relate each input text with a desired interpretation. Finally, some preliminary results are shown and some conclusions are depicted.

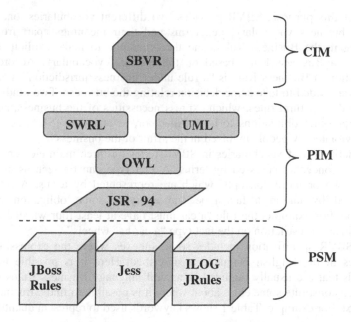

Fig. 1. Languages, standards and tools selected to build each level of the model driven architecture for K-Site Rules

2 Extracting Business Rules from Text

The process of extracting Business Rules from unrestricted text is supported by the SBVR standard and by the OWL language. The SBVR standard defined by the OMG group provides a way to document the semantics of business vocabulary, business facts and business rules. One part of this standard is devoted to the construction of several vocabularies, according to a hierarchical structure to be identified for the specific domain considered, to store concepts and relations among them. In the approach covered in this work, these vocabularies are going to be substituted by ontologies, either pre-existing (for example, the SUMO Finance ontology [8]) or specifically defined for the working domain. This representation will be later easily referenced by a SWRL expression of the Business Rule, which will also be expressed using OWL. On the other hand, tools to manage ontologies, like Protégé [6], can be used to maintain the vocabulary up to date. Although the notation provided by SBVR is not going to be directly applied, some clues provided by this standard will be applied to detect the presence of possible Business Rules in free text sentences.

2.1 Semantic Business Vocabulary and Rules, SBVR

This specification adopted by the OMG and defined by the Business Rules Group, BRG[1], tries to include semantics in the definition of the business and its governing

---

[1] http://www.businessrulesgroup.org

rules. For this purpose, SBVR provides two different vocabularies, one of them to describe business vocabulary, i.e., terms and their meanings apart from the ones appearing in the Business Rules, and the other one to make explicit the meaning related to a Business Rule, based on the business vocabulary. According to this specification, a Business Rule is "a rule under business jurisdiction", i.e., it exposes the criteria needed to take some decision relating the business. Two kinds of rules are considered: structural rules, which express necessities of the business, and operative rules, expressing obligations to be fulfilled, or, as stated in [5], "rules that can be directly violated by people involved in the affair of the business".

The definition of vocabularies in SBVR relies on three main elements: rules, fact types and concepts expressed by terms. A fact type can be seen as an association between two or more concepts, which are represented by terms. A rule is always constructed by taking a fact type and applying some obligation or necessity restriction. For example, the rule "a car must have at least four wheels" corresponds to an obligation restriction on the fact type "a car has wheels".

The SBVR specification includes some annexes where the expression of SBVR structures in the English language is described. Here it is possible to find some keywords that are usually included in natural language Business Rules expressions, which can constitute some clues about where it is possible to find a Business Rule in a given text. For example, Table 1 shows keywords used to represent quantifiers.

Table 1. Example of keywords for quantifiers identified in the SBVR specification

Keyword	Quantifier
Each	universal quantification
Some	existential quantification
at least one	existential quantification
at least n	at-least-n quantification
at most one	at-most-one quantification
at most n	at-most-n quantification
exactly one	exactly-one quantification
exactly n	exactly-n quantification
at least n and at most m	numeric range quantification
more than one	at-least-n quantification with n=2

Another language that has been used as a source of linguistic clues to detect Business Rules is RuleSpeak [1] that makes some suggestions about the best way to express rules to assure a correct interpretation and understanding by business agents. The target language in the approach presented in this work is Spanish, so these keywords and cues have been transformed into their counterparts for the Spanish language.

2.2 The Semantic Web Rule Language (SWRL)

The initiative to include semantics in the World Wide Web initiated by Tim Berners-Lee, the so called Semantic Web, has produced a language to annotate web sites with meaning. This language is called OWL, which is related to the concept of ontology ("a specification of a shared conceptualization", [4]), where concepts, along with their valid interpretations, and relations among them are somehow represented. This language is proposed in our work to represent the business vocabulary that constitutes the basis to build Business Rules. Another initiative, sponsored by the W3C, has defined a language, grounded on OWL, to represent Business Rules, this language is called SWRL and it is based on RuleML, a language to express rules using Horn clauses. This language has two syntaxes, one of them XML based and the other one RDF based. According to [5], a rule has two parts, an antecedent (body) and a consequent (head), both of them constituted by a set of atoms that could be empty. An atom can be an OWL instance, an OWL data value, an OWL description or data range, an OWL property or an OWL built-in relation. In this way, rules can be referenced to existing concepts in an OWL ontology.

This language will be used to construct a platform independent representation for Business Rules. As described in the previous section, a transformation from CIM to PIM must be provided, i.e., a way to translate from a natural language expression of the rule to an SWRL representation. This is the purpose of the system described in this paper, to help in the identification of Business Rules present in a free text and to express them using the SWRL language. In this way, the automation of the rest of the process to implement the given Business Rule will be assured.

2.3 Business Rule Recognition Process

Taking into account the mentioned technologies, the process to get a free text and returning a set of possible Business Rules and their representation in SWRL is shown in the following example.

In the framework of the ITECBAN project, a use case was defined to test involved technologies. The use case is a functional document describing the way some saving products, offered by a financial entity, should work. Suppose that the starting point is a functional analysis document where it is possible to read the sentence:

"Si un Producto Ahorro es Deposito Financiero entonces la divisa asociada solo puede ser euro." [If a Savings Product is a Financial Deposit, then the associated currency has to be the euro.]

Then, a process to detect if this sentence can describe a Business Rule is launched, a linguistic analysis is made, linguistic rules are applied to know if the sentence can contain a Business Rule and a typical if – then structure is detected. Besides, an ontology was specifically defined for this use case, and some of the concepts described in it were also detected in the sentence. As a result, the sentence is marked as a candidate to contain a Business Rule and the concepts residing in the ontology are also tagged. A final step is considered where a business analyst decides about the validity of the Business Rule and, if it is necessary, modifications to the rule sentence are made. Once the natural language content of the rule is available, a SWRL equivalent expression is automatically built to be able to continue with the Business Rule development process. Fig. 2 shows the final SWRL code for the rule given as an example.

```
<swrl:Imp rdf:ID="DivisaProductoAhorro">
    <swrl:head>
        <swrl:AtomList>
            <rdf:rest rdf:resource= "http://www.w3.org/1999/02/22-
                rdf-syntax-ns#nil"/>
            <rdf:first>
                <swrl:IndividualPropertyAtom>
                    <swrl:argument1>
                        <swrl:Variable rdf:ID="x"/>
                    </swrl:argument1>
                    <swrl:argument2 rdf:resource="#EURO"/>
                    <swrl:propertyPredicate
                        rdf:resource="#con_divisa"/>
                </swrl:IndividualPropertyAtom>
            </rdf:first>
        </swrl:AtomList>
    </swrl:head>
    <swrl:body>
        <swrl:AtomList>
            <rdf:first>
                <swrl:ClassAtom>
                    <swrl:argument1 rdf:resource="#x"/>
                    <swrl:classPredicate
                        rdf:resource="#ProductoAhorro"/>
                </swrl:ClassAtom>
            </rdf:first>
            <rdf:rest>
                <swrl:AtomList>
                    <rdf:first>
                        <swrl:ClassAtom>
                            <swrl:classPredicate rdf:resource=
                                "#DepositoFinanciero"/>
                            <swrl:argument1 rdf:resource="#x"/>
                        </swrl:ClassAtom>
                    </rdf:first>
                    <rdf:rest rdf:resource=
                        "http://www.w3.org/1999/02/22-rdf-syntax-
                        ns#nil"/>
                </swrl:AtomList>
            </rdf:rest>
        </swrl:AtomList>
    </swrl:body>
</swrl:Imp>
```

Fig. 2. Example rule expressed using SWRL

3 System Architecture

The system described in the previous section has been implemented according to the architecture depicted in Fig. 3. The linguistic analysis component is in charge of processing the input text to include morphological, syntactical and semantic information. The semantic data is based on compiled resources and on the ontology describing the application domain. The output of this analysis constitutes the input of the rule detection component, which is built on a set of linguistically motivated rules or patterns inspired in SBVR and RuleSpeak languages. The output of this rule detector component is supervised by a business analyst and then introduced in the rule transformation component, which provides an automatic translation from the natural language expression of the rule to its SWRL counterpart. It is worth mentioning that every concept referenced in the rule must be included in the ontology that supports the rule system.

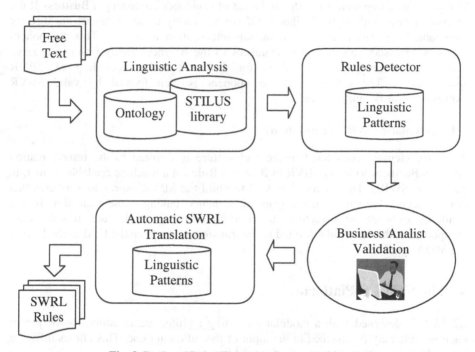

Fig. 3. Business Rules Extraction System Architecture

3.1 Linguistic Analysis

The linguistic analysis component is built based on STILUS®, a library for the morphologic, syntactic and semantic analysis of texts in Spanish. This software component, developed by DAEDALUS [2], is used to divide the input text in sentences, to provide the lemma for each word. To include information contained in the ontology, this lemma will be matched against the lexical realizations associated to each concept. On the other hand, each term in the document is related with the semantic tag present in the STILUS lexical base, if any. These data will be the basis for the rules detection component.

3.2 Rules Detector

This element of the architecture takes the linguistic information provided for the input text and applies a set of rules to decide if a sentence can contain a Business Rule or some clues to define one. This set of rules is based on some indications given by SBVR and RuleSpeak. A very simple example of these rules is '*if the sentence has an if – then structure then the probability for the sentence to contain a Business Rule is High*'.

3.3 Business Analyst Validation

The previous component has extracted a set of sentences containing a Business Rule, or part of one, with high likelihood. Of course, rarely these sentences are directly applicable as Business Rules, and human intervention is needed. This component includes necessary tools to show sentences to the Business Analyst and to retrieve valid natural language expressions for Business Rules according to SBVR specifications. The output of the component is then formed by valid SBVR expressions for Business Rules.

3.4 Automatic SWRL Translation

The final element considered in the architecture is devoted to the transformation between Business Rules in SBVR to Business Rules in a machine readable format, in particular, SWRL. The process followed to build the SBVR expressions assures that only concepts known by the organization, hence having some reflection in the available software infrastructure, are used to define rules. In fact, this software component is the one really devoted to the transformation from the CIM to the PIM in the MDA approach.

4   The STILUS Platform

STILUS is designed with a modular cascading architecture in which the output of each module may be attached to the input of the following one. This allows different versatile combinations to perform advanced text processing:

- **Text Segmentation:** segmentation is the process of dividing written text into words (usually called tokens) or other similar meaningful units, such as sentences or topics.
- **Part-of-speech tagger:** each word is assigned with its corresponding POS analysis. The morphological model is adopted from the ARIES platform [3]. The main idea is to automatically generate, for a given word, all possible root and derivative morphemes which can concatenate to each other according to the inflectional and derivational rules for Spanish. For example, root morphemes such as "hab-", "hub-", "hay-" are generated for the verb "haber", derivative morphemes "-o", "-a", "-os", "-as" for nominal gender and number inflection ("niñ-o-a-os-as"). In addition, the lexical database also includes other entries such as lexicalized words (those irregular forms that cannot be obtained with morphological inflection) and multiword expressions such as "a costa de", "Juan Carlos I", etc. Each unit has its own morphological information such as gender, number, person, verb tense, verb mode, word lemma, etc. The POS tagger uses this information to tag each word. The

lexical database is stored in a trie data structure that allows a very efficient read access.

- **Multiword recognition:** multiword units are identified by means of rules (such as dates or numbers) or linguistic resources (e.g. toponyms, film titles...)
- **POS disambiguation:** rules for morphosyntactic disambiguation that try to select the correct analysis for a given token among the proposed alternatives by the POS tagger. These rules cover linguistic patterns for specific words or combination of grammatical cases.
- **Syntactic parser:** this module turns a list of words into a syntactical tree with information about the type and function of each part of speech. This process consists of several steps including morphological and syntactical levels of analysis carried out in a bottom-up strategy. STILUS allows both a shallow parsing that simply identifies chunks (basic syntactic constituents of a phrase) and also a more complex analysis that computes chunks and the functional relations among them, thus tackling problems that have a semantic nature. The advantage of shallow parsing is in the case of ill-formed sentence, because the analyzer is still able to parse at least parts of the sentence.

4.1 Linguistic Rules

The linguistic knowledge is represented in a proprietary rule language specifically designed to abstract the linguists from the actual parser implementation and allowing them to focus on the linguistic phenomena.

The basic structure of a linguistic rule is:

```
IF
        conditions
THEN
        actions
END
```

"Conditions" include different test functions connected by boolean operators (AND, OR) and, if necessary, grouped into brackets. Table 2 shows different examples of test functions. For example, EXISTENTIAL_POS receives two arguments (token to evaluate and a regular expression for the POS tag) and is true if any of the word analysis fulfils the condition. In turn, UNIVERSAL_POS receives the same arguments but is true if all analysis fulfill the condition.

Table 2. Some examples of test functions

Function	Meaning
WORD(<pos>,<regexp>)	If the word matches the given regular expression
STARS_WITH_I (<pos>,<regexp>)	If the word starts with the given regular expression, case insensitive
EXISTENTIAL_POS (<pos>,<regexp>)	If any of the word analysis matches the given regular expression
UNIVERSAL_LEMMA (<pos>,<regexp>)	If all lemmas match the given regular expression

"Position" indicates the rule focus, i.e., the word(s) under inspection. Usually a generic position N and relative scrolling (... N-2, N-1, N+1, N+2, ...) are used. In this case, the rule focus will move throughout the sentence, looking for a context that fulfils the given conditions. In addition, there are other position functions, for example, CHILD_POSITION, which allows testing syntactic structures (trees) instead of flat sentences, or FIRST_EXISTENTIAL_POS_POSITION, which uses a predicate to look for the position of the first token that fulfils the given condition.

Last, "actions" comprise a set of operations that may be applied over the sentence. Some examples are shown in Table 3.

Table 3. Some examples of action functions

Function	Meaning
JOIN_SYNTAGM (<pos1>,<pos2>,<tag>)	Create a new part of speech joining the words from <pos1> to <pos2> and assigning POS tag <tag>.
SELECT_TAG (<pos>,<regexp>)	Disambiguate the given word filtering out the tags that do not fulfill the given regular expression
ERROR(<pos1>,<pos2>, <type>,<msg>)	Marks the context from <pos1> to <pos2> with an error

5 Preliminary Experiments

The system developed and described in Section 3 has been executed on the use case defined in the framework of the ITECBAN project. As already mentioned, this use case describes the way that some saving products must be managed when customers of a banking entity want to work with them. For this use case, the ITECBAN Business Rules team has created an ontology containing concepts defined by the bank. Furthermore, an input document with the functional analysis for the application, written in natural language, has been studied and Business Rules have been identified and written in an SBVR version for Spanish. The document is formed by a total of 216 sentences, from which forty eight Business Rules have been extracted. The input to the system is formed by a file with the functional analysis document, written in Spanish, and another file with the ontology (in OWL format) that has been defined for the use case. The output is composed by an annotation, in XML, of the input text, including a score that can be interpreted as the likelihood for the sentence to represent a Business Rule. This score ranges from 0, when the sentence does not include a Business Rule, to 10, when it is almost sure that the sentence can be used to define a Business Rule. At this point, there is a validation step with human intervention to confirm if there is a rule in the sentence and, probably, to re-write it in a clearer expression using the SBVR language. Fig. 4 shows two examples of the system output (before the human validation step).

Table 4 shows some results produced, comparing the total number of sentences, the number of sentences marked by the system as containing a Business Rule (or part of it) and the number of Business Rules produced by hand by the Business Rules team of the project.

```
<frase puntuacion='5'> Cualquiera de los padres/tutores puede
efectuar  cualquier  movimiento  que  admita  el  producto  sin
necesidad  de  contar  con  la  firma  del  resto  de
padres/tutores.</frase>

<frase puntuacion='8'> La titularidad y divisa de la  cuenta
asociada deberán ser idénticas a las del contrato que se
apertura, y no podrá cancelarse mientras no haya finalizado la
operación inicial y sucesivas, en su caso.</frase>
```

Fig. 4. Examples of the output of the Business Rules Detector system

Table 4. Preliminary results for the Business Rules Detector system

Total number of sentences	Number of handmade Business Rules	Number of sentences possibly containing a Business Rule
216	48	42

It can be surprising noticing that the number of sentences automatically detected to contain a Business Rule is smaller than the number of handmade Business Rules. This is normal because there is no a one to one relation between a sentence and a Business Rule, i.e., a sentence can be broken down into several Business Rules, and, of course, repeated Business Rules must be deleted.

Of course, this is only a preliminary evaluation and there must be and will be a further evaluation carried out with the business analysts and developers that will act as final users.

6 Conclusions

Natural language specification of products, services, procedures, regulations, etc. in business environments tends to use simple and unambiguous standard language conventions. In particular, typical sentences make use of an implicit world vision (an ontology) and express definitions, concepts and restrictions with a clear associated semantics. On the other side, there is a pressing need for companies to develop new products and services at the fast pace imposed by global markets. The only way to achieve this is by giving business experts a central role in the whole software-based product development life cycle. The tools developed in the framework of the ITECBAN project are intended to assist business analysts in the specification, implementation and deployment of business applications, with minimum support from IT specialists. Although this work is still in an early stage, and full experimentation and evaluation has to be completed, evidence has collected about the interest and viability of this approach in a real environment.

Acknowledgements

This work has been partially supported by the Spanish Center for Industry Technological Develpoment (CDTI, Ministry of Industry, Tourism and Trade), through the project

ITECBAN (Architecture for Core Banking Information Systems), INGENIO 2010 Programme, and the project "Software Process Management Platform: modeling, reuse and measurement" (TIN2004-07083). Other partners in ITECBAN are INDRA Sistemas, CajaMadrid, Sun Microsystems and Grid Systems.

References

1. Business Rule Solutions: BRS RuleSpeak® Practitioner's Kit. Business Rule Solutions, LLC. PDF (2001-2004), http://BRSolutions.com/p_rulespeak.php
2. DAEDALUS, Data, Decisions and Language, S.A, http://www.daedalus.es
3. González, J.C., Goñi, J.M., Nieto, A.F.: ARIES: a ready for use platform for engineering Spanish-processing tools. In: Digest of the Second Language Engineering Convention, London, October 1995, pp. 219–226 (1995)
4. Gruber, T.R.: A translation approach to portable ontology specification. Knowledge Acquisition 5(2), 199–220 (1993)
5. Horrocks, P.F., Patel-Schneider, H., Boley, S., Tabet, B., Grosof, M.: Dean: SWRL: A Semantic Web Rule Language Combining OWL and RuleML, draft 0.5 (November 2003), http://www.daml.org/2003/11/swrl/
6. Knublauch, H.: Ontology-Driven Software Development in the Context of the Semantic Web: An Example Scenario with Protégé/OWL. In: International Workshop on the Model-Driven Semantic Web, Monterey, CA (2004)
7. Miller, J., Mukerji, J.: MDA Guide Version 1.0.1, OMG (June (2003), http://www.omg.org/docs/omg/03-06-01.pdf
8. Niles, I., Pease, A.: Towards a Standard Upper Ontology. In: Welty, C., Smith, B. (eds.) Proceedings of the 2nd International Conference on Formal Ontology in Information Systems (FOIS-2001), Ogunquit, Maine, October 17-19 (2001)
9. OMG: Semantics of Business Vocabulary and Business Rules (SBVR), First Interim Specification (March (2006), http://www.omg.org
10. Selman, D.: Java Rule Engine API Specification JSR-94. Draft 1.0 (2002), http://jcp.org/en/jsr/detail?id=94

Paraphrasing OCL Expressions with SBVR

Raquel Pau[1] and Jordi Cabot[2]

[1] GTD Ingeniería de Sistemas y de Software
raquel.pau@gtd.es
[2] Estudis d'Informàtica, Multimedia i Telecomunicació, Universitat Oberta de Catalunya
jcabot@uoc.edu

Abstract. A conceptual schema (CS) should be explained to the stakeholders to validate that it is an appropriate representation of all knowledge of the domain. One of the best ways to explain the CS is to describe it by means of natural language expressions (*paraphrasing*). Even though paraphrasing has been studied for the most typical elements of a CS, current methods are, in general, unable to cope with the textual business rules that complement the CS. In this paper, we cover this gap by presenting a method that generates natural language explanations for business rules expressed in OCL (Object Constraint Language), the standard language to specify business rules on UML-based CSs. As an intermediate step, our method translates the OCL expression into a SBVR (Semantics of Business Vocabulary and Business Rules) representation.

1 Introduction

The specification of an information system must include a formal representation of the knowledge of the domain required by the system to perform its functions. In conceptual modelling, this representation is known as its conceptual schema (CS). A CS must include the definition of all relevant business rules of the domain.

To validate the correctness of the CS we must rely on the external stakeholders, as the experts in the domain. Nevertheless, to do so, customers must be able to understand the knowledge represented in the CS. To facilitate this task, designers should reexpress the CS in a way that they can easily understand. Typically, one of the most useful alternatives is to describe the elements of the CS by means of natural language expressions. This process is known as *paraphrasing*.

The OMG has recently proposed the SBVR (Semantics of Business Vocabulary and Business Rules [4]) specification as a way to facilitate the translation among a set of different languages. SBVR is specially suited for acting as an intermediate step in a CS to natural language transformation since it is conceptualized optimally for business people.

In this sense, the goal of this paper is to introduce an automatic translation from UML-based CSs to SBVR, where the initial UML schemas [5] are complemented with textual expressions in OCL (Object Constraint Language [3]) to define all business rules that cannot be graphically represented. In particular, since the own SBVR standard already sketches the translation of the graphical UML elements that may appear in the CS (classes, associations, attributes and so forth), in this paper we

E. Kapetanios, V. Sugumaran, M. Spiliopoulou (Eds.): NLDB 2008, LNCS 5039, pp. 311–316, 2008.
© Springer-Verlag Berlin Heidelberg 2008

will focus on the translation of its OCL business rules. Then, the obtained SBVR representation can be presented to the customers as a list of self-explaining natural language expressions using the predefined alternative notations to express SBVR concepts by means of English statements (either in the SBVR Structured English style or using the RuleSpeak approach [4]) included in the SBVR standard.

As far as we know, ours is the first proposal to provide such translation. To the best of our knowledge, only [1] is aimed at generating natural language sentences from OCL expressions. However, this approach provides an ad-hoc solution instead of relying in a standard format for defining the business rules, as SBVR.

The rest of the paper is structured as follows. Section 2 presents our preliminary OCL to SBVR translation. In Section 3, we generate natural language expressions from the resulting SBVR excerpts. Finally, Section 4 presents some conclusions.

2   From OCL to SBVR

OCL is a formal high-level language used to specify textual business rules that cannot be expressed using the graphical constructs provided by the UML.

In OCL, invariants (i.e. static constraints) are defined in the context of a specific type, called the *context type* of the constraint. The actual OCL expression stating the constraint condition is called the *body* of the constraint. The body is a boolean expression that must be satisfied by all instances of its context type. As an example, in constraint *ownersDrive* (Fig. 1) *Car* is the context type, the variable *self* refers to an instance of *Car* and the *exists* condition (the body) must hold for all possible values of *self* (i.e. for all instances of *Car*).

The aim of this section is to introduce a method for the translation of OCL constraints[1] into SBVR logical formulations. A logical SBVR formulation structures the meaning of a rule using several propositions. Different kinds of simple propositions are predefined, as quantifications, logical operations (conjunction, disjunction, equivalence...) and projections, which can be roughly defined as logical formulation of a condition that results in a set or a bag (multi set) of instances. Fig. 1 shows the logical formulation for the previous *ownersDrive* constraint.

As a first step, each constraint is expressed as a new SBVR logical formulation. In particular, they are translated as *necessities*. Given a constraint *c* with a body *b* and defined over a context type *t,* the equivalent SBVR representation for *c* would be:

The rule is a proposition meant by a necessity claim.
. The necessity claim embeds a universal quantification. (to ensure that all instances of t satisfy b)
.. The universal quantification introduces a variable. (the self variable)
... The variable ranges over the concept 't'. (self will be mapped to each different t instance)
.. The universal quantification scopes over ...'b condition'....

Then, the body *b* of the OCL constraint is translated into an appropriate SBVR representation using a set of translation patterns, where each pattern defines how to translate a different construct[2] that may appear in an OCL expression. The full OCL

---

[1] Note that not all kinds of OCL expressions can be directly mapped to SBVR concepts. Required extensions to SBVR to achieve a full translation are left as further work.
[2] To simplify the process, complex OCL constructs are first rewritten in terms of basic ones [2].

expression translation is performed by combining a set of appropriate patterns depending on the expression structure. We assume that the UML schema has already been translated to SBVR (e.g. following the guidelines provided in the own SBVR standard) before translating the OCL expressions themselves.

Translation of Basic OCL operators

The following patterns describe the translation of basic OCL operators. In the patterns, X and Y represent variables or constant literals of the appropriate type. The sentence *"previous translation"* refers to the result of translating a previous part of the OCL expression. *"Translation of X"* and *"translation of Y"* refer to the result of translating the OCL expressions corresponding to the X and Y variables. For each pattern we define the target OCL expression and its SBVR translation. The dots before each proposition indicate the proposition nesting level in complex expressions.

- X *implies* Y
  The previous translation introduces an implication.
  .The implication has an antecedent that is a.... translation of X ...
  .The implication has a consequent that is a ... translation of Y ...

- X *and/or/xor* Y
  The previous translation introduces a [conjunction | disjunction | exclusive disjunction].
  . The [conj./disj./excl. disj.] has a logical operand₁ that is a... translation of X ...
  . The [conj./disj./excl. disj.] has a logical operand₂ that is a... translation of Y ...

- *Not X*
  The previous translation introduces a logical negation.
  . The logical negation has a negand that is a... translation of X ...

- X *[< | > |=] Y*
  The previous translation introduces an atomic formulation
  . The atomic formulation is based on the fact type quantity₁ [is less than | is greater than| is] quantity₂
  . The atomic formulation has the first role binding.
  . . The first role binding is of the role quantity₁
  . . The first role binding binds to the ... translation of X ...
  . The atomic formulation has the second role binding.
  . . The second role binding is of the role quantity₂
  . . The second role binding binds to the... translation of Y ...

- $X.oclIsTypeOf(T)$
  The previous instantiation introduces an instantiation formulation.
  . The instantiation formulation considers the concept 'T'.
  . The instantiation formulation binds to the... translation of X ...

Translation of navigation expressions

In OCL, the *dot* notation applied over an object (or collection of objects) allows to navigate from that object to related objects in the system using the associations defined in the CS. That is, given an object o of type t_1 and a type t_2, related with t_1 through an association *assoc* with roles r_1 and r_2, respectively, the expression $o.r_2$ returns the set of objects of t_2 related with o. This expression is translated as follows:

The first universal quantification introduces the first variable (the initial object o)
. The first variable scopes over the concept 't₁'

The first universal quantification introduces the first variable (the initial object o)
. The first variable scopes over the concept 't_1'
The first universal quantification scopes over a second universal quantification
. The second universal quantification introduces the second variable
. . The second variable scopes over the concept 't_2'
. The second universal quantification is restricted by an atomic formulation
. . The atomic formulation is based on the fact type 'assoc'
. . The atomic formulation has the first role binding.
. . . The first role binding is of the role 'r_1'
. . . The first role binding binds to the first variable
. . The atomic formulation has the second role binding.
. . . The second role binding is of the role 'r_2'
. . . The second role binding binds to the second variable
. The second universal quantification scopes over ... translating the rest of the expression.

Note that, in SBVR, navigations are expressed by defining a second variable of type t_2 (to represent the set of objects retrieved by the navigation) and stating that this second variable returns the subset of instances of t_2 related with the first variable (of type t_1). Access to attributes or association classes is translated in the same way.

Translation of collection expressions
In OCL, a collection C of objects of type T can be manipulated through a predefined set of OCL operations and iterators. Each expression introduces a new quantification (that in some cases may be merged with the translation of the previous part of the OCL expression). Their basic translation is the following:

- *C->exists(v | boolean-condition-over-v)*
  The previous translation of C introduces an at-least-one quantification
  . The at-least-one quantification introduces a variable. (v)
  . . The variable ranges over the concept T
  . The at-least-one quantification scopes over the translation of boolean-condition-over-v

- *C->forAll (v | boolean-condition-over-v)*
  The previous translation of C introduces a universal quantification
  . The universal quantification introduces a variable. (v)
  . . The variable ranges over the concept T,
  . The universal quantification scopes over the translation of boolean-condition-over-v

- *C->select (v | boolean-condition-over-v)[3]*
  The previous translation of C introduces a universal quantification
  . The universal quantification introduces a variable
  . . The variable ranges over the concept T
  . The universal quantification scopes over the translation of boolean-condition-over-v

- *C-> iterate (v [acc:T'] | expression-with-v)*
  The previous translation of C introduces a universal quantification
  . The universal quantification introduces a variable (v)
  . . The variable ranges over the concept T
  . The universal quantification scopes over a second universal quantification
  . . The second universal quantification introduces a second variable (acc)
  . . . The second variable ranges over the concept T'
  . . The second universal quantification scopes over the translation of expression-with-v

---

[3] The translation of a *select* expression will differ depending on the context where it is applied. The one provided herein is useful when the *select* is the last OCL operator in the expression.

Original OCL expression:
```
context Car inv ownersDrive:
self.driver->exists(d|d=self.owner)
```

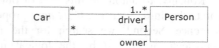

1) Semantic formulation in SBVR:

The rule is a proposition meant by a necessity claim. (ownersDrive)
. The necessity claim embeds a universal quantification.
.. The universal quantification introduces a first variable. (self)
... The first variable ranges over the concept 'car'.
.. The first universal quantification scopes over an at-least-one quantification. (exists)
... The at-least-one quantification introduces a second variable. (d, i.e.: self.driver)
.... The second variable ranges over the concept 'person'.
.... The at-least-one quantification is restricted by an atomic formulation.
..... The atomic formulation is based on the fact type 'driver of car'.
..... The atomic formulation has the first role binding.
...... The first role binding is of the role 'car'.
...... The first role binding binds to the first variable.
.... The atomic formulation has the second role binding.
...... The second role binding is of the role 'driver'.
...... The second role binding binds to the second variable.
.... The at-least-one quantification scopes over a third universal quantification.
..... The third universal quantification introduces a third variable. (self.owner)
...... The third variables ranges over the concept 'person'
...... The atomic formulation is restricted by the fact type 'owner of car'.
...... The atomic formulation has the first role binding.
.... The first role binding is of the role 'car'.
.... The first role binding binds to the first variable.
...... The atomic formulation has the second role binding.
.... The second role binding is of the role 'owner'.
.... The second role binding binds to the third variable
..... The third universal quantification scopes over an atomic formulation.
...... The atomic formulation is based on the fact type $thing_1 = thing_2$
...... The atomic formulation has the first role binding.
.... The first role binding is of the role '$thing_1$'.
.... The first role binding binds to the second variable.
...... The atomic formulation has the second role binding.
.... The second role binding is of the role '$thing_2$'.
.... The second role binding binds to the third variable

2) Structured English expression:

It is necessary that at least one driver *of* a car is the owner *of* the car

Fig. 1. Example of an: 1) OCL to SBVR and 2) SBVR to natural language translation. Next to each SBVR proposition, we show the subset of the OCL expression it refers to.

3 From SBVR to Natural Language

Once we have the SBVR-based representation of the OCL rules, the predefined mappings provided in the SBVR standard can be used to generate natural language descriptions of the rules in the *Structured English* or *RuleSpeak English* notations.

Roughly, to automate this part of our translation process we proceed as follows. First, the initial *'necessity claim'* proposition generates an *'It is necessary that'* sentence. Second, we search for the most inner quantification q in the SBVR representation. Let lf be the logical formulation that q scopes over. Then, the translation continues by translating lf as well as by (recursively) translating all variables involved in lf and q according to the appropriate patterns described in [4]. See Fig. 1 for an example.

4   Conclusions and Further Work

We have presented a method to facilitate the validation of business rules specified in the UML/OCL languages. Our method uses the SBVR standard to bridge the gap from UML/OCL to natural language, the only representation that most customers are able to understand, and thus, validate. We believe our paper can also be a first step towards a tighter integration of the business rules and UML/OCL communities.

As a further work we would like to propose a set of extensions to the SBVR standard to be able to translate all kinds of OCL expressions. We would also like to test our approach in industrial case studies and use the customers' feedback to refine and improve the quality of the final natural language expressions.

Acknowledgements

Thanks to Ruth Raventós for their useful comments to previous drafts of this paper. This work was partially supported by the Ministerio de Ciencia y Tecnologia and FEDER under project TIN2005-06053.

References

1. Burke, D.A., Johannisson, K.: Translating Formal Software Specifications to Natural Language: A Grammar-Based Approach. In: Blache, P., Stabler, E.P., Busquets, J.V., Moot, R. (eds.) LACL 2005. LNCS (LNAI), vol. 3492, pp. 51–66. Springer, Heidelberg (2005)
2. Cabot, J., Teniente, E.: Transformation Techniques for OCL Constraints. Science of Computer Programming 68(3), 152–168 (2007)
3. OMG: UML 2.0 OCL Specification. OMG Adopted Specification (ptc/03-10-14) (2003)
4. OMG: Semantics of Business Vocabulary and Rules (SBVR) Specification. OMG Adopted Specification (dtc/06-03-01) (2006)
5. OMG: UML 2.1.1 Superstructure Specification. OMG Adopted Specification (formal/07-02-03) (2007)

A General Architecture for Connecting NLP Frameworks and Desktop Clients Using Web Services

René Witte[1] and Thomas Gitzinger[2]

[1] Department of Computer Science and Software Engineering
Concordia University, Montréal, Canada
[2] Institute for Program Structures and Data Organization (IPD)
University of Karlsruhe, Germany

Abstract. Despite impressive advances in the development of generic NLP frameworks, content-specific text mining algorithms, and NLP services, little progress has been made in enhancing existing end-user clients with text analysis capabilities. To overcome this software engineering gap between desktop environments and text analysis frameworks, we developed an open service-oriented architecture, based on Semantic Web ontologies and W3C Web services, which makes it possible to easily integrate any NLP service into client applications.

1 Introduction

Research and development in natural language processing (NLP), text mining, and language technologies has made impressive progress over the last decade, developing innovative strategies for automatically analyzing and transforming large amounts of natural language data. Machine translation, question-answering, summarization, topic detection, cluster analysis, and information extraction solutions, although far from perfect, are nowadays available for a wide range of languages and in both domain-dependent and -independent forms [1].

Despite these advances, none of the newly developed technologies have materialized in the standard desktop tools commonly used by today's knowledge workers—such as email clients, software development environments (IDEs), or word processors. The vast majority of users still relies on manual retrieval of relevant information through an information retrieval tool (e.g., Google or Yahoo!) and manual processing of the (often millions of) results—forcing the user to interrupt his workflow by leaving his current client and performing all the "natural language processing" himself, before returning to his actual task.

This lack of integration is not caused by missing NLP services, nor can it be attributed to a lack of interest by end users: Rather, a gap exists between the domains of software engineers (developing end-user clients) and language engineers (implementing NLP services) that so far has been keeping these worlds apart. One approach seen in the past is the development of new (graphical) user interfaces for a particular analysis task from scratch. While this allows for

E. Kapetanios, V. Sugumaran, M. Spiliopoulou (Eds.): NLDB 2008, LNCS 5039, pp. 317–322, 2008.
© Springer-Verlag Berlin Heidelberg 2008

highly specialized clients, it also requires significant development efforts. Furthermore, today's users, who face constant "information overload," particularly need support within clients commonly used to develop and access natural language content—such as email clients or word processors. Reengineering all these existing clients just to deliver NLP services (like question-answering or summarization) is not a feasible approach.

The solution proposed here is an open service-oriented architecture (SOA) based on established standards and software engineering practices, such as W3C Web services with WSDL descriptions, SOAP client/server connections, and OWL ontologies for service metadata, which together facilitate an easy, flexible, and extensible integration of any kind of language service into any desktop client. The user is not concerned with the implementation or integration of these services, from his point of view he only sees context-sensitive *Semantic Assistants* relevant for his task at hand, which aid him in content analysis and development.

2 Related Work

Some previous work exists in building personalized information retrieval agents, e.g., for the Emacs text editor [2]. These approaches are typically focused on a particular kind of application (like emails or word processing), whereas our approach is general enough to define NLP services independently from the end-user application through an open, client/server, standards-based infrastructure.

The most widely found approach for bringing NLP to an end user is the development of a new interface (be it Web-based or a "fat client"). These applications, in turn, embed NLP frameworks for their analysis tasks, which can be achieved through the APIs offered by frameworks such as GATE [3] or UIMA [4]. The Bio-RAT system [5] targeted at biologists is an example for such a tool, embedding GATE to offer advanced literature retrieval and analysis services. In contrast with these approaches, we provide a service-oriented architecture to broker any kind of language analysis service in a network-transparent way. Our architecture can just as well be employed on a local PC as it can deliver focused analysis tools from a service provider.

Recent work has been done in defining Web services for integrating NLP components. In [6], a service-oriented architecture geared towards the field of terminology acquisition is presented. It wraps NLP components as Web services with clearly specified interface definitions and thus permits language engineers to easily create and alter concatenations of such components, also called processing configurations. This work can be seen as complimentary to our approach, since the granularity of our ontology model does not extend to individual NLP components, but rather complete NLP pipelines (applications build from components).

3 The General Integration Architecture

In this section, we present an overview of our architecture integrating NLP and (desktop) clients, delivering "Semantic Assistants" to a user.

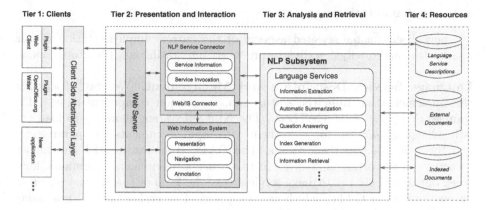

Fig. 1. Service-oriented architecture for integrating NLP and end-user clients

3.1 Architectural Overview

Our architecture is based on a multi-tier information system design (Fig. 1).

Tier 1: Clients.   This tier has the main purpose of providing access to the system. Typically, this will be an existing client (like a word processor or email client), extended to connect with our architecture through a plug-in interface. Besides facilitating access to the whole system, clients are also in part responsible for presentation, e.g., of language service results. In addition to the actual client applications, an abstraction layer is part of Tier 1. It shields the clients from the server and provides common functionality for NLP services.

Tier 2: Presentation and Interaction.   Tier 2 consists of a standard Web server and a module labeled "NLP Service Connector" in the diagram. One responsibility of this module is interaction, in that it handles the communication flow between the NLP framework and the web server. Moreover, it prepares language service results, by collecting output from the NLP services and transforming them into a format suitable for transmission to the client. Finally, the NLP Service Connector reads the descriptions of the language services, and therefore "knows" the vocabulary used to write these descriptions. In addition, our architecture can also provide services to other (Web-based) information systems, like a Wiki [7].

Tier 3: Analysis and Retrieval.   Tiers 1 and 2 are the ones the user has direct contact with. Tier 3 is only directly accessed by the NLP Service Connector. It contains the core functionality that we want, through Tiers 1 and 2, to bring to the end user. Here, the language services reside, and the NLP subsystem in whose environment they run (such as GATE [3] or UIMA [4]). NLP services can be added and removed from here as required by a language engineer.

Tier 4: Resources.   The final tier "Resources" contains the descriptions of the language services using an OWL-DL ontology. Furthermore, indexed documents as well as external documents (e.g., on the Web) count as resources.

3.2 Implementation

We now discuss some selected aspects of the current implementation of our architecture and briefly describe the process of integrating new (desktop) clients.

Language Service Description and Management. We start discussing the implementation from the status quo, a common component-based NLP framework—in our case, GATE [3]. The NLP subsystem allows us to load existing language services, provide them with input documents and parameters, run them, and access their results. The GATE framework's API also allows to access all the artifacts involved in this process (documents, grammars, lexicons, ontologies, etc.). Language services take the form of (persistent) *pipelines* or *applications* in GATE, which are composed of several sequential *processing resources* or *PRs*.

To describe the deployed language analysis services, we developed a comprehensive OWL ontology[1] that models all parameters required for finding applicable services, based on the user's task and language capabilities. This ontology is queried by the NLP Service Connector using SPARQL[2] in order to present applicable Semantic Assistants to the user. For example, we can query the service description metadata for all language pipelines that process input documents in German and deliver results in English. Additionally, the ontology contains all information needed to find, load, and execute a language service (including necessary parametrization), and to locate and retrieve the result(s) delivered.

Web Services. Thus far, we can search, load, parametrize, and execute language services. However, all input/output channels are still local to the context of the NLP framework's process. To make NLP services available in a distributed environment, we have to add network capabilities, which we achieve using *Web services*, a standard defined by the W3C:[3] *"A Web service is a software system designed to support interoperable machine-to-machine interaction over a network. It has an interface described in a machine-processable format (specifically WSDL[4]). Other systems interact with the Web service in a manner prescribed by its description using SOAP[5] messages, typically conveyed using HTTP with an XML serialization in conjunction with other Web-related standards."* In essence, a requester agent has to know the description of a Web service to know how to communicate with it, or, more accurately, with the provider agent implementing this Web Service. It can then start to exchange SOAP messages with it in order to make use of the functionality offered by the service. Provider agents are also referred to as Web service *endpoints*. Endpoints are referenceable resources to which Web service messages can be sent. Within our architecture (Fig. 1), the central piece delivering functionality from the NLP framework as a Web service

---

[1] OWL Web Ontology Language Guide, http://www.w3.org/TR/owl-guide/

[2] SPARQL Query Language for RDF, see http://www.w3.org/TR/rdf-sparql-query/

[3] Web Services Architecture, see http://www.w3.org/TR/ws-arch/

[4] Web Services Description Language (WSDL), see http://www.w3.org/TR/wsdl

[5] Simple Object Access Protocol, see http://www.w3.org/TR/soap/

endpoint is the NLP Service Connector. Our implementation makes use of the Web service code generation tools that are part of the Java 6 SDK and the Java API for XML-Based Web services (JAX-WS).[6]

The Client-Side Abstraction Layer (CSAL). We have just published a Web service endpoint, which means that the server of our architecture is in place. On the client side, our *client-side abstraction layer* (CSAL) is responsible for the communication with the server. This CSAL offers the necessary functionality for clients to detect and invoke brokered language services. The implementation essentially provides a proxy object (of class SemanticServiceBroker), through which a client can transparently call Web services. A code example, where an application obtains such a proxy object and invokes the getAvailableServices method on it to find available language analysis services, is shown below:

```
// Create a factory object
SemanticServiceBrokerService service = new SemanticServiceBrokerService();
// Get a proxy object, which locally represents the service endpoint (= port)
SemanticServiceBroker broker = service.getSemanticServiceBrokerPort();
// Proxy object is ready to use. Get a list of available language services.
ServiceInfoForClientArray sia = broker.getAvailableServices();
```

Client Integration. After describing the individual parts of our architecture's implementation, we now show how they interact from the point of view of a system integrator adding NLP services to a client application. The technical details depend on the client's implementation language: If it is implemented in Java (or offers a Java plug-in framework), it can be connected to our architecture simply by importing the CSAL archive, creating a SemanticServiceBrokerService factory, and calling Web services through a generated proxy object. After these steps, a Java-enabled client application can ask for a list of available language services, as well as invoke a selected service. The code example shown above demonstrates that the client developer can quite easily integrate his application with our architecture, and does not have to worry about performing remote procedure calls or writing network code.

A client application developer who cannot use the CSAL Java archive still has access to the WSDL description of our Web service. If there are automatic client code generation tools available for the programming language of his choice, the developer can use these to create CSAL-like code, which can then be integrated into or imported by his application.

3.3 Application Example: Word Processing Integration

Word processor applications are one of the primary tools of choice for many users when it comes to creating or editing content. Thus, they are an obvious candidate for our architecture, bringing advanced NLP support directly to end users in form of "Semantic Assistants." We selected the open source OpenOffice.org[7] application *Writer* for integration. We developed a plug-in for *Writer*

---

[6] Java API for XML-Based Web Services (JAX-WS), see https://jax-ws.dev.java.net/
[7] Open source office suite OpenOffice.org, see http://www.openoffice.org/

that seamlessly integrates NLP services, like summarization, index generation, or question-answering, into the GUI [8]. These services are discovered using their OWL description and presented to the user in additional dialog windows; the services can be executed on either the current document (e.g., for summarization) or a (highlighted) selection (e.g., for question-answering). Results are delivered depending on the produced output format, e.g., as a new *Writer* document or by opening a Web browser window.

4   Conclusions

In this paper, we described the design, implementation, and deployment of a service-oriented architecture that integrates end-user (desktop) clients and natural language processing frameworks. Existing applications, or newly developed ones, can be enhanced by so-called Semantic Assistants mediated through our architecture, e.g., information extracting systems, summarizers, and sophisticated question-answering systems. Existing NLP services can be added to the architecture without the need to change any code, thereby immediately becoming available to any connected client.

Our work fills an important gap in the emerging research area intersecting NLP and software engineering that has been neglected by existing approaches. The developed architecture [9] will be made available under an open source license, which we hope will foster a vibrant ecosphere of client plug-ins that finally allow to bring the hard won advances in NLP research to a mass audience.

References

1. Feldman, R., Sanger, J.: The Text Mining Handbook: Advanced Approaches in Analyzing Unstructured Data. Cambridge University Press, Cambridge (2006)
2. Rhodes, B.J., Maes, P.: Just-in-time Information Retrieval Agents. IBM Syst. J. 39(3–4), 685–704 (2000)
3. Cunningham, H., Maynard, D., Bontcheva, K., Tablan, V.: GATE: A Framework and Graphical Development Environment for Robust NLP Tools and Applications. In: Proc.of the 40th Anniversary Meeting of the ACL (2002), http://gate.ac.uk
4. Ferrucci, D., Lally, A.: UIMA: An Architectural Approach to Unstructured Information Processing in the Corporate Research Environment. Natural Language Engineering 10(3-4), 327–348 (2004)
5. Corney, D.P., Buxton, B.F., Langdon, W.B., Jones, D.T.: BioRAT: Extracting Biological Information from Full-Length Papers. Bioinformatics 20(17), 3206–3213 (2004)
6. Cerbah, F., Daille, B.: A Service Oriented Architecture for Adaptable Terminology Acquisition. In: Proc.NLDB (2007)
7. Witte, R., Gitzinger, T.: Connecting Wikis and Natural Language Processing Systems. In: Proc.of the 2007 Intl.Symp.on Wikis (WikiSym 2007) (2007)
8. Gitzinger, T., Witte, R.: Enhancing the OpenOffice.org Word Processor with Natural Language Processing Capabilities. In: Natural Language Processing resources, algorithms and tools for authoring aids, Marrakech, Morocco (June 1, 2008)
9. Semantic Assistants Architecture, http://semanticsoftware.info

Conceptual Modelling and Ontologies
Related Posters

Conceptual Model Generation from Requirements Model: A Natural Language Processing Approach

Azucena Montes[1], Hasdai Pacheco[1], Hugo Estrada[1], and Oscar Pastor[2]

[1] National Research and Technological Development Centre (CENIDET)
Interior Internado Palmira s/n, Col. Palmira. Cuernavaca, Morelos, Mexico
{amr,ehsanchez06c,hestrada}@cenidet.edu.mx
[2] Technical University of Valencia, Camino de Vera s/n, Valencia, Spain
opastor@dsic.upv.es

1 Introduction

In the conceptual modeling stage of the object-oriented development process, the requirements model is analyzed in order to establish the static structure (conceptual model) and dynamic structure (sequence diagrams, state diagrams) of the future system. The analysis is done manually by a requirements engineer, based upon his experience, the system's domain and certain rules of extraction of classes, objects, methods and properties. However, this task could get complicated when large systems are developed. For this reason, a method to automatically generate a conceptual model from the system's textual descriptions of the use case scenarios in Spanish language is presented. Techniques of natural language processing (NLP) and conceptual graphs are used as the basis of the method. The advantage of the proposed method, in comparison to other approaches [1, 2, 3], is that it makes exhaustive use of natural language techniques to face the text analysis. This allows us to consider text with a certain level of ambiguity and to cover relevant linguistic aspects, like composition of nouns, and verbal periphrases.

2 Proposed Method

The proposed method uses NLP techniques for the identification of linguistic elements [4, 5] that correspond to the basic elements of a object-oriented conceptual model (classes, methods, relationships, properties). The input to the method (Fig. 1) is a set of use case scenarios. The **syntactic analysis** stage consists of the text labeling and its decomposition in simple sentences, while the **semantic analysis** is responsible of eliminating the ambiguities found on the text (name composition, verbal periphrases).

Once an ambiguity-free text has been obtained, the **creation of an intermediate representation model** proceeds, that reflects the relationships between the linguistic elements. The intermediate representation model is a conceptual graph with morphological, lexical and statistical information.

The final stage of the method consists of traversing the conceptual graph and transforming the nodes into their equivalent representation in the object-oriented **conceptual model**. The output of the method is an OASIS code for the class diagram of the future system.

E. Kapetanios, V. Sugumaran, M. Spiliopoulou (Eds.): NLDB 2008, LNCS 5039, pp. 325–326, 2008.

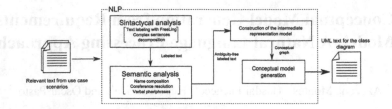

Fig. 1. Transformation method of requirements models into conceptual models

3 Conclusions

There is a need for mechanisms that help the analysts to systematically produce a conceptual model that appropriately represents the expected functionality of the software system. In this context, most of the current research works in the area use very restricted language to write the use case scenarios. The approach proposed in this work address some of the complex linguistic problems found in textual descriptions: composition of nouns and verbal periphrases for Spanish Language. With this work we are taking a further step in the process to define a complete software production process that starts with natural language definition of the system requirements and finishes with the generation of the information system.

References

1. Díaz, I., Sánchez, J., Pastor, O.: Metamorfosis: A Framework for the Functional Requirements Analysis. In: Proceedings of the Workshop on Requirements Engineering, Portugal, pp. 233–244 (2005)
2. Overmyer, S.P., Lavoie, B., Rambow, O.: Conceptual Modeling through Linguistic Analysis using LIDA. In: Proceedings of the 23rd International Conference on Software Engineering, pp. 401–410. IEEE, Los Alamitos (2001)
3. Harmain, H.M., Gaizauskas, R.: CM-Builder: An Automated NL-Based CASE Tool. In: Proceedings of the 15th IEEE International Conference on Automated Software Engineering, pp. 45–53. IEEE, Los Alamitos (2000)
4. Chen, P.: Entity-Relationship Diagrams and English Sentence Structure. In: Proceedings of the 1st International Conference on the Entity-Relationship Approach to Systems Analysis and Design, pp. 13–14 (1980)
5. Lyons, J.: Sémantique Linguistique. Larousse (1990)

Bivalent Verbs and Their Pragmatic Particularities

B. Patrut, I. Furdu, and M. Patrut

University of Bacau
bogdan@edusoft.ro, ifurdu@ub.ro, monica@edusoft.ro

Abstract. In this paper we present some pragmatic particularities of verbs describing actions or feelings that can appear between two human beings.

1 Introduction

In information extraction an important task is relation extraction, whose goal is to detect and characterize semantic relations between entities in a text [1,2]. Within this paper we investigate the pragmatic particularities of one class of verbs representing actions or feelings that can appear between two human beings. This class can serve as a model for other aspects related to human communication.

Let us consider "John loves Mary. William likes Mary. Which of them does Mary take for a walk?" (assuming Mary knows their feelings). From the information we have, we are tempted to say that John is the chosen one. Why? Because everybody knows that to love is "more" than to like. We also know that if Mary goes for a walk with John, then John takes Mary for a walk too. Even more, Mary will (probably) wish to walk with someone who loves or likes her. For finding the answer, let us consider a generalization: we know that XV_1Z and YV_2Z. We are asked to find out if XV_3Z exclusive-or XV_3Y. We have considered that X, Y and Z are persons, V_1, V_2, V_3 are verbs expressing relations between persons and that we can't have XV_3Z and YV_3Z simultaneously.

2 Bivalent Verbs and Characteristics

There are some verbs (in every language), about which we can say they are *bivalent interpersonal verbs*, when we refer to the number of persons who take part to the event. Generally speaking, we can talk about any verb's valence, which designates the number of actors who take part to the event. A verb that designates an action referring to persons, a feeling or a human state, is characterized, along with usual linguistic features, by its valence. The valence represents the number of the persons which the verb refers to; for example, we can say that "to be (ill, in love etc)" has the valence 1, "to love somebody" has the valence 2, "to lecture" has a valence higher than or equal to 3. Thus, a bivalent verb would be a verb with valence 2. Taking into discussion phrases of the type "a V r" that represent events in which *a* is the agent (the subject), *v* is the verb (predication) and *r* is the receiver (a prepositional object, the person who benefits of the result of the V's action), we can introduce certain notions.

E. Kapetanios, V. Sugumaran, M. Spiliopoulou (Eds.): NLDB 2008, LNCS 5039, pp. 327–328, 2008.
© Springer-Verlag Berlin Heidelberg 2008

A verb is *symmetric* when from the phrase XVY automatically results that YVX; if from formulations like XVY and YVZ results that XVZ too, the verb V is *transitive*. Persons have states, are sensitive and somebody might hurt them, might treat them well or might be indifferent to them; if from the action XVY, Y is the hurt person, we say that V is a *negative-receiver* verb. Similarly, we define the *positive-receiver* and *neutral-receiver* verbs; in a similar way we can talk about the sign of V, referring to the agent, with reference to the state of X person. So, a verb can be *positive-agent, negative-agent or neutral-agent)*; Two verbs can have the same meaning, no matter the context, or can have the same meaning only in certain contexts. If we have V_1, V_2 with similarities between them and if Y is more influenced in XV_1Y than in XV_2Y, then we say that V_1 is more powerful than V_2. So, the bivalent verbs can be represented in classes of equivalence. The verbs from a certain relative-synonymy class can be in relations of implication inside their class and over the verbs from other classes (fig. 1).

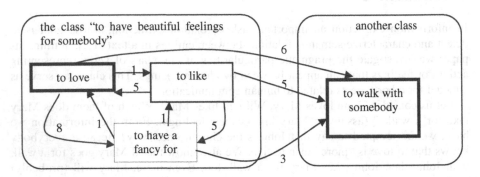

Fig. 1. Relations of implication from within a class and towards a verb from another class

3 Conclusions

Now it is clear why Mary will take John for a walk: "to love" is more powerful than "to like" and the value of its relation of implication over the other verbs is more powerful (6>5) and "to walk with somebody" is symmetric. We developed an algorithm is a method of reasoning for solving the generalized problem, or for inferring new knowledge, with the help of the human user.

References

1. Bunescu, R., Mooney, R. J.: A shortest path dependency kernel for relation extraction, in Proceedings of HLT/EMNLP (2005)
2. Jiang, J., Cheng, X. Z.: A Systematic Exploration of the Feature Space for Relation Extraction Proceedings of Human Language Technologies: The Annual Conference of the North American Chapter of the Association for Computational Linguistics, 113-120, (2007)

Using Ontologies and Relatedness Metrics for Semantic Document Analysis on the Web

Antonio Picariello and Antonio M. Rinaldi

Universitá di Napoli Federico II - Dipartimento di Informatica e Sistemistica
80125 Via Claudio, 21 - Napoli, Italy
{picus,amrinald}@unina.it

1 Motivations

The problem of defining new methodologies for Knowledge Representation has a great influence on information technology and, from a general point of view, on cognitive sciences. In the last years some new approach strictly related to the above matter has been proposed and some of them are based on ontologies to reduce conceptual or terminological mess and to have a common view of the same information. We here propose an ontological model to represent information and we implement it in a content based information system for scoring documents w.r.t. a given topic an ad hoc metric; we use the Web as our context of interest.

2 The Framework

The aim of our paper is to define and implement a model for knowledge representation using a conceptualization as much as possible close to the way in which the concepts are organized and expressed in human language and use it for improving the knowledge representation accuracy for document analysis. Our model is based on concepts, words, properties and constraints arranged in a class hierarchy, resulting from the syntactic category for concepts and words and from the semantic or lexical type for the properties. The implementation of the ontology is obtained by means of a Semantic Network (i.e. DSN), dynamically built using a dictionary based on WordNet [1]. The DSN is built starting from the domain keyword that represents the context of interest for the user. We then consider all the component synsets and construct a hierarchy, only based on the hyponymy property; the last level of our hierarchy corresponds to the last level of WordNet's one. After this first step we enrich our hierarchy considering all the other kinds of relationships in WordNet. In our framework, after the DSN building step, we compare it with the analyzed documents. The intersection between DSN and documents give us a list with the relevant terms related to the represented knowledge domain. All terms are linked by the properties from the DSN. We define a system grading ables to assign a *vote* to the documents on the basis of their syntactic and semantic content taking into account the *centrality* of a term in a given context using the polysemy property, a

E. Kapetanios, V. Sugumaran, M. Spiliopoulou (Eds.): NLDB 2008, LNCS 5039, pp. 329–330, 2008.

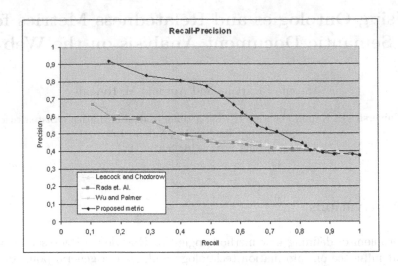

Fig. 1. Recall-Precision curve

syntactic semantic component using a statistical approach (SSG: Syntactic Semantic Grade) and a semantic contribution based on a combination of the path length (l) between pairs of terms and the depth (d) of their subsumer (i.e. the first common ancestor), expressed as number of hops; the correlation between the terms is the semantic relatedness and it is computed through a nonlinear function (SeG: Semantic Grade). We implemented our model and the proposed information extraction and analysis process in a system for document analysis using the Web as context of interest. To evaluate the performances of the proposed system from a quantitative point of view, we have carried out some experiments in order to evaluate the precision of results using a *test set collection* created by means of interaction with the directory service of the search engine Yahoo (www.yahoo.com). The directory service provides the category referred to each Web pages. In this way we can have a *relevance assessment* in order to compare the results. We compare our system results with other metrics used to measure the semantic relatedness between words. We focus our attention on three metrics well-known in literature: *Rada et. al.*, *Wu and Palmer* and *Leacock and Chodorow* [2]. In Figure 1 we can observe a good improvement respect to the other metrics putting into evidence great precision values for low recall one.

References

1. Miller, G.A.: Wordnet: a lexical database for english. Commun. ACM 38(11), 39–41 (1995)
2. Budanitsky, A.: Lexical semantic relatedness and its application in natural language processing. Technical report, Department of Computer Science, University of Toronto (1999)

Information Retrieval Related Posters

The *TSRM* Approach in the Document Retrieval Application*

Efthymios Drymonas, Kalliopi Zervanou, and Euripides G.M. Petrakis

Intelligent Systems Laboratory, Electronic and Computer Engineering Dpt.,
Technical University of Crete (TUC), Chania, Crete, Greece
max@softnet.tuc.gr, {kelly,petrakis}@intelligence.tuc.gr
http://www.intelligence.tuc.gr

Abstract. We propose *TSRM*, an alternative document retrieval model
relying on multi-word, rather than mere single key-word domain terms,
typically applied in traditional IR.

Keywords: term extraction, term similarity, information retrieval.

1 The *TSRM* Approach

TSRM is built upon the idea of computing the similarity among multi-word
terms using internal (lexical) and external (contextual) criteria, while taking
into consideration term variation (morphological and syntactic). *TSRM* can be
viewed as a three phase process:

A. Corpus processing:	1. Corpus Pre-Processing
	2. Term Extraction (C/NC value)
	3. Term Variants Detection (FASTR)
	4. Term Similarity Estimation
B. Query processing:	1. Query Pre-Processing
	2. Query Term Extraction (C/NC value)
	3. Query Term Expansion by term variants (FASTR)
C. Document Retrieval:	1. Similarity Computation (TSRM/A or TSRM/B)
	2. Document Ranking

The C/NC-value method [1] is a domain-independent and hybrid (linguis-
tic/statistical) method for the extraction of multi-word and nested terms. The
candidate noun phrase termhood is represented by NC-value. FASTR [2] identifies term variants based on a set of morpho-syntactic rules. Based on the Nenadic
et al. [3] study, term similarity (TS) in *TSRM*, is defined as a linear combination
measure of two similarity criteria referred to as lexical similarity and contextual
similarity. In *TSRM/A* document similarity is calculated as

$$Similarity(d, q) = \frac{\sum_i \sum_j q_i d_j TS(i, j)}{\sum_i \sum_j q_i d_j},$$ (1)

---

* This work was supported by project TOWL (FP6-Strep, No. 26896) of the EU.

E. Kapetanios, V. Sugumaran, M. Spiliopoulou (Eds.): NLDB 2008, LNCS 5039, pp. 333–334, 2008.

where i and j are terms in the query q and the document d respectively. The weight of a term (q_i, d_j respectively) is computed as the NC-value of the term. Eq. 1 takes into account dependencies between non-identical terms. In **TSRM/B** document similarity can be computed as:

$$Similarity(d,q) = \frac{1}{2}\{\frac{\sum_{i \in q} idf_i \max_j TS(i,j)}{\sum_{i \in q} idf_i} + \frac{\sum_{j \in d} idf_j \max_i TS(j,i)}{\sum_{j \in d} idf_j}\} \quad (2)$$

TSRM has been tested on OHSUMED, a standard TREC collection. The results in Fig. 1 demonstrate that *TSRM*, with both formulae for document matching, outperform Vector Space Model (VSM) in most cases (i.e., VSM performs well only for the first few answers).

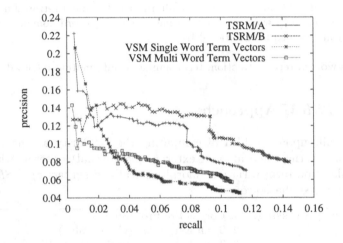

Fig. 1. Precision-recall diagram for retrieval on OHSUMED using *TSRM* and VSM

2   Conclusions

We have discussed on the potential improvements to traditional IR models related to document representation and conceptual topic retrieval and proposed *TSRM* to demonstrate our ideas.

References

1. Frantzi, K., Ananiadou, S., Mima, H.: Automatic recognition of multi-word terms: The C-Value/NC-value Method. Intl. J. of Digital Libraries 3(2), 117–132 (2000)
2. Jacquemin, C., Klavans, J., Tzoukermann, E.: Expansion of multi-word terms for indexing and retrieval using morphology and syntax. In: 35th Annual Meeting of the Assoc. for Comp. Linguistics, Spain, pp. 24–31 (1997)
3. Nenadic, G., Spasic, I., Ananiadou, S.: Automatic Discovery of Term Similarities Using Pattern Mining. Intl. J. of Terminology 10(1), 55–80 (2004)

Enhanced Services for Targeted Information Retrieval by Event Extraction and Data Mining

Felix Jungermann and Katharina Morik

Artificial Intelligence Unit, TU Dortmund
http://www-ai.cs.tu-dortmund.de

Abstract. We present a framework combining information retrieval with machine learning and (pre-)processing for named entity recognition in order to extract events from a large document collection. The extracted events become input to a data mining component which delivers the final output to specific user's questions. Our case study is the public collection of minutes of plenary sessions of the German parliament and of petitions to the German parliament.

Keywords: Web-based Information Services, Knowledge Extraction, Application Framework.

1   Targeted Information Retrieval

The integration of Named Entity Recognition as slots into Event Templates allows to automatically gather data directly from texts for further analyses. Our integrated system (respect [1] and figure) consists of the following components:

IR-Component. This component extracts all the plenary session documents and printed papers and stores them for later use. Additionally the websites of the members of parliament are extracted in order to fill event templates containing personal information. In addition we build up an index of all the documents using the open-source indexing environment *lucene.* Furthermore, we extract the information of every printed paper into an event-like template for accessing the information easily.

IE-Component. The IE-component is necessary for other processes for which it captures trigger-words or patterns. Additionally, it delivers relation extraction. We have developed a plugin for information extraction (IE) for the open-source software RapidMiner[1] [2]. Using the IE-plugin one can access every textual document, extracted further. Then, one can use various pre-processing-operators. After the pre-processing steps, one can use multiple machine learning methods. We use *conditional random fields* for sequence-labeling in order to extract entities, relations or events.

---

[1] Formerly known as YALE.

E. Kapetanios, V. Sugumaran, M. Spiliopoulou (Eds.): NLDB 2008, LNCS 5039, pp. 335–336, 2008.
© Springer-Verlag Berlin Heidelberg 2008

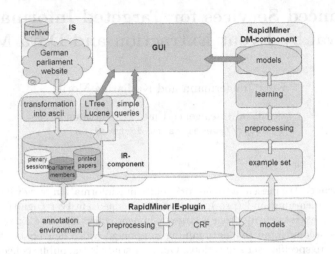

DM-Component. The innovation of our system is the opportunity to use extracted events as input for data mining experiments. It is nice to get questions answered like *'How many requests are recommended to be rejected?'*, but data mining goes beyond that. It offers the opportunity to get to know why or under which circumstances a request has been rejected. All found requests are converted into examples as an input for a data mining task. These examples can be processed – internally by RapidMiner – by various machine learning techniques.

2   Evaluation

We made an exemplary event extraction experiment to extract all *requests* and their *recommendations*. Accordingly we extracted some additional information to each *request* like the *signing party*, the *number of request*, and some more. The machine learning task was to predict the *recommendation* for each *request*. Our framework extracted 1.935 *requests*. 1.225 of them have a *recommendation*. The automatic extraction of the kind of *recommendation* (label) ended up in 794 *rejections*, 251 *acceptances*, 44 *deposes* and 166 *unidentified recommendations*.

A ten-fold-cross-validation over the requests ended up in *67,1 % accuracy* to predict the label basically on the basis of the *number of supporting parties* and the *particular party*.

References

1. Jungermann, F., Morik, K.: Enhanced services for targeted information retrieval by event extraction and data mining. Technical Report 04/2008, Sonderforschungsbereich 475, University of Dortmund (2008)
2. Mierswa, I., Wurst, M., Klinkenberg, R., Scholz, M., Euler, T.: YALE: Rapid Prototyping for Complex Data Mining Tasks. In: Eliassi-Rad, T., Ungar, L.H., Craven, M., Gunopulos, D. (eds.) Proceedings of the 12th ACM SIGKDD International Conference on Knowledge Discovery and Data Mining (KDD 2006), pp. 935–940. ACM Press, New York (2006)

Querying and Question Answering
Related Posters

Improving Question Answering Tasks by Textual Entailment Recognition[*]

Óscar Ferrández, Rafael Muñoz, and Manuel Palomar

Natural Language Processing and Information Systems Group
Department of Computing Languages and Systems, University of Alicante
{ofe,rafael,mpalomar}@dlsi.ua.es

Abstract. This paper explores a suitable way to integrate a Textual Entailment (TE) system, which detects unidirectional semantic inferences, into Question Answering (QA) tasks. We propose using TE as an answer validation engine to improve QA systems, and we evaluate its performance using the *Answer Validation Exercise* framework. Results point out that our TE system can improve the QA task considerably.

1 The Method

Textual Entailment (TE) has been proposed recently as a generic framework for modeling language variability. Competitions such as the *Recognising Textual Entailment Challenges* (RTE) [1] and the *Answer Validation Exercise* (AVE[1]) track within the *Cross–Language Evaluation Forum*[2] describe appropriate frameworks for evaluating the use of TE approaches within QA environments.

The proposed approach features lexical and semantic inferences computed as features for a SVM algorithm. It is an extension of our previous system presented in [2] adding the improvements discussed throughout the paper.

The lexical perspective is covered by the computation of a wide variety of measures[3] such as the Needleman-Wunsch algorithm, Smith-Waterman algorithm, Jaro distance, Euclidean distance, Dice's coefficient and Cosine similarity.

The semantic inferences are obtained from the WordNet taxonomy. We have used the Java WordNet Similarity Library (JWSL[4]), which implements some of the most commons semantic similarity measures. We have developed a procedure that obtains the maximum similarity among all measures. Also, we propose another inference based on the detection and correspondence of Named Entities (NEs). In [3] the authors successfully build their system only using the knowledge supplied by the recognition of NEs. In our case, we use NERUA [4] to

---

[*] This research has been partially subsidized by the Spanish Government under project TIN2006-15265-C06-01 and by the QALL-ME consortium, 6th Framework Research Programme of the European Union (EU), FP6-IST-033860.
[1] http://nlp.uned.es/clef-qa/ave/
[2] http://www.clef.campoaign.org
[3] http://www.dcs.shef.ac.uk/~sam/simmetrics.html
[4] http://grid.deis.unical.it/similarity/

E. Kapetanios, V. Sugumaran, M. Spiliopoulou (Eds.): NLDB 2008, LNCS 5039, pp. 339–340, 2008.

recognise Spanish and English entities, and we integrated it into the system as a preprocessing tool in order to detect correspondences between the entities of the texts.

2 Evaluation and Discussion

To evaluate the benefits that the system contributes to QA systems, the AVE framework is probably the best evaluation environment. Initially, the system was designed to deal with English corpora, but owing to AVE also provides corpora in Spanish, we were able to prove the system's portability to Spanish. A baseline setting all the pairs as positive were generated. Table 1 shows the results as well as the place that our system would have reached in the AVE ranking.

Table 1. Results obtained over the AVE evaluation framework

System	dev. data			test data				rank
	Prec.	Recall	F	Prec.	Recall	F	QA acc.	
$Baseline_{alltrue}$	0.116	1	0.208	0.108	1	0.194	–	–
Lexical	0.281	0.792	0.414	0.277	0.857	0.419	0.2089	3rd
Lex+JWSL	0.266	0.8	0.399	0.286	0.952	0.44	0.2238	3rd
Lex+JWSL+ENT	0.266	0.8	0.399	0.323	0.952	**0.482**	0.2388	2nd
over Spanish Corpora								
$Baseline_{alltrue}$	0.145	1	0.254	0.23	1	0.374	–	–
Lexical	0.293	0.717	0.416	0.382	0.96	0.546	0.4823	1st
Lex+ENT	0.293	0.717	0.416	0.423	0.991	**0.593**	0.4176	1st

Results show that accurate precision is reached from a lexical point of view, but although the lexical analysis is of paramount importance in the entailment decision making process, without deep semantic analysis the TE task can not be tackled completely. However, the addition of these units would definitely increase the response time preventing the system from being used in real–time tasks.

References

1. Giampiccolo, D., Magnini, B., Dagan, I., Dolan, B.: The Third PASCAL Recognizing Textual Entailment Challenge. In: Proceedings of the ACL-PASCAL Workshop on Textual Entailment and Paraphrasing, Prague, June 2007, pp. 1–9 (2007)
2. Ferrández, O., Micol, D., Muñoz, R., Palomar, M.: A Perspective-Based Approach for Solving Textual Entailment Recognition. In: Proceedings of the ACL-PASCAL Workshop on Textual Entailment and Paraphrasing, Prague, June 2007, pp. 66–71 (2007)
3. Rodrigo, A., Peñas, A., Verdejo, F.: UNED at Answer Validation Exercise 2007. In: Working Notes of the CLEF 2007 Workshop, Budapest, Hungary (September 2007)
4. Kozareva, Z., Ferrández, O., Montoyo, A., Muñoz, R.: Combining data-driven systems for improving Named Entity Recognition. Data and Knowledge Engineering 61(3), 449–466 (2007)

Supporting Named Entity Recognition and Syntactic Analysis with Full-Text Queries

Luísa Coheur, Ana Guimarães, and Nuno Mamede

L²F/INESC-ID Lisboa
Rua Alves Redol, 9, 1000-029 Lisboa, Portugal
{lcoheur,arog,nuno.mamede}@l2f.inesc-id.pt

Abstract. JaTeDigo is a natural language interface (in Portuguese) to a cinema database that has to deal with a vocabulary of more than 2500000 movies, actors and staff names. As our tools were not able to deal with such a huge amount of information, we decided to profit from full-text queries to the database to support named entity recognition (NER) and syntactic analysis of questions. This paper describes this methodology and evaluates it within JaTeDigo.

1 Introduction

JaTaDigo is a Natural Language Interface (in Portuguese) to a cinema database. Its database has information from IMDB, OSCAR.com and PTGate[1].

Regarding question interpretation, JaTeDigo runs over a natural language processing chain, responsible for a morpho-syntactic analysis and for a semantic interpretation based on domain terms. Due to the huge quantity of names that can be asked about, two problems arose: first, it was impossible to load that information in the dictionaries and run the system in a reasonable time; second, as the part-of-speech tagger suggests labels for unknown words – and names can be entered both in Portuguese and English –, the morpho-syntactic analysis became erratic, and it was impossible to build syntactic rules to capture those sequences.

Considering other NLIDB, the last decade brought some interesting and promising NLIDB/QA systems such as BusTUC [1] or Geo Query [2]. However, typically these NLIDB run over small databases, and they do not accept named entities in other languages than the query language. As so, after several experiments, and before going into sophisticated QA techniques, we decided to use full-text queries and see the results. In this paper we explore this hypothesis. Further details about JaTeDigo can be found in [3].

2 Overview of the Methodology

1. The question – possibly without some words, such as interrogative pronouns – is submitted to the database in a full-text query;

---

[1] http://www.imdb.com/, http://www.oscars.org/awardsdatabase/ and http://www.cinema.ptgate.pt, respectively.

E. Kapetanios, V. Sugumaran, M. Spiliopoulou (Eds.): NLDB 2008, LNCS 5039, pp. 341–342, 2008.
© Springer-Verlag Berlin Heidelberg 2008

2. The resulting first 100 rows – value obtained empirically – are compared with the question sequences and named entities are captured;
3. A disambiguation step is performed, if there is more that one entity with the same name;
4. A local grammar is created and added to the main grammar;
5. A morpho-syntactic and a semantic analysis are performed over the question and if during morpho-syntactic analysis the part-of-speech tagger suggests wrong tags, the local grammar overrides them;
6. Syntactic and semantic analysis are performed, an SQL query is built and the answer is retrieved from the database.

3 Evaluation

For evaluation proposes, 20 users performed 198 questions. From these, 41 got no response and 157 were answered. From the 157, 7 got wrong answers. Although there were many causes for these errors, if names were correct and complete, NER based in full-text queries was 100% accurate. It should be noticed however that if names are misspelled or incomplete, results can be unanswered questions or – worse – wrong answers.

4 Conclusions and Future Work

In this paper, we have presented a methodology to support NER and a syntactic analysis, based on full-text queries. The questions is submitted to the database in a full-text query, named entities are identified from the returned results and a local grammar is created, overriding possible errors from the part-of-speech tagger. This method provides recognition of named entities with little effort and although a simple solution, it can be useful to NLIDB developers that are using tools that cannot deal with large vocabularies.

Acknowledgment

This work was funded by PRIME National Project TECNOVOZ number 03/165.

References

1. Amble, T.: BusTUC - a natural language bus route oracle. In: Proceedings of the 6th Applied Natural Language Processing Conference, Seattle, Washington, USA, Association for Computational Linguistics, pp. 1–6 (2000)
2. Kate, R.J., Wong, Y.W., Mooney, R.J.: Learning to transform natural to formal languages. In: AAAI, pp. 1062–1068 (2005)
3. Guimarães, R.: Játedigo- uma interface em língua natural para uma base de dados de cinema. Master's thesis, Instituto Superior Técnico (2007)

Document Processing and Text Mining Related Posters

Multilingual Feature-Driven Opinion Extraction and Summarization from Customer Reviews

Alexandra Balahur and Andrés Montoyo

Department of Software and Computing Systems,
University of Alicante, Spain
{abalahur,montoyo}@dlsi.ua.es

Abstract. This paper presents a feature-driven opinion summarization method for customer reviews on the web based on identifying general features (characteristics) describing any product, product specific features and feature attributes (adjectives grading the characteristics). Feature attributes are assigned a polarity using on the one hand a previously annotated corpus and on the other hand by applying Support Vector Machines Sequential Minimal Optimization[1] machine learning with the Normalized Google Distance[2]. Reviews are statistically summarized around product features using the polarity of the feature attributes they are described by.

Motivation and Contribution. Mining user reviews on the web is helpful to potential buyers in the process of decision making, but the large quantity of data available and the language barrier problems can only be solved using a system that automatically analyzes and extracts the values of the features for a given product, independent of the language the customer review is written in, and presents user with concise results. What follows is a description of such a system that presently works on Spanish and English. Our approach concentrates on two main problems. The first one is discovering the features that will be quantified, implicitly or explicitly appearing in text. We define a set of product independent features and deduce a set of product dependent features, which we complete using the synonymy, meronymy and hyponymy relations in WordNet[1], and a set of ConceptNet[2] relations. Feature attributes are discovered by applying the *"has attributes"* relation in WordNet to product features and adding the concepts found in the ConceptNet relations *PropertyOf* and *CapableOf*. Using EuroWordNet we map the concepts discovered in English to Spanish. Bigram features are determined running Pedersens Ngram Statistics Package[3] with target co-occurrences of features on a corpus of 200 reviews.

Implementation. The implementation includes two stages: *preprocessing* and *main processing*. In the *preprocessing stage*, the documents retrieved containing the product name in different languages are filtered with Lextek[4] and split

---

[1] http://wordnet.princeton.edu
[2] http://web.media.mit.edu/ hugo/conceptnet/
[3] www.d.umn.edu/ tpederse/nsp.html
[4] http://www.lextek.com

E. Kapetanios, V. Sugumaran, M. Spiliopoulou (Eds.): NLDB 2008, LNCS 5039, pp. 345–346, 2008.

according to the language they are written in - English or Spanish. We determine the product category and extract the product specific features and feature attributes accordingly. The *main processing* is done in parallel for English and Spanish, using specialized tools for anaphora resolution, sentences splitting and Named Entity Recognition. We extract the sentences from texts containing at least one feature or feature attribute and parse them. We also define a set of context polarity shifters (negation, modifiers and modal operators). We extract triplets of the form (feature, attributeFeature, valueOfModifier), by determining all attribute features that relate to the features in the text. Feature attributes appearing alone are considered to implicitly evoke the feature that it is associated with in the feature collection previously built for the product.

Assignment of Polarity to Feature Attributes. In order to assign polarity to each of the identified feature attributes of a product, we employ SMO SVM machine learning and the Normalized Google Distance (NGD). The set of anchors contains the terms {*featureName, happy, unsatisfied, nice, small, buy*}. Further on, we build the classes of positive and negative examples for each of the feature attributes considered. We use the anchor words to convert each of the 30 training words to 6-dimensional training vectors defined as $v(j,i) = NGD(wi,aj)$, where aj with j ranging from 1 to 6 are the anchors and wi, with i from 1 to 30 are the words from the positive and negative categories. After obtaining the total 180 values for the vectors, we use SMO SVM to learn to distinguish the product specific nuances. For each of the new feature attributes we wish to classify, we calculate a new value of the vector $vNew(j,word)=NGD(word, aj)$, with j ranging from 1 to 6 and classify it using the same anchors and the trained SVM model. The precision of classification was beween 0.72 and 0.80 and the kappa value was above 0.45.

Summarization of Feature Polarity. We statistically summarize the polarity of the features, as ratio between the number of positive or negative quantifications to feature attributes and the total number of quantifications made in the considered reviews to that specific feature respectively, in the form (feature, percentagePositiveOpinions, percentageNegativeOpinions).

Evaluation. For the evaluation of the system, we annotated a corpus of 50 customer reviews for each language at the level of feature attributes. The results obtained have the following values for Accuracy, Precision and Recall respectively: English - 0.82, 0.80, 0.79, Spanish - 0.80, 0.78, 0.79, Combined - 0.81, 0.79, 0.79, with baselines for English - 0.21, 0.20, 0.40 and Spanish - 0.19, 0.20, 0.40. We can notice how the use of NGD helped the system acquire significant and correct new knowledge about the polarity of feature attributes.

References

1. Platt, J.: Sequential minimal optimization: A fast algorithm for training support vector machines. Microsoft Research Technical Report MSR-TR-98-14 (1998)
2. Cilibrasi, D., Vitanyi, P.: Automatic Meaning Discovery Using Google. IEEE Journal of Transactions on Knowledge and Data Engineering (2006)

Lexical and Semantic Methods in Inner Text Topic Segmentation: A Comparison between C99 and Transeg

Alexandre Labadié and Violaine Prince

LIRMM
161 rue Ada
34392 Montpellier Cedex 5, France
{labadie,prince}@lirmm.fr
http://www.lirmm.fr

Abstract. This paper present a semantic and syntactic distance based method in topic text segmentation and compare it to a very well known text segmentation algorithm: c99. To do so we ran the two algorithms on a corpus of twenty two French political discourses and compared their results.

Keywords: Text segmentation, topic change, c99.

1 C99 and Transeg

In this paper we compare Transeg, a text segmentation method based on distances between text segments and a fairly deep syntactic and semantic analysis, to c99 [1]. Our goal is to assess the importance of syntactic and semantic information in text segmentation.

1.1 C99

Developed by Choi [1], c99 is text segmentation algorithm strongly based on the lexical cohesion principle. It is, at this time, one the best algorithms in the text segmentation domain.

1.2 Transeg: A Distance Based Method

We have developed a distance based text segmentation specifically designated to find topic variations inside the text called Transeg.

The first step of our approach is to convert each text sentence into a semantic vector obtained using the French language parser SYGFRAN [2].

Using this sentence representation, we try to find transition zones inside the text using a protocol described in [3].

In our first implementation of this method, we used the angular distance to compute the transition score. In this paper we use an extended version of the concordance distance first proposed by [4]. This improvement has enhanced the discriminant capabilities of our method.

E. Kapetanios, V. Sugumaran, M. Spiliopoulou (Eds.): NLDB 2008, LNCS 5039, pp. 347–349, 2008.

2   Experiment and Result on French Political Discourses

In order to compare the two methods, we tried them on a set of twenty two French political discourses and we measured their scores in text topic boundaries detection.

We chose a set of French political discourses for two main reasons: (1) As they were identified by experts, internal boundaries looked less artificial than just beginnings of concatenated texts. (2) The topical structure of a political discourse should be more visible than other more mundane texts.

From an original corpus of more than 300, 000 sentences of a questionable quality we extracted and cleaned 22 discourses totalizing 1, 895 sentences and 54, 551 words.

We set up a run of both Transeg and the LSA augmented c99 (Choi's algorithm) on each discourse separately. To be sure that there is not any implementation error, we used the 1.3 binary release that can be downloaded on Choi's personal Linguaware Internet page (http://www.lingware.co.uk/homepage/ freddy.choi/software/software.htm).

To evaluate the results of both methods, we used the DEFT'06 **tolerant recall and precision** ([5]).

Table 1. c99 and Transeg topic segmentation results

	Words	Sentences	Transeg			c99		
			Precision	Recall	FScore	Precision	Recall	FScore
Text 3	2767	92	42.86	85.71	**28.57**	20	14.29	8.33
Text 6	5348	212	8.7	18.18	5.88	20	18.18	**9.52**
Text 9	1789	53	75	100	**42.86**	25	16.67	10
Text 19	678	26	33.33	33.33	16.67	50	66.67	**28.57**
Text 22	1618	40	60	75	**33.33**	100	25	20

First of all, we see that results (table 1) are not spectacular. *FScore* is a very strict measure, even when softened by using tolerant recall and precision.

Transeg has a better *FScore* on 16 on the 22 documents composing the corpus. On these 16 texts, our recall is always better or equal to c99 and our *FScore*. Transeg has also the best *FScore* of both runs with 42.86 on text 9. C99 has a better *FScore* on 6 texts. Anyway, we should notice that c99 has comparatively good precision on most of the texts. Thus, when examining texts where c99 is better we see that they fit into two categories: (1) Texts with few boundaries. C99 seems to be very effective on short texts with just one inner topic boundary. (2) Enumerations. Text 6 for example, which is quite big, is a record of the government spokesman where he enumerates dealt subjects during the weekly minister reunion.

According to the experiment results, Transeg seems to be more effective at finding inner text segments than c99. It seems to be efficient on longer documents, with multiple and related topics. Whereas a lexically based method is efficient on either short texts with very few topics, or enumerations and/ or concatenation of unrelated topics or subjects.

References

1. Choi, F.Y.Y., Wiemer-Hastings, P., Moore, J.: Latent semantic analysis for text segmentation. In: Proceedings of EMNLP, pp. 109–117 (2001)
2. Chauché, J.: Un outil multidimensionnel de l'analyse du discours. In: Proceedings of Coling 1984, vol. 1, pp. 11–15 (1984)
3. Prince, V., Labadié, A.: Text segmentation based on document understanding for information retrieval. In: Kedad, Z., Lammari, N., Métais, E., Meziane, F., Rezgui, Y. (eds.) NLDB 2007. LNCS, vol. 4592, pp. 295–304. Springer, Heidelberg (2007)
4. Chauché, J., Prince, V., Jaillet, S., Teisseire, M.: Classification automatique de textes á partir de leur analyse syntaxico-sémantique. In: Proceedings of TALN 2003, pp. 55–65 (2003)
5. Azé, J., Heitz, T., Mela, A., Mezaour, A., Peinl, P., Roche, M.: Présentation de deft 2006 (defi fouille de textes). In: Proceedings of DEFT 2006, vol. 1, pp. 3–12 (2006)

An Application of NLP and Audiovisual Content Analysis for Integration of Multimodal Databases of Current Events

Alberto Messina

Università degli Studi di Torino, Dip. di Informatica,
Corso Svizzera 185, I-10149 Torino, Italy
messina@di.unito.it

Abstract. The nowadays explosion of pervasive services based on the Internet makes information about current events easy and rapidly available to any user. A critical point is concerned with the possibility for the users to have a comprehensive, trustworthy and multi-source informative dossiers on their events of interest. This paper presents a completely unsupervised architecture for the automatic detection and integration of informative sources about current events based on natural language processing and audiovisual content analysis. The system provides a multimodal RSS feed integrating both audiovisual and textual contributions.

1 Aggregating Information Items in Hybrid Spaces

Modern times are characterised by the annihilation of spatiotemporal barriers between occurrence of events and the awareness of them by the public. This revolutionary scenario brings in its own new issues, e.g. users wanting to have a complete view on *a certain event* struggle with the variegated sources of information all dealing with the same subject but under concurrent perspectives.

This innovative model requires innovative multimodal aggregation services. Intelligent information aggregation and organisation has been a hot topic in the world wide web developers community in the recent years, and specifically in the area of news aggregation, e.g. see [1]. Analysis and indexing of audiovisual material is traditionally treated as a separate topic of research. In our context, programme segmentation is the key subject and particularly all techniques concerning news item detection and indexing, e.g. [2]. Despite the liveliness of the two described areas, very little has happened to combine efforts towards a complete multimodal framework, able to exploit the strength of both information delivery media. We present a novel solution in the context of intelligent information aggreagation (IIA), in which we tried to tackle and address these new issues. We propose a hybrid clustering algorithm based on the intuitive principle of *semantic relevance* between information items belonging to a primary (C_1) and an ancillary (C_2) space, which are heterogeneous each other. We then define an empirical entailment measurement S in the space C_1 by measuring the set intersection on C_2 of the semantic relevance function.

E. Kapetanios, V. Sugumaran, M. Spiliopoulou (Eds.): NLDB 2008, LNCS 5039, pp. 350–351, 2008.
© Springer-Verlag Berlin Heidelberg 2008

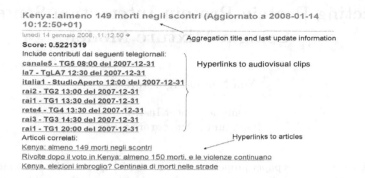

Fig. 1. Example of published multimodal aggregation as RSS item

Concretely, we used the set of itelevision newscasts channels as ancillary space (C_2), and the newspapers sites as primary space (C_1). Seven major national channels were acquired and digitised, and news items were automatically detected using a segmentation method based on audiovisual content analysis. The spoken content of detected news items was transcribed to text by a speech to text engine and indexed by an open source full text search engine (Lucene). We used RSS feeds published by the online newspapers as surrogates of the full articles, using the title and description sentences as a summary of the content.

RSS information items were elaborated by a part of speech tagger which extracts elements of the title and description sentences which are important from the *linguistic* point of view, which were used to build full text queries, run on the news items index provided by Lucene, thus implementing the semantic relevance function.

Cross-space aggregations are collected in a multimodal RSS service, each item of which points at both the semantically relevant articles and news items as depicted in Figure 1. The overall accuracy has been measured by the following indicators: a)actual relevance of the aggregation detection (AR): what part the detected aggregations represent a semantically coherent aggregation?; b)precision of the induced aggregation (PIA): what part of the RSS items of an aggregation are relevant w.r.t. the semantics of it?; c) precision of the cross-space aggregation (PCA): what part of the aggregated audiovisual clips concerning a detected current event are relevant to it? The values of the three indicators measured on a database of 162 detected aggregations, corresponding to two weeks of system run, were respectively 0.77, 0.97, and 0.96.

References

1. Das, A.S., Datar, M., Garg, A., Rajaram, S.: Google news personalization: scalable online collaborative filtering. In: Proc. of the 16th international conference on World Wide Web, pp. 271–280 (2007)
2. De Santo, M., Percannella, G., Sansone, C., Vento, M.: Unsupervised news video segmentation by combined audio-video analysis. In: MRCS, pp. 273–281 (2006)

Detecting Protein-Protein Interaction Sentences Using a Mixture Model

Yun Niu and Igor Jurisica

Ontario Cancer Institute
{yun,juris}@cs.toronto.edu

Abstract. This paper proposes a mixture model approach for automatic detection of **protein-protein interaction sentences**. We combine rule-based and SVM algorithms to achieve both high precision and recall. Experimental evaluation shows that the mixture model outperforms each individual model. Particularly, recall is substantially improved using the mixture model compared to SVM with only a small reduction in precision. We apply this approach to annotate interactions in I^2D database of protein-protein interactions with references from Medline.

1 Introduction

Despite the success of automatic protein-protein interaction (PPI) detection using information extraction algorithms, these approaches often suffer from low coverage. Our evaluation of automatic PPI detection shows that *sentence-level PPI information* is a crucial factor that could contribute to improving coverage of interaction relation detection[1]. *Sentence-level PPI information* indicates whether a sentence expresses at least one interaction. In this paper, we propose to use a mixture model to combine a rule-based model with support vector machine algorithm to identify **interaction sentence** in text.

2 Methods and Results

A rule-based approach. In automatic PPI extraction research, some words have been recognized as important cues that suggest existence of PPIs in text. We developed a simple rule-based approach to assign a positive label to a sentence containing at least one keyword (i.e., an **interaction sentence**). Our keyword set is developed based on the list in [2]. The final keyword collection comprises each word in the list, combined with all its morphological variations.

Support Vector Machines (SVM) approach. We develop a supevised machine learning approach using unigrams as features to combine lexical cues to determine whether a sentence is an **interaction sentence**. In the experiment, we use SVM^{light} implementation of SVMs [3] with a linear kernel.

A mixture model. In order to improve the overall performance, we developed a mixture model to take advantage of the high recall of the rule-based approach, and the high precision of SVM. In the mixture model, the machine learning and

E. Kapetanios, V. Sugumaran, M. Spiliopoulou (Eds.): NLDB 2008, LNCS 5039, pp. 352–354, 2008.

Table 1. Comparison of different models in detecting **interaction sentence**

Approach	Precision (%)	Recall (%)	F-score (%)
keyword-based	40.5	81.1	56.4
SVM (unigrams + sigmoid)	59.7	61.3	60.5
mixture model	57.1	71.0	63.3

the keyword-based approaches has its own input to the final label of a data point (since output of SVM is distance value, we train a sigmoid model [4] to map SVM output into probability). Our goal of combining the two approaches is to construct a consensus partition that summarizes the information from the two inputs. Let y_i, $i = 1, \ldots, n$ be the label of data point x_i, $i = 1, \ldots, n$. The labels y_i are modeled by a mixture of two component densities. The probability of getting y_i for a data point x_i is: $P(y_i|x_i) = \sum_{k=1}^{2} \alpha_k P_k(y_i|x_i)$. We estimate the coefficients α_k using the EM algorithm in the experiment.

Evaluation. The mixture model is evaluated on the *AImed* data set [5][1], performing a 10-fold cross validation. Achieved performance is reported in Table 1. In comparison with the keyword-based approach, F-score of the mixture model is improved by about 7% with an increase of about 17% in precision and a cost of 10% in recall. The mixture model substantially improved recall compared to SVM, with only small reduction in precision of about 2%. The results show that the mixture model outperforms both approaches. In addition, a mixture model has more flexibility in controlling precision and recall of the system by varying the mixture coefficients α.

3   Conclusion

We proposed a mixture model approach that combines a rule-based strategy with a supervised model of SVM to achieve higher coverage on detecting **interaction sentence** with high precision. The results show that the mixture model outperforms each model. Particularly, recall is substantially improved using the mixture model compared to SVM with only a small decrease in precision. The work is under way to annotate all PPIs in the protein-protein interaction database I^2D (http://ophid.utoronto.ca/i2d).

References

1. Niu, Y., Otasek, D., Jurisica, I.: A feature-based approach to detect protein protein interactions from unstructured text (submitted)
2. Plake, C., Hakenberg, J., Leser, U.: Optimizing syntax patterns for discovering protein-protein interactions. In: Proceedings of the 2005 ACM symposium on Applied computing, pp. 195–201 (2005)

---

[1] The evaluation data set is derived by labeling a sentence in *AImed* that describes at least one interaction as positive and otherwise is it labeled as negative instance.

3. SVM^{light} Support Vector Machine, http://svmlight.joachims.org/
4. Platt, C.J.: Advances in large margin classifiers. MIT Press, Cambridge (1999)
5. Bunescu, C.R., Mooney, J.R.: Subsequence kernels for relation extraction. In: Proceedings of the Nineteenth Annual Conference on Neural Information Processing Systems (2005)

Using Semantic Features to Improve Task Identification in Email Messages

Shinjae Yoo[1], Donna Gates[1], Lori Levin[1], Simon Fung[1],
Sachin Agarwal[1], and Michael Freed[2]

[1] Language Technology Institute
5000 Forbes Ave, Pittsburgh, PA 15213
{sjyoo,dmg,lsl,sfung,sachina}@cs.cmu.edu
[2] SRI International
333 Ravenswood Ave., Menlo Park, CA 94025
freed@ai.sri.com

Abstract. Automated identification of tasks in email messages can be very useful to busy email users. What constitutes a task varies across individuals and must be learned for each user. However, training data for this purpose tends to be scarce. This paper addresses the lack of training data using domain-specific semantic features in document representation for reducing vocabulary mismatches and enhancing the discriminative power of trained classifiers when the number of training examples is relatively small.

Keywords: classification, semantic features, construction grammar, ontology.

1 Introduction

Lack of training data and idiosyncratic, highly ambiguous user definitions of tasks make email task classification very challenging. A simple bag-of-words approach is problematic mainly due to vocabulary mismatches. Some such mismatches can be resolved by proper text preprocessing such as stemming or tokenization. However, the differentiation of word senses and the generalization of word meanings for synonyms are not easily solved. Bloehdorn and Hortho [2] explored the use of ontologies (semantic hierarchies) such as WordNet and claimed that classification performance improved on a standard benchmark dataset such as Reuters 21578. However, their baseline performance was lower than the baseline reported by Li and Yang [2] where no such additional Natural Language Processing features were used.

Our hypothesis is that classification performance can be improved by using domain specific semantic features. In this paper, we investigate the use of semantic features to overcome the vocabulary mismatch problem using a general ontology and domain specific knowledge engineering. When we extracted the ontology information using domain specific grammars indicating the appropriate semantic ancestor (e.g., *conference-session* versus the too general *thing* or too specific *tutorial* concepts), we observed improvement over naively extracting ontology information.

E. Kapetanios, V. Sugumaran, M. Spiliopoulou (Eds.): NLDB 2008, LNCS 5039, pp. 355–357, 2008.
© Springer-Verlag Berlin Heidelberg 2008

2 Natural Language Feature Extraction

Our approach uses rules to extract features from natural language text. We use syntactic information from dependency parses produced by Connexor Machinese Syntax[4] and semantic concepts from the RADAR's Scone[6] ontology system. Sets of construction rules (inspired by Construction Grammar [3]) map Connexor's syntactic structures onto semantic frames. The construction rules encode patterns that are matched against the syntactic parse to add general pragmatic information (e,g., speech-acts: *request-information, request-action, reject-proposal*), semantic information (e.g., semantic roles: *agent, patient, location*) and domain specific features (e.g., *conference-session, briefing, meeting*) to the final semantic frame. The inventory of domain specific features was determined manually. At run time, features are extracted using ConAn rules that combine Scone concepts with speech-act information and/or semantic roles. These domain features are tied to domain specific classes or sub-tasks (e.g., *information, web, meetings, food, conference sessions*, etc...) identified by the classifier.

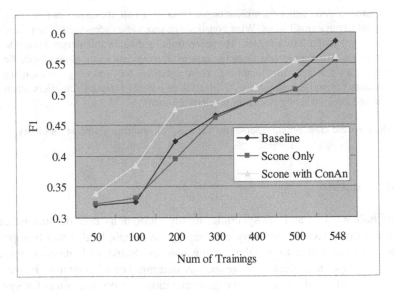

Fig. 1. Macro F1 of SVM

3 Evaluation and Results

We tested our semantic features using SVM [1] on 548 training emails and evaluated on 107 unseen emails. The training emails were hand-labeled with eight previously defined categories (task types). We tested two conditions: first, naively extracting Scone concepts for each word and its semantic parent(s) and grandparent(s) and then, second, extracting semantic features to find the most appropriate domain specific ancestor(s) using construction rules. Naively extracted Scone concepts hurt system performance. However, the construction grammar driven semantic features did well on training data of up to 500 emails. The ontology, construction analyzer and classifier modules described in this paper are being developed as a part of RADAR [5], a system that will act as a personal assistant in such tasks as classifying e-mail, updating web information, scheduling rooms and scheduling meetings.

Acknowledgement

The material described in this paper is based upon work supported by the Defense Advanced Research Projects Agency (DARPA) under contract No. NBCHD030010. Any opinions, findings and conclusions or recommendations expressed in this material are those of the author(s) and do not necessarily reflect the views of the Defense Advanced Research Projects Agency (DARPA).

References

1. Li, F., Yang, Y.: A loss function analysis for classification methods in text categorization. In: ICML 2003 (2003)
2. Bloehdorn, S., Hortho, A.: Boosting for Text Classification with Semantic Features. In: Mobasher, B., Nasraoui, O., Liu, B., Masand, B. (eds.) WebKDD 2004. LNCS (LNAI), vol. 3932, pp. 149–166. Springer, Heidelberg (2006)
3. Goldberg, A.E.: Constructions: a new theoretical approach to language. Trends in Cognitive Sciences 7(5), 219–224 (2003)
4. Tapanainen, P., Järvinen, T.: A non projective dependency parser. In: Fifth Conference on Applied Natural Language Processing, pp. 64–71 (1997)
5. Freed, M., Carbonell, J., Gordon, G., Hayes, J., Myers, B., Siewiorek, D., Smith, S., Steinfeld, A., Tomasic, A.: RADAR: A Personal Assistant That Learns to Reduce Email Overload. In: AAAI 2008 (2008)
6. Fahlmann, S.: Scone User's Manual (2006),
   http://www.cs.cmu.edu/~sef/scone/

Text Pre-processing for Document Clustering

Seemab Latif and Mary McGee Wood

School of Computer Science
The University of Manchester, M13 9PL United Kingdom
{latifs,mary}@cs.man.ac.uk

Abstract. The processing performance of vector-based document clustering methods is improved if automatic summarisation is used in addition to established forms of text pre-processing.

Keywords: Document Clustering, Automatic Summarisation, Document Pre-processing.

1 Pre-processing Documents for Clustering with Summarisation

The Vector Space Model [1] commonly used for document clustering is badly affected by the very high-dimensional vectors which represent longer texts [2]. Established techniques for reducing the dimensionality of vectors are largely limited to stopword removal and stemming. In this experiment we looked at the effect of automatic summarisation as a pre-processing technique on the overall processing efficiency of document clustering algorithms.

2 Experimental Data and Design

To test the effect of summarisation on the performance of document clustering algorithms, we used some 600 documents averaging around 1000 words in length, student answers from three exams held at the University of Manchester using the ABC (Assess By Computer) software [3], [4]. Half length unsupervised summaries (i.e. the keywords used were generated automatically by analyzing the source texts) were used. Three clustering algorithms were used: K-means, agglomerative, and particle swarm. Each clustering algorithm was used to divide each dataset (i.e. the student answers to a specific exam question) into three clusters. For each algorithm, we compared its processing time on the full texts with the sum of its performance on the summaries added to the summarisation time.

3 Experimental Results

There is a reduction of 35%, 34% and 53% in the number of terms, pre-processing time and similarity matrix calculation time respectively after applying summarisation as a pre-processing step. Figure 1 shows that the pre-processing time has

E. Kapetanios, V. Sugumaran, M. Spiliopoulou (Eds.): NLDB 2008, LNCS 5039, pp. 358–359, 2008.
© Springer-Verlag Berlin Heidelberg 2008

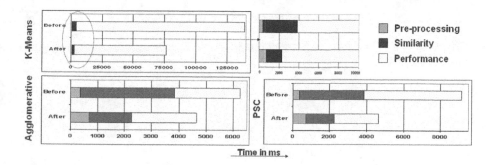

Fig. 1. K-Means, Agglomerative and PSC Time Before and After Summarisation

increased more then double, but the overall time performance of the three algorithms has decreased. The time needed to cluster the documents using the K-Means and PS has been reduced to half and that of the Agglomerative is reduced by 25%. The reason agglomerative time has little effect from summarisation is that it is highly dependent on its implementation.

4 Conclusion

Automatic summarisation has been applied as a text pre-processing step for document clustering to improve overall processing time. Our experiment showed that there is approximately 50% and 27% reduction in the computational time of k-means and PSC algorithms and agglomerative clustering respectively. The next step is to evaluate the summary-based clusters for their quality, cohesion and content.

Acknowledgements. This work is part of the ABC project. The first author would like to acknowledge funding by the National University of Sciences and Technology/ Government of Pakistan. We are grateful for this support.

References

1. Salton, G.: The SMART Retrieval System: Experiments in Automatic Document Processing. Prentice-Hall, Englewood Cliffs (1971)
2. Chang, H.-C., Hsu, C.-C.: Using topic keyword clusters for automatic document clustering. In: 3rd International Conference on Information Technology and Applications, pp. 419–424. IEEE Computer Society, Los Alamitos (2005)
3. Sargeant, J., Wood, M.M., Anderson, S.: A human-computer collaborative approach to the marking of free text answers. In: 8th International Conference on Computer Aided Assessment, Loughborough, UK, pp. 361–370 (2004)
4. Wood, M.M., Jones, C., Sargeant, J., Reed, P.: Light-weight clustering techniques for short text answers in HCC CAA. In: 10th International Conference on Computer Aided Assessment, Loughborough, UK, pp. 291–305 (2006)

Software (Requirements) Engineering and Specification Related Posters

Trade Oriented Enterprise Content Management: A Semantic and Collaborative Prototype Dedicated to the "Quality, Hygiene, Safety and Environment" Domain

Christophe Thovex[1] and Francky Trichet[2]

[1] Aker Yards France – avenue Bourdelle
44600 – Saint Nazaire
christophe.thovex@akeryards.fr, cthovex@free.fr
[2] LINA – Laboratoire d'Informatique de Nantes Atlantique (UMR-CNRS 6241)
Team Knowledge and Decision – University of Nantes
Polytech'Nantes - La Chantrerie - rue Christian Pauc BP 50609
francky.trichet@univ-nantes.fr

Abstract. With the opening of office document formats such as ODF and Open XML, the content management business is evolving more and more towards an issue of knowledge management. For industry, the main challenge is to model a software architecture incorporating trades intelligence of documentary content to knowledge management. While some sectors acquire systems dedicated to the management of semantic knowledge (e.g. press or high-tech industrial), in most cases, these systems are still isolated from the most widespread office context, *i.e.* Open Office, Star Office and MS Office. Interoperability with wells of structured information in Information Systems (IS) remains very limited. And yet, this is an important lever of streamlining and improving the productive value of IS. By developing interoperability, it is human capital that is expected to grow, with the ease to cooperate. By now, the field "Quality, Safety, Hygiene and Environment" is a vehicle for major progress of the European industry. The prototype we have developed in the context of the shipbuilding industry is based on the integration of a semantic cross-lingual component for managing QHSE trades knowledge within a collaborative ECM.

Keywords: semantic enterprise content management, multilingual and cross lingual ontologies, learning process, knowledge browsing, Open XML, SKOS, SPARQL.

Aker Yards SA is a French shipbuilding industry of the Scandinavian group Aker Yards. It includes many guilds, multilingual users of the information system. The inner need for collaborative software tools is important and, in a complex architecture where the end user workstation is an applicative common-core, content management corporate office is a strong element. An Enterprise Content Management (ECM) system becomes collaborative as soon as it allows workgroups to cooperate by sharing information around a theme (activity, project). The collaborative Aker Yards environment, *SharePoint Server 2007* (SPS), makes the most of Open XML, normalized document format according W3C XML, XSD, XSLT recommendations and the WebDAV protocol (access control, check-in/check-out, versions). It integrates multilingual indexation and research services

E. Kapetanios, V. Sugumaran, M. Spiliopoulou (Eds.): NLDB 2008, LNCS 5039, pp. 363–364, 2008.
© Springer-Verlag Berlin Heidelberg 2008

dealing with resources, internal or external to the system, at the syntactic level. It does not deal with the office information (unstructured data, structured and semi-structured) at the semantic level.

A collaborative and semantic ECM prototype, dedicated to the "Quality, Hygiene, Safety and Environment" area, has been developed. This prototype has been defined from an existing knowledge-based application which includes approximately 22,000 files and 14,000 folders (identifying a corpora of 4.7 Giga) related to QHSE norms and standard used for Aker Yards industry. The technologies used to extract the content of this corpora and to convert it into Open XML are J2EE, VB .NET and C#. An ASP.NET Web Part component, dedicated to semantic querying based on ontologies, has been implemented. Among the semantic languages from the W3C, Simple Knowledge Organization System (SKOS) was selected for its relevance to the context. Indeed, its level of abstraction, intermediary between RDF-RDFS and OWL, facilitates access to taxonomies and lightweight ontologies.

Two lightweight ontologies have been identified and connected together in a single SKOS file. The first one, which is an application ontology, combines the concepts which are strictly proper to the application. The second one includes the concepts used by the application that are not specific to it. This is, conceptually, the intersection of the application ontology and a domain ontology of the shipbuilding industry.

The triplet SKOS-SPARQL-ASP.NET has been adopted as a solid foundation for defining the architecture properly. A SPS Web Part (servlet) has been developed in C#, using the SPARQL and SPS 2007 API. SPARQL could even use on the fly volatile rules system construction.

Semantic and cross-lingual functions were finally conceived upstream of indexed syntactic research, integrated to an environment they interface:

1. The Web Part component receives the end-user keywords in SPS 2007 ;
2. It extracts, from the SKOS definition, the key concepts semantic mapping;
3. It provides a refined list of keywords to the standard syntactic search system;
4. The SPS standard search service presents the refined results to the end-user.

By a satisfied bundle of examples, this prototype clearly demonstrates the validity of a semantic and cross-lingual search engine in the area of shipbuilding industry, and specifically in the "Quality, Health, Safety and Environment" area.

For the European shipbuilding industry, semantics is necessary for the productive organization of cross-lingual trades referential. Our work can be considered as the first step of an ambitious project which aims at federating the five major European manufacturers on the development of a common platform for the semantic management of documents and knowledge of the shipbuilding industry.

Doctoral Symposium Papers

Mapping Natural Language into SQL in a NLIDB*

Alessandra Giordani

University of Trento, Povo di Trento, Italy
agiordani@disi.unitn.it

Abstract. Since the 1970's, there has been growing interest in understanding and answering human language questions. Despite this, little progress has been made in developing an interface that any untrained user can use to query very large databases using natural language. In this research, the design of a novel system is discussed. Tree-like structures are built for every question and each query, and a tree kernel, which represents trees in terms of their substructures, is used to define feature spaces. A machine learning algorithm is proposed that takes pairs of trees as training input and derives the unknown final SQL query by matching propositional and relational substructures.

Keywords: NLIDB, Natural Language Question Answering, Tree Kernel.

1 Introduction

The task of natural language understanding and answering has been extensively studied in the past thirty years [2]. The great success obtained by search engines in the last decade confirms that users continuously need to retrieve a lot of information in an easy and efficient way. This information is usually stored in databases, split across tables and retrieved using machine-readable instructions or filling in fields of ad-hoc forms. Obviously, these procedures are not easy, fast or pleasant enough even for an expert user.

The aim of this research is to design and to develop a novel system for answering natural language questions against a relational database, mapping them into SQL queries. To address this problem, we start from building syntactic propositional trees from natural language questions and relational trees from SQL queries. The first step towards our goal is to classify similar propositional and relational trees. This is done using a tree kernel, which computes the number of common substructures between two trees, generating our feature spaces. The next step is exploiting the classification to answer natural language questions, generating SQL codes that are ranked based on the number of matching substructures.

The paper is organized as follows. Section 2 reviews some state-of-the-art systems that address the problem of question answering. Section 3 gives a definition of the problem and an example that motivates the need for a novel system. Section 4 illustrates the proposed solution and the research methodology.

---

* Ph.D. advisor: Periklis Andritsos, University of Trento, Italy. periklis@disi.unitn.it

E. Kapetanios, V. Sugumaran, M. Spiliopoulou (Eds.): NLDB 2008, LNCS 5039, pp. 367–371, 2008.
© Springer-Verlag Berlin Heidelberg 2008

2 Related Work

Despite the numerous attempts on answering natural language questions made in the past thirty years, it has turned out to be much more difficult to build useful natural language interfaces than what was originally expected. State-of-the-art approaches to address that problem can be classified into three categories: spoken language understanding, keyword-based searching and natural language question answering. In this section we will review only recent systems that enable automatic translation from natural language to SQL. (For a review of other systems, please refer elsewhere [2].)

Table 1 shows a list of natural language interfaces to database (NLIDBs) together with important features, advantages and disadvantages of such approaches.

Table 1. Systems' review. Dom.I./Db.I. stand for domain/database independence.

Name	Automatical Approach	Intervention Required	Dom.I.	Db.I.
EasyAsk	Costruct contextual dictionary	-	✓	✓
NaLIX	Disambiguate using ontology	Rephrase questions	✓	✗
SQ-Hal	Use relatioships between tables	Enter relationships/synonyms	✗	✗
English Query	Extract few basic relationships	Model refinement	✓	✗
ELF	Extract most table relationships	-	✓	✓
MASQUE/SQL	Express semantic meaning	Configure domain words	✗	✗
Tiscover	Detect laguage. Check spell	Construct dictionaries	✗	✗
Edite	Exploit mapping tables	Construct dictionaries	✗	✗
EzQ	Express semantic meaning	Construct dictionary/Rephrasing	✗	✗
Precise	Solve problem with maxflow	Rephrase questions	✓	✗
MaNaLa	Map prop. trees into rel. trees	-	✓	✓

EasyAsk [10] is a commercial application that offers both keyword and natural language search over relational databases. NaLIX [6] enables casual user to query XML database, relying on schema-free xQuery. A system that, in contrast to the others, requires the user to know the underlying schema, is the open-source application SQ-HAL [13]. This approach defeats the aim of building an NLIDB for naïve users that is domain and database independent (which means that the interface can be used with different domains and databases without intervention).

English Query [11] and English Language Front-End [12] are two domain independent commercial systems: they automatically create a model collecting objects and extracting relationships from the database.

Among systems that need to be configured for each domain and database there are: MASQUE/SQL [1], a modified version of the famous MASQUE system; Tiscover [3] and Edite [4], two tourism multilingual interfaces that require to construct a domain-specific dictionary for every language they accept; the Spanish NLIDB [5] used in EzQ [7], which implements a domain customization module to maintain dictionaries.

The system described in this paper, MaNaLa (Mapping Natural Language), is independent from both database and domain: it automatically detects relationships exploiting database metadata and expresses semantic meaning mapping NL into SQL.

The NLIDB Precise [8] is 100% accurate when answering "semantically tractable" questions. This class is a small subset of all questions that would be practically legal. The goal of our system is to answer a wider class of questions, possibly outperforming the optimum result of Precise.

3 Motivation and Problem Definition

A central issue in natural language processing is understanding the user intent when she poses a question to a system, identifying concepts and relationships between constituents and resolving ambiguities. In an NLIDB the answer should be retrieved from a database, translating the natural language question into an SQL query.

Suppose that a user types the following natural language question:

WHICH STATES BORDER TEXAS?   (NL1)

The answer to this question can be found in the Geoquery database [8]. With respect to that database, a translation of question NL1 into a SQL query can be retrieved visiting the relational tree SQL1 (Fig. 1) in pre-order. Knowing this structure, the user would be able to straightforwardly substitute values of similar concepts (e.g. Texas VS Iowa) to create new SQL queries (see relational tree VIEW1 in Fig. 1). Additionally, one can pose a composite question, like the following:

WHICH CAPITALS ARE IN STATES BORDERING IOWA?   (NL2)

The figure below represents the NL questions with syntactic propositional parse trees and SQL queries with relational parse trees. It shows that NL2 shares some similarities with NL1, which are reflected also by corresponding relational translated trees: SQL2 embeds VIEW1, whose structure is identical to the one of SQL1.

The problem of question answering can thus be regarded as a matching problem. Given pairs of trees, we can compute if and to what extent they are related using a tree kernel. The kernel that we consider in this research represents tree-like structures in terms of their substructures and defines feature spaces by detecting common substructures. This methodology is described in detail in the next section.

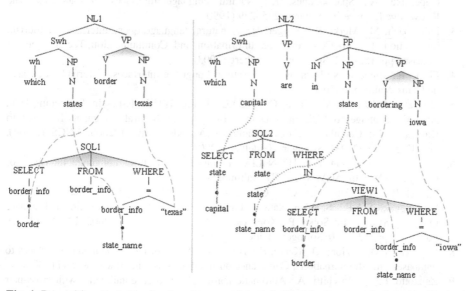

Fig. 1. Propositional/relational pairs of trees and the relationships between their nodes. Dashed lines and triangles reflect the similarities between the structure of SQL2 and the substructure VIEW1 of SQL1. Dotted lines indicate additional relationships between NL2 and SQL2.

4 Research Methodology

Looking at the figure above, we note that SQL representations reflect a semantic structure isomorphous to human conceptual structures. We can exploit this, learning how to map NL *questions* into SQL *queries*. The first step is learning to classify correct pairs *<question, query>* (a pair is correct if a valid transformation of the question leads to the query) based on their parse trees. For this purpose, we add some relational information to trees such that constituents of the natural language sentence and concepts of the relational tree are linked together. Then, given a set of positive (correct) and negative examples, we train the classifier.

To measure the syntactic similarity between pairs of trees we use a tree kernel, which computes the number of common substructures (e.g. the structure highlighted with a grey triangle in Fig.1 is shared by both subtrees of questions NL1 and NL2).

As pointed out elsewhere [9] when we have pairs of trees whose nodes encode relational information, we can use both intra-pair and cross-pair similarities to derive more accurate models for automatic learning from examples. The literature has shown that the use of such a kernel function to evaluate the maximum similarity between two pairs of sentences is effective and efficient even on very large datasets.

References

1. Androutsopoulos, I., Ritchie, G., Thanisch, P.: MASQUE/SQL - An Efficient and Portable Natural Language Query Interface for Relational Databases. In: 6th Int. Conf. on Industrial & Engineering Applications of Artificial Intelligence and Expert Systems (1993)
2. Copestake, A., Sparck-Jones, K.: Natural language interfaces to databases. The Knowledge Engineering Review, 225–249 (1990)
3. Dittenbach, M., Merkl, W., Berger, H.: A natural language query interface for tourism information. In: Int. Conference on Information and Communication Technologies in Tourism, pp. 152–162. Springer, Heidelberg (2003)
4. Filipe, P., Mamede, N.: Databases and Natural Language Interfaces. V Jornadas de Bases de Datos. Valladolid (2000)
5. Gonzalez, B.J.J., Pazos, R.R.A., Cruz, C.I.C., Fraire, H.H.J.: Issues in Translating from Natural Language to SQL in a Domain-Independent Natural Language Interface to Databases. In: Gelbukh, A., Reyes-Garcia, C.A. (eds.) MICAI 2006. LNCS (LNAI), vol. 4293, pp. 922–931. Springer, Heidelberg (2006)
6. Li, Y., Yang, H., Jagadish, H.: NaLIX: an Interactive Natural Language Interface for Querying XML. In: ACM SIGMOD, Baltimore (2005)
7. Pazos, R.R.A., Gelbukh, A.F., González, B.J.J., Alarcón, R.E., Mendoza, M.A., Domínguez, S.A.P.: Spanish Natural Language Interface for a Relational Database Querying System. In: Sojka, P., Kopeček, I., Pala, K. (eds.) TSD 2002. LNCS (LNAI), vol. 2448, pp. 123–130. Springer, Heidelberg (2002)
8. Popescu, A.M., Etzioni, O., Kautz, H.: Towards a theory of natural language interfaces to databases. In: 8th International Conference on Intelligent user interfaces. Miami (2003)
9. Zanzotto, F.M., Moschitti, A.: Automatic learning of textual entailments with cross-pair similarities. In: 21st International Conference on Computational Linguistics and 44th Annual Meeting of the Association for Computational Linguistics, Sydney (2006)

10. EasyAsk (a.k.a. English Wizard), http://www.easyask.com
11. English Query, http://msdn2.microsoft.com/en-us/sqlserver/
12. English Language Front End (VB-ELF), http://www.elfsoft.com
13. SQ-HAL, http://www.csse.monash.edu.au/hons/projects/2000/
    Supun.Ruwanpura/

Improving Data Integration through Disambiguation Techniques*

Laura Po

Dipartimento di Ingegneria dell'Informazione
Universitá di Modena e Reggio Emilia
laura.po@unimore.it

Abstract. In this paper Word Sense Disambiguation (WSD) issue in the context of data integration is outlined and an Approximate Word Sense Disambiguation approach (AWSD) is proposed for the automatic lexical annotation of structured and semi-structured data sources.

1   Introduction

The focus of data integration systems is on producing a comprehensive global schema successfully integrating data from heterogeneous data sources (heterogeneous in format and in structure) [8,2]. The amount of data to be integrated can be scattered at many sources and there may not be available domain experts to perform the integration process. For these reasons and for saving time and human intervention, the integration process should be as much automated as possible. Thus, in recent years, many different data-integration tools have been improved with methods for automatic discovery of mappings and thus matching among schemata.

The biggest difficulty in schema matching lays on being able to discover the right relationships among schemata from different sources. Usually, data sources are organized by developers, according to different categorization. Therefore, it is necessary to understand the modelling logic behind structuring information (i.e the *structural relationships* among schema elements). Further, it is important to deal with the problem of how the data are "labelled"; it is often hard to understand the meaning behind the names denoting schemata elements. Annotation becomes, thus, crucial to understand the meaning of schemata.

Annotation is the inclusion of extra information on a data source. The annotation process can be performed in relation to a reference, like ontology or vocabulary. During the lexical annotation process (i.e. annotation w.r.t. a vocabulary or thesaurus) the concepts and attributes of a data sources (which in the following will be called generally terms) are annotated according to a lexical reference database (WordNet[1] in this implementation, but the method is independent on this choice), in which the terms are interconnected by *lexical relationships* that can be syntactic or semantic.

The followed approach faces the disambiguation problem in the field of integrating structured and semi-structured data sources schemata. These data have got some special

---

* Advisor: Prof. Sonia Bergamaschi.
[1] See http://wordnet.princeton.edu for more information on WordNet.

E. Kapetanios, V. Sugumaran, M. Spiliopoulou (Eds.): NLDB 2008, LNCS 5039, pp. 372–375, 2008.

features that distinguish them from text data. On structured or semi-structured data sources there is not as a wide context as in a text source; in addition, there are less words that concur to the definition of concepts (classes, attributes, properties). The contexts on these data can be defined considering the structural and semantic relationships over a source, like sets of attributes belonging to the same class, or classes connected by aggregations or hierarchies.

As integration system, MOMIS-Ontology Builder [2] has been used; MOMIS is a framework able to create a domain ontology representing a set of selected data sources, described with a standard W3C language wherein concepts and attributes are annotated according to the lexical reference database.

2 Related Work

A semantic approach to schema matching has been proposed in [5]. The method calculates mappings between schema elements by computing semantic relationships. The semantic relations are determined by analysing the meaning which is codified in the elements and the structures of schemas. Labels at nodes are translated into propositional formulas which explicitly codify the label's intended meaning. The use of the semantic relations in this approach is similar to what will be proposed here.

Some other authors, dealing with ontology matching, have proposed a method to solve semantic ambiguity in order to filter the appropriate mappings between different ontologies [6]. Using the semantic similarity measures, the mappings found by an ontology matching tool can be filtered, so the precision of the system improves. The limit of this method is that it does not disambiguate the label of the ontology classes, while it evaluates the possible meanings only.

3 An Approximate Word Sense Disambiguation Approach

As it is described in [7] the WSD task involves two steps: (1) the determination of all the different meanings for every word under consideration; and (2) a means to assign to each occurrence of a word its appropriate meaning. The most recent works on WSD rely on predefined meanings for step (1), including: a list of meanings such as those found in dictionaries or thesauri.

The use of a well-known and shared thesaurus (in this case WordNet) provides a reliable set of meanings and allows to share with others the result of the disambiguation process. Moreover, the fundamental peculiarity of a thesaurus is the presence of a wide network of relationships between words and meanings.

The disadvantage in using a thesaurus is that it does not cover, with same detail, different domains of knowledge. Some terms may not be present or, conversely, other terms may have many associated and related meanings. These considerations and the first tests made led to the need of expanding the thesaurus with more specific terms. On the other hand, when a term has many associated and related meanings, we need to overcome the usual disambiguation approach and relate the term to multiple meanings: i.e. to union of the meanings associated to it. Even Resnik and Yarowsky [9] ratify that there are common cases where several fine-grained meanings may be correct.

The proposed AWSD approach may associate more than one meaning to a term and this, thus, differs from the traditional approaches.

Different methods to disambiguate structured and semi-structured data sources have been developed and tested (two in CWSD [3] and one in MELIS [1]). The results of the cited methods are good, even if not totally satisfying as they do not disambiguate all the terms in a data integration scenario. In [3] it is shown that the combination of methods is an effective way of improving the WSD process performance. The problem focuses on how the different method have to be combined.

Instead of concentrating in determining a unique best meaning of a term, AWSD associates a set of related meanings to a term, each one with its own reliability degree. Our idea is supported by Renisk and Yarowsky that have introduced [9] that the problem of disambiguation is not confined to search for the best meaning. They thought it is significant that a method reduces all possible meanings associated to a term, and that, within this set assigns a probability to the correct meanings.

Let us suppose that a disambiguation method, called A, provides two possible meanings for a term t (t#1 and t#2[2]) while a second method, called B, gives t#2 and t#3. In a combined approach, both methods are applied, and, eventually, the list of all its associated meanings with their probabilities is proposed. The probabilities will be associated to the methods on the basis of their reliability. So if a method is trusted, the method will have a high probability. Moreover, if a WSD method is iterative, it could obtain different meanings with changing probabilities during iteration.

In the previous example, let us suppose that the first method is more reliable than the second one and that the meanings extracted by both methods are different but equiprobable. Let us suppose, method A has 70% probability, while method B has 30%. The final result will be as shown in table 1.

So far the probability associated with a method is defined by the user who can interpret the experimental results provided by methods (for example, in [3], CWSD combines a structural disambiguation algorithm (SD), with a WordNet Domains based disambiguation algorithm (WND); the SD method has a higher precision then WND, then the user can associate to SD method a higher probability).

Table 1. Example of the AWSD output

term	meaning	probability - method A	probability - method B	probability - ASWD
t	t#1	50%	-	35%
t	t#2	50%	50%	50%
t	t#3	-	50%	15%

Choosing more meanings for a term means that the number of discovered lexical relationships connecting a term to other meanings in the thesaurus increase.

Let us assume that it is necessary to analyze the relationships among the terms t, s and r. Filtering the result by a threshold, t#1 and t#2 are maintained as the correct synset for the term t. If there is a lexical relationship between t#1 and s#2,

---

[2] t#2 is the meaning associated to the second sense of the word occurring in the label "t".

and another lexical relationship between t#2 and r#1, choosing both t#1 and t#2 for disambiguate the term t means that two relationship among the three terms are maintained.

The set of lexical relationships together with the set of structural relationships in a dynamic integration environment are the input of the mapping process. Enriching the input of the mapping tool means to improve the discover mappings and so to refine the global schema.

Moreover widening the use of probability to the discover relationships allows computing probabilistic mappings. Finding probabilistic mapping is the basis of managing uncertainty in data integration system. And this leads to new method of query answering like suggested in [4].

4 Conclusion

In a dynamic integration environment, annotation has to be performed automatically, i.e. without human intervention. This leads to uncertain disambiguation results i.e. more than one meaning is associated to a term with a certain probability. In this paper an approximate approach is proposed that associates to each term its meanings with a certain probability. In a lexical database the meanings are related with lexical relationships. By exploiting lexical relationships among the meanings of the terms extracted by AWSD, we can evaluate a similarity degree between the terms and compute probabilistic mappings.

References

1. Bergamaschi, S., Bouquet, P., Giacomuzzi, D., Guerra, F., Po, L., Vincini, M.: An incremental method for the lexical annotation of domain ontologies. Int. J. Semantic Web Inf. Syst. 3(3), 57–80 (2007)
2. Bergamaschi, S., Castano, S., Vincini, M.: Semantic integration of semistructured and structured data sources. SIGMOD Record 28(1), 54–59 (1999)
3. Bergamaschi, S., Po, L., Sala, A., Sorrentino, S.: Data source annotation in data integration systems. In: Proceedings of the Fifth International Workshop on Databases, Information Systems and Peer-to-Peer Computing (DBISP2P) (2007)
4. Dong, X.L., Halevy, A.Y., Yu, C.: Data integration with uncertainty. In: VLDB, pp. 687–698 (2007)
5. Giunchiglia, F., Shvaiko, P., Yatskevich, M.: Semantic schema matching. In: OTM Conferences, vol. (1), pp. 347–365 (2005)
6. Gracia, J., Lopez, V., D'Aquin, M., Sabou, M., Motta, E., Mena, E.: Solving semantic ambiguity to improve semantic web based ontology matching. In: Proceedings of the Workshop on Ontology Matching (OM 2007) at ISWC/ASWC 2007, Busan, South Korea (2007)
7. Ide, N., Véronis, J.: Introduction to the special issue on word sense disambiguation: The state of the art. Computational Linguistics 24(1), 1–40 (1998)
8. Lenzerini, M.: Data integration: A theoretical perspective. In: Popa, L. (ed.) PODS, pp. 233–246. ACM, New York (2002)
9. Resnik, P., Yarowsky, D.: Distinguishing systems and distinguishing senses: new evaluation methods for word sense disambiguation. Natural Language Engineering 5(2), 113–133 (2000)

An Ontology-Based Focused Crawler

Lefteris Kozanidis

Computer Engineering and Informatics Department, Patras University, 26500 Greece
kozanid@ceid.upatras.gr

Abstract. In this paper we present a novel approach for building a focused crawler. The goal of our crawler is to effectively identify web pages that relate to a set of pre-defined topics and download them regardless of their web topology or connectivity with other popular pages on the web. The main challenges that we address in our study are: (i) how to effectively identify the pages' topical content before these are fully downloaded and processed and (ii) how to obtain a well-balanced set of training examples that the crawler will regularly consult in its subsequent web visits.

1 Introduction

Web search engines are the entry point to the web for millions of people. In order for a search engine to be successful towards offering web users with the piece of information they are looking for, it is vital that the engine's index is as full and qualitative as possible. Considering the billions of pages that are available on the web and the evolution of the web population, it becomes evident that building a search engine to fit all purposes might be a tedious task. To alleviate both search engine users and search engine developers from the burden of going over large volumes of data, researchers have proposed the implementation of vertical search engines; each containing specialized indices and thus serving particular user needs.

In this paper, we address the problem of building vertical search engines and we focus on the first step towards this pursuit, which is the implementation of a focused web crawler. The motivation of our study is to design a web crawler that will be able to focus its web visits on particular pages that are of interest to the vertical search engine users. The main problem that we have to overcome is to equip our crawler with a decision making mechanism upon which it will operate for judging the usefulness of the pages it comes across during its web visits. More formally, consider a web crawler that navigates in the web space and looks for resources to download, simply by relying on their links. In the simplest case the crawler will download every resource it comes across regardless of the resource's content or usefulness to the search engine users. In case though the crawler is intended to feed a vertical engine's index it must focus every web visit on the page's content so as to ensure that only qualitative and interesting pages will be downloaded. Based on the above, we realize that building a focused crawler is a complex yet vital task if we want to equip search engines with valuable indices that will be able to serve specific user requests and facilitate web users experience successful web searches. To accomplish the above goal we propose a novel approach for the design and implementation of a focused

E. Kapetanios, V. Sugumaran, M. Spiliopoulou (Eds.): NLDB 2008, LNCS 5039, pp. 376–379, 2008.

crawler that leverages a number of resources and tools from the NLP community in order to compile a rich knowledge base that the crawler will consult for making decisions about whether to download a given page, what is the download priority of every page it comes across and what should be the update policy that the crawler will need to adopt for ensuring that the topical focused index remains fresh. Before delving into the details of our method, let's first outline the current knowledge on the field and present the most widely known attempts of other researches who study the focused crawling problem. In Section 3, we present our proposed methodology, we describe the individual components of our focused crawling module and we outline our plans for future work.

2 Related Work

Researchers have studied the focused crawling problem in the past and proposed a number of algorithms. For instance in [5] the authors propose the Fish Search algorithm, which treats every URL as fish whose survivability depends on visited page relevance and server speed. In [7] the Shark-Search algorithm improves Fish-Search as it uses vector space model in order to calculate the similarity between visited page and query. [3] suggested the exploitation of a document taxonomy for building classification models and proposed two different rules for link expansion, namely the Hard Focus and the Soft focus rules. In [4] two separate models are used to compute the page relevance and the URL visiting order, i.e. the online training by examples for the URL ranking and the source page features for deriving the pages' relevance. Later [6] introduced a new focused crawling approach based on a graph–oriented knowledge representation and showed that their method outperforms typical breath–first keyword or taxonomic based approaches. [1] developed a rule-based crawler which uses simple rules derived from inter class linkage patterns to decide its next move. Moreover their rule-based crawler support tunnelling and try to learn which off topic pages can eventually lead to high quality on-topic pages. This approach improves the crawler in [2] in both in harvest rate and coverage.

3 Our Proposed Focused Crawling Methodology

To design a topical focused crawler, we rely on two distinct yet complementary modules: (i) a classifier that integrates a topical ontology in order to firstly detect the topical content of a web page and thereafter compute the relevance of the page to the ontology topic in which it has been assigned and (ii) a passage extraction algorithm that given a classified web page, it semantically process its content in order to extract the text nugget that is semantically closest to the page's topical content. Based on both the topic relevance values and the topic similar extracts of a large number of web pages, we can built a knowledge base of training examples that the crawler will periodically consult in its web walkthroughs in order to judge whether a new (i.e. yet un-visited) page should be downloaded and if so in which priority order. Figure 1 illustrates the general architecture of our focused crawler. As the figure illustrates the pre-requisites of our focused crawler are: a topical ontology, a classification module, a passage extraction algorithm and a generic crawling infrastructure. Let's now turn our attention to the description of how the above systems can be put together in order to turn a generic crawler into a topical focused one. Given a set of web pages, we preprocess them in order to derive a set of thematic keywords from their contents. For keywords extraction one could employ any of the well-known techniques in the literature. However, in our current study we experiment with the lexical chaining approach [2], which explores the WordNet lexical ontology [9] in order to generate sequences of semantically related term, the so-called thematic terms.

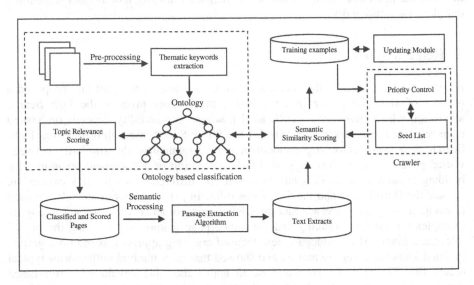

Fig. 1. Focused Crawler Architecture

Having extracted a number of keywords from the page's contents, we employ a topical ontology and we map the pages' keywords to their corresponding ontology nodes in order to compute an appropriate ontology topic for representing the page's thematic content. Again the ontology that could be used for web pages' classification might be any ontology one would like to use. In our work, we are experimenting with the TODE topical ontology [8] that has proved to be richly encoded and useful for web pages' classification. Based on the ontology nodes that match the page's thematic keywords we can compute the degree to which every ontology topic is to likely represent the page's content as the fraction of the page's thematic terms that match an ontology topic T, formally given by

$$\text{Topic Relevance}(p) = \frac{|\text{thematic keywords in p matching T}|}{|\text{thematic keywords in p}|} \quad (1)$$

Based on the above formula we estimate the probability with which every page belongs to an ontology topic. Following web pages' classification to the ontology topics, we employ a passage extraction algorithm in order to extract from the pages contents the text nugget that is semantically closest to the identified topics. The passage extraction algorithm segments the pages' textual content into passages of 50 consecutive terms and relies on the semantic similarity values that the terms in every passage exhibit to the identified page's topic. In particular, to estimate the semantic similarity between a given ontology topic and the terms in the passage of a page that has been assigned to this topic we work as follows. We map all terms inside every page to the ontology and we apply the Wu and Palmer [10] similarity measure, which computes the degree to which passage terms semantically relate to the ontology concepts that represent topical categories. Having computed paired similarity values between passage terms and the ontology topics we take the average similarity values between a passage and a topic. Thereafter from all the page's passages we rely on the one that has the highest overall similarity to the most relevant page topic and we use it as a training example of the crawler's knowledge base. This way, we generate a number of example passages that semantically correlate to specific ontology topics. These examples are given as input to a core crawling module together with a seed list of URLs to be downloaded. Every time the crawler reaches to a page absent from the seed list, it downloads a 10% portion of its content in which it looks for patterns that match the crawler's training examples. If such patterns are found the page is downloaded. Depending on the semantic correlation that the detected patterns exhibit to the pre-defined topics the crawler's priority policy can be fine-tuned so as to ensure index freshness, quality and completeness. Currently, we are experimenting with the crawler's effectiveness in performing complete topical crawls and we are testing its accuracy in building priority queues that are topic specific and resources affordable. In the future, we plan to evaluate our crawler's effectiveness by considering other ontological resources and NLP modules and by comparing its performance to the performance of existing focused crawlers.

References

1. Altingovde, I.S., Ulusoy, O.: Exploiting interclass rules for focused crawling. In: Intelligent Systems, pp. 66–73. IEEE, Los Alamitos (2004)
2. Barzilay, R.: Lexical chains for text summarization. Master's Thesis (1997)
3. Chakrabarti, S., van den Berg, M., Dom, B.: Focused Crawling: A new approach to Topic-Specific Web Resource Discovery. Computer Networks 31(11-16) (1999)
4. Chakrabarti, S., Punera, K., Subramanyam, M.: Accelerated focused crawling through online relevance feedback. In: WWW 2002, Honolulu, Hawaii, USA, May 7-11, 2002. ACM, New York (2002)
5. De Bra, P., Houben, G., Kornatzky, Y., Post, R.: Information Retrieval in Distributed Hypertexts. In: Proceedings of the 4th RIAO Conference, New York, pp. 481–491 (1994)
6. Ehrig, M., Maedche, A.: Ontology-focused Crawling of Web Documents. In: ACM Symposium on Applied computing (2003)
7. Hersovici, M., Jacovi, M., Maarek, Y., Pelleg, D., Shtalhaim, M., Ur, S.: The shark-search algorithm - An application: Tailored Web site mapping. In: Proc. 7th Intl. World-Wide Web Conference, Brisbane, Australia (April 1998)
8. Stamou, S., Krikos, V., Kokosis, P., Ntoulas, A., Christodoulakis, D.: Web directory construction using lexical chains. In: 10th NLDB Conference, pp. 138–149 (2005)
9. WordNet, http://www.cogsci.princeton.edu/~wn
10. Wu, X., Palmer, M.: Web semantics and lexical selection. In: 32nd ACL Meeting (1994)

Impact of Term-Indexing for
Arabic Document Retrieval

Siham Boulaknadel[1,2]

[1] LINA FRE CNRS 2729 Université de Nantes
2 rue la Houssinière BP 92208 44322 Nantes cedex 03, France
[2] GSCM Université Mohammed V
BP 1014 Agdal Rabat-Maroc
siham.boulaknadel@univ-nantes.fr

Abstract. In this paper, we adapt the standard method for multi-word term extraction for Arabic language. We define the linguistic specifications and develop a term extraction tool. We experiment the term extraction program for document retrieval in a specific domain, evaluate two kinds of multi-word term weighting functions considering either the corpus or the document, and demonstrate the efficiency of multi-word term indexing for both weighting up to 5.8% of average precision.

1 Introduction

With the Internet evolution, several challenges appear in Information retrieval (IR) that could be classified in three main types: information volume management, presence of multiple supports and, finally, web multilingual character. In this last case, importance of other languages than English caused the development of automatics tools and techniques in order to allow their data-processing treatment. In comparison with English and more generally with Semitic languages, Arabic language presents distinctive features, namely agglutination and vocalisation.

Our purpose is to demonstrate the efficiency of multi-word term (MWT) indexing in IR for Arabic language applying on a scientific domain. To extract MWTs from corpora, we adopt the standard approach that combined grammatical patterns and statistical score [1]. We defined the linguistic specification of MWTs for Arabic language. Then, we develop a term extraction program and evaluate several statistical measures in order to filter the extracted term-like units for keeping the most representative of domain specific corpus. Then, we demonstrate the efficiency of using MWT indexing compare to single term indexing in an Arabic Information Retrieval System using the TREC 2001 evaluation protocol and evaluating two kinds of multi-word term weighting functions considering either the corpus or the document.

E. Kapetanios, V. Sugumaran, M. Spiliopoulou (Eds.): NLDB 2008, LNCS 5039, pp. 380–383, 2008.

2   Linguistic Specifications of Arabic Multi-word Terms

2.1   MWT Patterns

MWTs are noun phrases that share limited morphosyntactic structures such as N ADJ, N1 N2, where N stands for noun and ADJ for adjective. To attest the terminological character of a MWT-like sequence, two solutions are possible: the one is to use a terminological database that includes Arabic terms like AGROVOC [1], and the other is to check if the English translation belongs to a terminological database. For this procedure, we used Eurodicautom [2]. Some examples met in our corpus are presented in Table 1. For the description of Arabic terms, we applied Buckwalter's transliteration system [3] which transliterates Arabic alphabet into Latin alphabet.

Table 1. MWT patterns

Pattern	Sub-pattern	Arabic MWT	English translation
N ADJ		Altlwv AlkmyAAy	chemical pollution
N1 N2		tlwv AlmAA	water pollution
	N1 b N2	Altlwv b AlrsAs	pollution with lead
N1 PREP N2	N1 l N2	AltErD l AlAmrAD	exposure to diseases
	N1 mn N2	Altxls mn AlnfAyAt	waste disposal

3   Multi-word Term Extraction

3.1   Method

The term extraction process is performed in two major steps: the selection of MWT-like units, and the ranking of MWT-like units by means of statistical techniques. To filter the MWT-like units and their variants, we used their part-of-speech that has been assigned by the diab's tagger [2]. Association measures [3] [4] [5] are used in order to rank MWT-like strings that have been collecting in the first step.

3.2   Experimentation and Evaluation

3.2.1   Corpus

As it does not exist specialised domain corpora, we built a specialised corpus: the texts are taken from the environment domain and are extracted from the Web sites "Al-Khat Alakhdar"[4] and "Akhbar Albiae"[5]. The corpus contains 1.062 documents, 475.148 words from which 54.705 are distinct.

---

[1] www.fao.org/agrovoc/

[2] http://ec.europa.eu/eurodicautom/Controller

[3] http://www.qamus.org/transliteration.htm

[4] http://www.greenline.com.kw

[5] http://www.4eco.com

3.2.2 Termhood Evaluation

We decide to evaluate termhood [6] precision. We measured the precision of each association measure by examining the first 100 MWT candidates. The results are shown in Table 2. We notice that the LLR, MI^3 and t-score measures, that are based on the significance of association measure, outperform the MI^3 measure. This result shows that the statistical null-hypothesis of the independent assumption is a better means for ranking and identifying MWTs, in comparison to the probability parameters approximation methods using the degree of association measures.

Table 2. Termhood precision of MWTs

Measure	Precision (%)
T-score	57%
LLR	85%
MI^3	26%

4   IR System and Evaluation

We used the document collection presented in Section 3.2.1. We worked out 30 queries following the TREC format but only including the title field.

4.1 IR System

The IR system can be divided into three parts:

1. **Diacritic removing and transcription.** Most of the texts of our corpus does not contain diacritic. However we detect some weak vowels appearing in a few words. We eliminate them before applying the Buckwalter's transliteration system.
2. **Tokenisation and POS tagging** We split the corpus into sentences and the word are isolated and morphologically analysed using the Diab's tagger[2].
3. **Text indexing.** We perform free indexing using the MWT extraction program presented in Section 3.

4.2 Results

The efficiency of multi-word term indexing is evaluated by comparing the results obtained with three strategies:

1. **SWT** the traditional single term indexing (SWT)
2. **MWT_C** the MWT indexing using the LLR ranking.
3. **MWT_D** the MWT indexing using a Okapi-BM25 weighting score [7].

The results in Table 3 show 5.8% enhancement in average precision for MWT_D and 3.6% for MWT_C. Whatever is the strategy, corpus or document weight computation, integrating multi-word terms in the indexing process increases the performance compared to the best results obtained by single terms indexing.

Table 3. 11- point average precision

	Avg Prec	Diff
SWT	26,1 %	
MWT_C	29,7%	+3,6%
MWT_D	31,9%	+ 5,8%

5   Conclusion

In this paper, we present an Arabic multi-word term (MWT) indexing for document retrieval for a specific domain. We define the linguistic specifications of
Arabic MWT and develop a term extraction tool. Selected terms are then used
for indexing with an Arabic Information Retrieval System. We show that combining MWT indexing with Okapi BM-25 weighting can improve the information
retrieval system performances, specially at low recall. We conclude that multi-
word term indexing could be used to support other IR enhancements, such as
automatic feedback of the top-returned documents to expand the initial query.

References

1. Cabré, M.T., Bagot, R.E., Platresi, J.V.: Automatic term detection: A review of
   current systems. In: Bourigault, D., Jacquemin, C., L'Homme, M.C. (eds.) Recent
   Advances in Computational Terminology. Natural Language Processing, vol. 2, pp.
   53–88. John Benjamins, Amsterdam (2001)
2. Diab, M., Hacioglu, K., Jurafsky, D.: Automatic tagging of arabic text: From raw
   text to base phrase chunks. In: Proceedings of HLT-NAACL 2004, Boston, pp. 149–
   152 (2004)
3. Dunning, T.: Accurate methods for the statistics of surprise and coincidence. Computational Linguistics 19(1), 61–74 (1994)
4. Kenneth, W.C., Hanks, P.: Word association norms, mutual information, and lexicography. In: Proceedings of the 27th. Annual Meeting of the Association for Computational Linguistics, Vancouver, B.C, pp. 76–83 (1989)
5. Church, K., Gale, W., Hanks, P., Hindle, D.: Using statistics in lexical analysis. In:
   Lexical Acquisition: Exploiting On-Line Resources to Build a Lexicon, pp. 115–164.
   U. Zernik (1991)
6. Kageura, K., Umino, B.: Methods of automatic term recognition: a review. Terminology 3(2), 259–289 (1996)
7. Darwish, K.: Probabilistic Methods for Searching OCR-Degraded Arabic Text. PhD
   thesis, University of Maryland, Maryland, USA (2003)

Table 3. 11-point average precision

	AvgP [illegible]	Diff [illegible]
[illegible]	[illegible]	[illegible]
[illegible]	[illegible]	[illegible]
[illegible]	[illegible]	[illegible]

5 Conclusion

In this paper, we present a [illegible] multi-word term (MWT) indexing for document retrieval. For a specific domain, we define the linguistic specifications of Arabic MWT and developed a term extraction [illegible]. These terms are then used [illegible] Arabic [illegible] Information Retrieval System. We show that combining MWT indexing with Okapi BM-25 [illegible] can improve the information retrieval system performance, especially [illegible] recall. We conclude that multi-word term [illegible] could be used [illegible] other IR enhancements, such as automatic feedback of the top-retrieved documents to expand the initial query.

References

1. [illegible] Automatic term detection: A review of current systems. In: Bourigault, D., Jacquemin, C., L'Homme, M.C. (eds.) Recent Advances in Computational Terminology. Natural Language Processing, vol. 2, pp. [illegible]. John Benjamins, Amsterdam (2001)
2. [illegible] Automatic [illegible] of arabic [illegible] text [illegible] (2004)
3. Dunning, T.: Accurate methods for the statistics of surprise and coincidence. Computational Linguistics [illegible] (1993)
4. [illegible] Hanks, P.: Word association norms, mutual information, and lexicography. In: Proceedings of the 27th Annual Meeting of the Association for Computational Linguistics, Vancouver, B.C., pp. 76–83 (1989)
5. [illegible] Gale, W., [illegible] Hindle, D.: Using statistics in lexical analysis. In: [illegible] pp. 115–164 [illegible]
6. [illegible] Methods of automatic term recognition [illegible]
7. [illegible]

Author Index

Lecture Notes in Computer Science

Sublibrary 3: Information Systems and Application, incl. Internet/Web and HCI

For information about Vols. 1–4658
please contact your bookseller or Springer

Vol. 4832: M. Weske, M.-S. Hacid, C. Godart (Eds.), Web Information Systems Engineering – WISE 2007 Workshops. XV, 518 pages. 2007.

Vol. 4831: B. Benatallah, F. Casati, D. Georgakopoulos, C. Bartolini, W. Sadiq, C. Godart (Eds.), Web Information Systems Engineering – WISE 2007. XVI, 675 pages. 2007.

Vol. 4825: K. Aberer, K.-S. Choi, N. Noy, D. Allemang, K.-I. Lee, L. Nixon, J. Golbeck, P. Mika, D. Maynard, R. Mizoguchi, G. Schreiber, P. Cudré-Mauroux (Eds.), The Semantic Web. XXVII, 973 pages. 2007.

Vol. 4823: H. Leung, F. Li, R. Lau, Q. Li (Eds.), Advances in Web Based Learning – ICWL 2007. XIV, 654 pages. 2008.

Vol. 4822: D.H.-L. Goh, T.H. Cao, I.T. Sølvberg, E. Rasmussen (Eds.), Asian Digital Libraries. XVII, 519 pages. 2007.

Vol. 4820: T.G. Wyeld, S. Kenderdine, M. Docherty (Eds.), Virtual Systems and Multimedia. XII, 215 pages. 2008.

Vol. 4816: B. Falcidieno, M. Spagnuolo, Y. Avrithis, I. Kompatsiaris, P. Buitelaar (Eds.), Semantic Multimedia. XII, 306 pages. 2007.

Vol. 4813: I. Oakley, S.A. Brewster (Eds.), Haptic and Audio Interaction Design. XIV, 145 pages. 2007.

Vol. 4810: H.H.-S. Ip, O.C. Au, H. Leung, M.-T. Sun, W.-Y. Ma, S.-M. Hu (Eds.), Advances in Multimedia Information Processing – PCM 2007. XXI, 834 pages. 2007.

Vol. 4809: M.K. Denko, C.-s. Shih, K.-C. Li, S.-L. Tsao, Q.-A. Zeng, S.H. Park, Y.-B. Ko, S.-H. Hung, J.-H. Park (Eds.), Emerging Directions in Embedded and Ubiquitous Computing. XXXV, 823 pages. 2007.

Vol. 4808: T.-W. Kuo, E. Sha, M. Guo, L.T. Yang, Z. Shao (Eds.), Embedded and Ubiquitous Computing. XXI, 769 pages. 2007.

Vol. 4806: R. Meersman, Z. Tari, P. Herrero (Eds.), On the Move to Meaningful Internet Systems 2007: OTM 2007 Workshops, Part II. XXXIV, 611 pages. 2007.

Vol. 4805: R. Meersman, Z. Tari, P. Herrero (Eds.), On the Move to Meaningful Internet Systems 2007: OTM 2007 Workshops, Part I. XXXIV, 757 pages. 2007.

Vol. 4804: R. Meersman, Z. Tari (Eds.), On the Move to Meaningful Internet Systems 2007: CoopIS, DOA, ODBASE, GADA, and IS, Part II. XXIX, 683 pages. 2007.

Vol. 4803: R. Meersman, Z. Tari (Eds.), On the Move to Meaningful Internet Systems 2007: CoopIS, DOA, ODBASE, GADA, and IS, Part I. XXIX, 1173 pages. 2007.

Vol. 4802: J.-L. Hainaut, E.A. Rundensteiner, M. Kirchberg, M. Bertolotto, M. Brochhausen, Y.-P.P. Chen, S.S.-S. Cherfi, M. Doerr, H. Han, S. Hartmann, J. Parsons, G. Poels, C. Rolland, J. Trujillo, E. Yu, E. Zimányie (Eds.), Advances in Conceptual Modeling – Foundations and Applications. XIX, 420 pages. 2007.

Vol. 4801: C. Parent, K.-D. Schewe, V.C. Storey, B. Thalheim (Eds.), Conceptual Modeling - ER 2007. XVI, 616 pages. 2007.

Vol. 4797: M. Arenas, M.I. Schwartzbach (Eds.), Database Programming Languages. VIII, 261 pages. 2007.

Vol. 4796: M. Lew, N. Sebe, T.S. Huang, E.M. Bakker (Eds.), Human–Computer Interaction. X, 157 pages. 2007.

Vol. 4794: B. Schiele, A.K. Dey, H. Gellersen, B. de Ruyter, M. Tscheligi, R. Wichert, E. Aarts, A. Buchmann (Eds.), Ambient Intelligence. XV, 375 pages. 2007.

Vol. 4777: S. Bhalla (Ed.), Databases in Networked Information Systems. X, 329 pages. 2007.

Vol. 4761: R. Obermaisser, Y. Nah, P. Puschner, F.J. Rammig (Eds.), Software Technologies for Embedded and Ubiquitous Systems. XIV, 563 pages. 2007.

Vol. 4747: S. Džeroski, J. Struyf (Eds.), Knowledge Discovery in Inductive Databases. X, 301 pages. 2007.

Vol. 4744: Y. de Kort, W. IJsselsteijn, C. Midden, B. Eggen, B.J. Fogg (Eds.), Persuasive Technology. XIV, 316 pages. 2007.

Vol. 4740: L. Ma, M. Rauterberg, R. Nakatsu (Eds.), Entertainment Computing – ICEC 2007. XXX, 480 pages. 2007.

Vol. 4730: C. Peters, P. Clough, F.C. Gey, J. Karlgren, B. Magnini, D.W. Oard, M. de Rijke, M. Stempfhuber (Eds.), Evaluation of Multilingual and Multi-modal Information Retrieval. XXIV, 998 pages. 2007.

Vol. 4723: M. R. Berthold, J. Shawe-Taylor, N. Lavrač (Eds.), Advances in Intelligent Data Analysis VII. XIV, 380 pages. 2007.

Vol. 4721: W. Jonker, M. Petković (Eds.), Secure Data Management. X, 213 pages. 2007.

Vol. 4718: J. Hightower, B. Schiele, T. Strang (Eds.), Location- and Context-Awareness. X, 297 pages. 2007.

Vol. 4717: J. Krumm, G.D. Abowd, A. Seneviratne, T. Strang (Eds.), UbiComp 2007: Ubiquitous Computing. XIX, 520 pages. 2007.

Vol. 4715: J.M. Haake, S.F. Ochoa, A. Cechich (Eds.), Groupware: Design, Implementation, and Use. XIII, 355 pages. 2007.

Vol. 4714: G. Alonso, P. Dadam, M. Rosemann (Eds.), Business Process Management. XIII, 418 pages. 2007.

Vol. 4704: D. Barbosa, A. Bonifati, Z. Bellahsène, E. Hunt, R. Unland (Eds.), Database and XML Technologies. X, 141 pages. 2007.

Vol. 4690: Y. Ioannidis, B. Novikov, B. Rachev (Eds.), Advances in Databases and Information Systems. XIII, 377 pages. 2007.

Vol. 4675: L. Kovács, N. Fuhr, C. Meghini (Eds.), Research and Advanced Technology for Digital Libraries. XVII, 585 pages. 2007.

Vol. 4674: Y. Luo (Ed.), Cooperative Design, Visualization, and Engineering. XIII, 431 pages. 2007.

Vol. 4663: C. Baranauskas, P. Palanque, J. Abascal, S.D.J. Barbosa (Eds.), Human-Computer Interaction – INTERACT 2007, Part II. XXXIII, 735 pages. 2007.

Vol. 4662: C. Baranauskas, P. Palanque, J. Abascal, S.D.J. Barbosa (Eds.), Human-Computer Interaction – INTERACT 2007, Part I. XXXIII, 637 pages. 2007.